Second Edition

VIOLENT OFFENDERS

Theory, Research, Policy, and Practice

EDITED BY

Matt DeLisi, PhD
Professor and Coordinator
Criminal Justice Studies
Iowa State University
Ames, Iowa

Peter J. Conis, PhD
Professor
Department of Sociology
Des Moines Area Community College
Boone, Iowa

JONES & BARTLETT
LEARNING

World Headquarters
Jones & Bartlett Learning
5 Wall Street
Burlington, MA 01803
978-443-5000
info@jblearning.com
www.jblearning.com

Jones & Bartlett Learning
Canada
6339 Ormindale Way
Mississauga, Ontario L5V 1J2
Canada

Jones & Bartlett Learning
International
Barb House, Barb Mews
London W6 7PA
United Kingdom

Jones & Bartlett Learning books and products are available through most bookstores and online booksellers. To contact Jones & Bartlett Learning directly, call 800-832-0034, fax 978-443-8000, or visit our website, www.jblearning.com.

Substantial discounts on bulk quantities of Jones & Bartlett Learning publications are available to corporations, professional associations, and other qualified organizations. For details and specific discount information, contact the special sales department at Jones & Bartlett Learning via the above contact information or send an email to specialsales@jblearning.com.

Production Credits:
Publisher, Higher Education: Cathleen Sether
Acquisitions Editor: Sean Connelly
Editorial Assistant: Caitlin Murphy
Production Manager: Jenny L. Corriveau
Associate Production Editor: Sara Fowles
Associate Marketing Manager: Lindsay White
Manufacturing and Inventory Control Supervisor: Amy Bacus
Composition: Laserwords Private Limited, Chennai, India
Cover Design: Kristin E. Parker
Rights and Permissions Manager: Katherine Crighton
Permissions and Photo Researcher: Amy Mendosa
Cover Image: © Dmitriy Gool/Dreamstime.com
Printing and Binding: Malloy, Inc.
Cover Printing: Malloy, Inc.

Library of Congress Cataloging-in-Publication Data
DeLisi, Matt.
Violent offenders: theory, research, policy, and practice / Matthew DeLisi and Peter Conis. — 2nd ed.
p. cm.
Includes author and subject index.
ISBN-13: 978-0-7637-9790-4 (pbk.)
ISBN-10: 0-7637-9790-1 (pbk.)
1. Violent offenders. 2. Violent crimes—Research. 3. Criminal psychology. 4. Criminology. I. Conis, Peter John. II. Title.
HV6133.D45 2011
364.15—dc22

2011008290

6048

Printed in the United States of America
15 14 13 12 11 10 9 8 7 6 5 4 3 2 1

Contents

Preface

What is it about bad guys (i.e., violent offenders) that gets people so interested in criminal justice? Why is it that the more extreme, the more reprehensible, and the more dramatic the antisocial behavior, the more likely that it will spawn a television series, movie, or book? Why are our criminology and criminal justice undergraduates so eager to become forensic psychologists, criminal profilers, FBI Special Agents, or crime scene investigators? The answer is obvious: They want to catch the bad guys.

For some, the interest in criminal justice and especially the interest in the most serious and violent offenders is simply intrigue. Others view the emphasis on violent criminals as a fad, or entertainment, or sensationalistic fodder for unrefined minds. Still others view violent offenders as a diversion from the real menace to society, white-collar criminals. None of these perceptions is right.

That we are so fascinated by the bad guys speaks to the fundamental goodness of people. Violent criminal offenders aggravate our moral sense. Those who murder, or rape, or rob, or molest, violate fundamental codes of conduct that are so basic to human existence that we are appalled. Violent offenders serve as a measuring stick to evaluate our own conduct and to try to understand the reasons why some people commit grievous forms of antisocial behavior. With violent offenders, we are at once interested and unnerved, fascinated and repulsed, outraged and sympathetic.

Bad guys get lots of attention for another important reason: They commit the bulk of bad acts that occur in a society. This is an important scientific point. The 10% of criminals who are most active account for easily more than half of all crimes, and even higher proportions of violent crimes, such as murder, rape, robbery, assault, and kidnapping, etc. Increasingly, understanding the bad guys is what criminology is about.

We wanted to create a book that could present the state-of-the-art on violent offenders. As former criminal justice practitioners, we are stridently prejudiced in favor of applied, practitioner expertise and scholarly perspectives that stem from hands-on research experience with real offenders. We want students to understand the breadth of violent offenders, including gangs, serial killers, sex offenders, career criminals, and related topics. We believe we have succeeded, and have made significant improvements to the second edition.

Violent Offenders: Theory, Research, Policy, and Practice, Second Edition, offers the insights of a list of contributors who are international experts on violent offenders. Some of these contributors are academics, some are criminal justice practitioners, and some are both. The second edition contains 10 new chapters of material from accomplished scholars with backgrounds in criminology and criminal justice, sociology, psychology, public affairs, political science, and interdisciplinary programs.

The book is divided into two sections. Part I, Theory and Research, explores the theoretical and disciplinary foundations of the study of violent behavior, spanning the disciplines of sociology, psychology, biology, and neuroscience. This section contains five exciting chapters of new material to broaden the conceptual and disciplinary reach of the study of violent offenders. These chapters offer an exploration of the overlap between violent offending and violent victimization (Chapter 2), the community-level correlates of crime and violence from a sociological perspective (Chapter 3), violent acts that occur in cyberspace (Chapter 5), the place of violence in human societies from an evolutionary psychology perspective (Chapter 8), and a focused look at female sexual offending (Chapter 12).

Part II, Policy and Practice, explores public policy and practical applications. This section describes the various ways that criminal justice systems respond to violent offenders from the insightful perspectives of people who work among violent offenders on a daily basis. This section also contains five new chapters of material. These include an overview of focused deterrence strategies to reduce gang and gun-related violence (Chapter 15), a mixed-method investigation of patrol officers' perspectives on the processing of violent offenders by the criminal justice system (Chapter 16), a look at alternative adjudication methods for juvenile and adult offenders with behavioral and psychiatric problems (Chapter 19), an overview of the supervision of violent offenders in the federal criminal justice system (Chapter 20), and coverage of the reentry of violent offenders into society after serving time in confinement facilities (Chapter 25).

Violent Offenders: Theory, Research, Policy, and Practice, Second Edition, spans the gamut of the criminal justice system and its attempts to treat, supervise, and punish offenders who commit the most violent forms of crime. Although the study of violent offenders is "heavy" material, the scholar–practitioners who contributed to this edition demonstrate that the human services and criminal justice systems are working diligently to prevent, control, and understand violence in our society.

Matt DeLisi
Peter J. Conis

Reviewers

We would like to thank the following reviewers for their contributions to the text:

Monic P. Behnken, Iowa State University

Paige H. Gordier, Lake Superior State University

Alan Kraft, Seminole State College of Florida

Contributors

Eric Beauregard, PhD
School of Criminology
Simon Fraser University
Burnaby, British Columbia, Canada

Kevin M. Beaver, PhD
College of Criminology and Criminal Justice
Florida State University
Tallahassee, Florida

Monic P. Behnken, PhD, JD
Criminal Justice Studies
Iowa State University
Ames, Iowa

Mark T. Berg, PhD
Department of Criminal Justice
Indiana University
Bloomington, Indiana

Anthony A. Braga, PhD
School of Criminal Justice
Rutgers University
New Brunswick, New Jersey
Program in Criminal Justice Policy and Management
Harvard University
Cambridge, Massachusetts

Jonathan W. Caudill, PhD
Department of Political Science
California State University, Chico
Chico, California

Heith Copes, PhD
Department of Justice Sciences
University of Alabama
Birmingham, Alabama

Peter J. Conis, PhD
Department of Sociology
Des Moines Area Community College
Boone, Iowa

Mark D. Cunningham, PhD, ABPP
Clinical and Forensic Psychologist
Lewisville, Texas

Matt DeLisi, PhD
Criminal Justice Studies
Iowa State University
Ames, Iowa

Alan J. Drury, PhD candidate
Criminal Justice Studies
Iowa State University
United States Probation Office
Des Moines, Iowa

David P. Farrington, PhD
Institute of Criminology
University of Cambridge
Cambridge, United Kingdom

Danielle A. Harris, PhD
Department of Justice Studies
San Jose State University
San Jose, California

Jay Healey, PhD candidate
School of Criminology
Simon Fraser University
Burnaby, British Columbia, Canada

ANDY HOCHSTETLER, PHD
Criminal Justice Studies
Iowa State University
Ames, Iowa

DONI LYNN HOMISH, PHD
State University of New York
Buffalo, New York

FARNAZ KAIGHOBADI, PHD
Department of Psychology
Florida Atlantic University
Davie, Florida

SGT. FRANK KARDASZ, PHD
Arizona ICAC Task Force
Phoenix Police Department
Phoenix, Arizona

BENOIT LECLERC, PHD
Pinel Institute of Montreal
University of Montreal
Montreal, Canada

MATTHEW R. LEE, PHD
Department of Sociology
Louisiana State University
Baton Rouge, Louisiana

SHELLEY JOHNSON LISTWAN, PHD
Department of Criminal Justice and Criminology
University of North Carolina
Charlotte, North Carolina

ROLF LOEBER, PHD
Department of Psychiatry
University of Pittsburgh
Pittsburgh, Pennsylvania

Patrick Lussier, PhD
School of Criminology
Simon Fraser University
Burnaby, British Columbia, Canada

Jean Marie McGloin, PhD
Department of Criminology and Criminal Justice
University of Maryland
College Park, Maryland

Brooke N. Miller, MS
Program in Criminology
The University of Texas
Dallas, Texas

Robert G. Morris II, PhD
Program in Criminology
The University of Texas
Dallas, Texas

Graham C. Ousey, PhD
Department of Sociology
College of William and Mary
Williamsburg, Virginia

Jean Proulx, PhD
Pinel Institute of Montreal
University of Montreal
Montreal, Canada

Shelley L. Reese, MSA
Dallas County Sheriff's Department
Waukee, Iowa

Roxann M. Ryan, PhD, JD
Iowa Department of Public Safety
Des Moines, Iowa

Todd K. Shackelford, PhD
Department of Psychology
Oakland University
Rochester, Michigan

Jennifer Schwartz, PhD
Department of Sociology
Washington State University
Pullman, Washington

Rebecca Stallings, PhD
University of Pittsburgh Medical Center
Pittsburgh, Pennsylvania

Denise Timmins, JD
Iowa Assistant Attorney General
Iowa Attorney General's Office
Des Moines, Iowa

Chad R. Trulson, PhD
Department of Criminal Justice
University of North Texas
Denton, Texas

Michael G. Vaughn, PhD
School of Social Work
Saint Louis University
Saint Louis, Missouri

Glenn D. Walters, PhD
Federal Correctional Institution—Schuylkill
Minersville, Pennsylvania

John Paul Wright, PhD
Department of Criminal Justice
University of Cincinnati
Cincinnati, Ohio

PART I

Theory and Research

CHAPTER 1

The Importance of Violent Offenders to Criminology

Matt DeLisi
Iowa State University

Peter J. Conis
Des Moines Area Community College

Kevin M. Beaver
Florida State University

> *"There exists, it is true, a group of criminals, born for evil, against whom all social cures break as against rock."*
>
> —Cesare Lombroso (1911, p. 447)

> *"For some, the term 'career criminal' is a label that will serve to further stigmatize and exacerbate the risk factors that chronic offenders experience. For others, this moniker is the mark of Cain."*
>
> —Matt DeLisi (2005, p. 118)

For most of the 20th century, criminology floundered because it focused on normal processes and situations that were purported to cause people to be delinquent. Strain, anomie, stress, poverty, living in a bad neighborhood, discrimination, and hanging out with friends who enjoyed breaking the law were some of the dominant explanations of crime. Over time, especially during the 1960s and 1970s, the causes of crime were even attributed to functional institutions of society. For instance, the criminal justice system was blamed for causing crime or enabling recidivism because the police, judicial, and correctional systems negatively intervened in the lives of criminals. Sociological, specious, and liberal, criminology had a credibility problem because its major theories of crime lacked the ring of truth. They were too academic.

Then, between 1985 and 1993, three major works appeared that saved criminology: James Q. Wilson and Richard Herrnstein's *Crime and Human Nature* (1985), Michael Gottfredson and Travis Hirschi's *A General Theory of Crime* (1990), and Terrie Moffitt's developmental taxonomy theory (1993). Individually and collectively, these works have made a towering impact on the study of criminal behavior.

Wilson and Herrnstein's work is probably known as much for the star power of the authors as for its substantive impact. Both are best-selling authors, both are distinguished academics who crossed over into the role of public intellectual (as least in terms of the media coverage of their works), and both are viewed as conservative, which in academic circles is controversial (DeLisi, 2003a). More than anything else, *Crime and Human Nature* wrestled the study of crime from the stranglehold of sociology by incorporating ideas from economics, psychology, and biology. Borrowing from economics, Wilson and Herrnstein articulated the idea that crime is fundamentally a matter of choice. As such, rational choice theory and the thought processes of individual actors are essential in understanding why some people use violence against others. Wilson and Herrnstein expanded this idea by using psychological theories to show that choosing to commit crime is not simply the outcome of rational calculus, but rather a decision that is molded and influenced by an array of factors, such as family background, social class, environmental influences, and prior experience and education. From biology, Wilson and Herrnstein (1985) drew the following idea:

> *[T]he existence of biological predispositions means that circumstances that activate behavior in one person will not do so in another, that social forces cannot deter criminal behavior in 100% of the population, and that the distribution of crime within and across societies may, to some extent, reflect underlying distributions of constitutional factors. Crime cannot be understood without taking into account predispositions and their biological roots (p. 103).*

With its insistence on the individual and its friendliness to biology, *Crime and Human Nature* shook the discipline of criminology.

Gottfredson and Hirschi's *A General Theory of Crime* (1990), which advanced the idea that low self-control is the indispensable predictor of crime and violence, received a similar reception because it occupied a similar niche. Like Wilson and Herrnstein, Gottfredson and Hirschi are very accomplished and influential academics who, because they focused on individual-level factors to explain crime, went against the grain of the discipline. *A General Theory of Crime* has single-handedly dominated criminological journals as a source since 1990, racking up more than 3,800 citations by early 2011. By comparison, Hirschi's other classic work, *Causes of Delinquency*, published in 1969, had approximately 4500 citations by early 2011.

Terrie Moffitt's developmental taxonomy theory first appeared as an article published in *Psychological Review* in 1993. It has been massively influential, as evidenced by the more than 3500 citations of this work by early 2011. Her theory advances two offender prototypes. The first is a normative group (adolescence-limited) of offenders who tend to engage in low levels of crime consisting of benign offenses committed during adolescence. Their antisocial behavior is the result of the growing pains experienced during the uneven and often tumultuous transitions from childhood to adulthood. The second is a pathological group (life-course persistent) of offenders whose antisocial behavior is chronic, frequent, serious, and violent. Their antisocial behavior is the product of interactions between neuropsychological deficits and adverse home environments. Biological factors—specifically neurological and genetic factors—are also implicated in the behavior of life-course persistent offenders. Perhaps the scholar who conducts the most purely scientific criminological research, Moffitt has advanced understanding of the interaction between nature and nurture in explaining crime and violence (Caspi & Moffitt, 2006; Moffitt, 2005; Moffitt, Caspi, & Rutter, 2005).

CURRENT FOCUS

This chapter explores three general contributions of the works described in the preceding section. First, as a result of the popularity, appeal, and empirical strength of these researchers, there has been a renewed emphasis on the individual as the correct locus or place to explain criminal and violent behavior. Second, there has been increased recognition of criminality as an important way—perhaps *the* important way—to understand violent criminal behavior. Third, there has been an increased appreciation for the diversity of criminals from an array of disciplinary perspectives. As a consequence of these new reformulations, criminology has become a more believable and scientifically confident discipline because it disproportionately focuses scholarly attention on the most serious offenders—namely, the violent offenders who pose the greatest threat to society.

REDUCTIONISM AND VIOLENCE

Individuals—not communities, regions, nations, or cultures—commit murder, rape, robbery, and assault. Thus, to understand violent offenders, it is imperative to understand individual-level factors, such as personality, temperament, self-control, temper, or psychopathy, that influence behavior. Boiling scientific explanations down to a more fundamental unit of analysis is a process known as reductionism. To elucidate the motivations and behaviors of violent offenders, then, it becomes necessary to understand the characteristics that appear to typify violent offenders, not just the statuses they might occupy. Of course, environmental or sociological factors also matter, but they have indirect effects on behavior because they are processed by individuals. This is an important distinction. Many phenomena presumed to correlate with crime and violent behavior, such as poverty, employment status, neighborhood of residency, and ethnicity, really owe their "relationship" to crime to individual-level factors. As Gottfredson and Hirschi (1990) suggested, "[it] is hard to overstate the magnitude of this problem in criminology because of the tendency of people with low self-control to avoid attachment to or involvement in all social institutions—a tendency that produces a negative correlation between institutional experience and delinquency. This gives all institutions credit for negative effects on crime, credit they may not deserve" (pp. 167-168).

An interest in individuals as the unit of analysis to explain crime was a major contribution of Wilson and Herrnstein. The very title of their work, *Crime and Human Nature*, speaks to the pressing need to acknowledge constitutional features of people that powerfully affect how they choose to behave. According to Wilson and Herrnstein (1985), "whatever factors contribute to crime . . . must all affect the behavior of individuals, if they are to affect crime. If people differ in their tendency to commit crime, we must express those differences in terms of how some array of factors affects their individual decisions" (p. 42).

What are the characteristics of violent individuals? Numerous profiles emerge from the criminological literature. Even as children and adolescents, violent offenders are noteworthy for their callousness and emotional traits that reflect low empathetic concern for others, little guilt or anxiety over misdeeds, and limited emotional range (Frick & Hare, 2002). Kathleen Heide (1995) studied adolescent patricide offenders (children who murder their parents) and found that many had been severely abused, some were severely mentally disturbed or psychotic, and some were dangerously antisocial or sociopathic. Donald Lynam and colleagues (2004) examined the case histories of violent offenders from a New Zealand birth cohort and found that they were suspicious, alienated, callous, cruel, unempathetic, and prone to overreact to stress. Violent offenders were also different in their attitudes toward constraint, which included levels of reflectivity, caution, ability to plan, and excitement seeking. Margo Watt and her colleagues (2000) reported that incarcerated violent offenders have low moral intelligence evidenced by low empathy for others, low self-control, self-centeredness, poor moral reasoning, and weak socialization. Finally, in a review of the predictors of recidivism, Wagdy Loza (2003) documented a laundry list of personality characteristics of violent offenders, such as incapacity to feel sympathy, inability to learn from experience, narcissistic traits, inner rage, paranoia, external locus of control, self-indulgence, and a globally irresponsible approach to life.

Although this is but a small sampling of perspectives, criminologists have copiously documented the individual-level characteristics of violent offenders and core personality traits that appear repeatedly in the literature. First and foremost, violent offenders are often unmoved by normal human emotions and exhibit little connectedness to others. Their lack of empathy facilitates their ability to harm others. As a quick illustration, children in the Pittsburgh Youth Study who were evaluated as lacking a sense of guilt about antisocial behavior were between 480% and 630% more likely than normal children to present with multiple behavioral problems (Loeber et al., 2002). A smattering of other traits, such as impulsiveness, irresponsibility, external locus of control, and narcissism, similarly compromise the ability of violent offenders to fit in well with others in traditional social settings. Moreover, these constitutional factors are present throughout the life-course, even in early childhood. When criminogenic traits become readily apparent early in life, and when they accompany violent or antisocial behaviors, it is likely that their etiology stems from the offenders themselves.

CRIMINALITY: THE MARK OF CAIN

The following scale items are used to measure psychopathic traits in children:

- Does not show feelings or emotions
- Helpful if someone is hurt, upset, or feeling ill
- Feels bad or guilty when he or she does something wrong
- Has at least one good friend
- Considerate of other people's feelings
- Kind to younger children
- Is concerned about how well he or she does at school
- Often has temper tantrums or a hot temper
- Generally obedient, usually does what adults request
- Often fights with other children or bullies them

- Often lies or cheats
- Steals from home, school, or elsewhere

A study conducted by Essi Viding, James Blair, Terrie Moffitt, and Robert Plomin (2005) found significant genetic risk for psychopathy in 7-year-olds. The data collected by these researchers are staggering. Among children with psychopathic tendencies, 81% of extreme antisocial behavior was explained by genetics, and 0% was explained by shared environmental factors. Among identical twins, nearly 70% of extreme callous-unemotional traits were explained by genes (Viding et al., 2005). The findings offer two frightening implications. First, extremely violent behavior develops in some people when they are very young children; second, it strikes from within.

These conclusions speak to the notion of criminality as a negative form of personality. Personality is the relatively enduring, distinctive, integrated, and functional set of psychological characteristics that result from people's temperaments interacting with their cultural and developmental experiences (Walsh & Ellis, 2007). Criminality, which can be understood as the raw material or potential to engage in crime and violence that every person has within, ranges along a continuum from very low to very high. Criminality is not normally distributed; most people have low criminality, whereas a small number of people have very high criminality. Criminality is a global characteristic in that it affects more than just criminal behavior. Its elemental components, for example, also affect school performance, sociability, work performance, and "bad habits," such as smoking, gambling, drinking, and other risky activities. These are consequences of criminality, but what constitutes it?

Arguably the most popular example of criminality is the low self-control construct developed by Gottfredson and Hirschi (1990), who suggest:

> *[P]eople who lack self-control will tend to be impulsive, insensitive, physical (as opposed to mental), risk-taking, short-sighted, and nonverbal, and they will tend therefore to engage in criminal, and analogous acts. Since these traits can be identified prior to the age of responsibility for crime, since there is considerable tendency for these traits to come together in the same people, and since the traits tend to persist through life, it seems reasonable to consider them as comprising a stable construct useful in the explanation of crime (pp. 90-91).*

In other words, theories of criminality attempt to identify the characteristics and behaviors that describe the "types of people" most likely to commit crime and violence.

For example, temper is the dimension that has the most obvious connection to violence and is useful in understanding how core components of a person's being influence that individual's behavior. While everyone has a temper to some degree, most people are able to stifle their tempers and muffle their impulses or desires to respond negatively or aggressively to another person. Most people stifle not only the physical urges but also the verbal urges that are aroused by our tempers. Instead of violently attacking someone who has made us angry, we turn the other cheek, ignore, or otherwise defuse the situation. Individuals with hair-trigger tempers, such as violent offenders, do not behave this way, however. They are unable and unwilling to check themselves, and as a result, even seemingly trivial affronts or confrontations are grounds for an argument, fight, or worse. Everyone has known a person with a bad temper. And everyone knows that such a person is significantly more likely to get into trouble because of a single, perhaps innate, personality trait (Beaver & Wright, 2005; Wright & Beaver, 2005). This is the logic of self-control theory.

The self-control approach to criminality has enjoyed tremendous research support (Gottfredson, 2006). Alex Piquero and his colleagues (2007) examined the link between self-control and criminal careers using a large sample of youths from a New Zealand birth cohort. They found that those youths with the lowest self-control were also the most frequent, serious, and persistent criminals. Using data from a Dutch sample of 1531 persons involved in traffic accidents, Marianne Junger and her colleagues (2001) found that persons who displayed risky traffic behavior (a component of low self-control) that resulted in an accident were also likely to engage in various forms of criminal behavior. Specifically, they were 160% more likely to have an arrest for a violent crime, 150% more likely to have an arrest for property crime, and 430% more likely to have an arrest for additional traffic crimes (Junger et al., 2001). Using a sample of extreme offenders with a minimum of 30 arrests, Matt DeLisi (2001) found that offenders with low self-control are likely to accumulate many arrests for the most serious forms of violence. In addition, poorly controlled offenders continue to engage in violence while incarcerated and continually run afoul of the criminal justice system (DeLisi, 2001a, 2001b; DeLisi & Berg, 2006). Irrespective of the context, low self-control increases the likelihood for trouble, violence, and other deleterious outcomes.

Although self-control is a compelling basis for understanding criminality, it lacks the scientific punch of Terrie Moffitt's description of the processes that give rise to the most serious, violent offenders (known as life-course persistent offenders, and discussed later in this chapter). Briefly, according to Moffitt, two types of neuropsychological defects—those affecting the verbal and executive functions—engender an assortment of antisocial behaviors. Verbal functions include reading ability, receptive listening, problem-solving skill, memory, speech articulation, and writing; in short, they constitute humans' verbal intelligence. Executive functions relate to behavioral and personality characteristics, such as inattention, hyperactivity, and impulsivity. Children with neuropsychological deficits in these areas are restless, fidgety, destructive, and noncompliant, and they can be violent. They are also exceedingly difficult to parent. Children with neuropsychological deficits coupled with adverse home environments are especially at risk (Moffitt, 1990). Indeed, neurodevelopmental and family risk factors at age 3 predict life-course persistent offending in adulthood (Moffitt & Caspi, 2001). Whereas Gottfredson and Hirschi invoke sociology to describe the etiology of crime and violence, Moffitt uses a multidisciplinary perspective that spans psychology and neuroscience.

Irrespective of one's theoretical orientation, criminology has only recently paid close attention to the criminality that likely distinguishes mundane from deeply problematic offenders. The results of such research are leading to new insights into the nature of criminality.

THE DIVERSITY OF VIOLENT CRIMINALS

Visualize the following two offenders: Offender 1 raped and murdered an adult female who was a stranger; offender 2 raped and murdered his own mother. It is more than the victimology in these cases that is different. While the first offender is obviously a violent person, he pales in comparison to the second offender. The second offender's conduct is considerably more reprehensible, taboo, and repellent than the first offender's conduct, because it violates social relationships between offender and victim that do not exist in the first case. Given these differences, it is likely that different theoretical explanations would be used to create a profile of the two offenders. Again, while one is simply violent, the other is not only violent but also draped in psychopathology.

Similarly, criminal justice practitioners can appreciate the diversity of violent offenders. For instance, cases involving domestic assaulters, bar fighters, and gang bangers often have a different "feel" because these offenders' violence is contextually different or even contextually driven. Whether a murderer kills by firearm, knife, or manual strangulation matters, because there are likely compelling differences in the offender's psychology to lend insight into his or her behavior.

A psychological phenomenon that has historically been helpful in understanding violent offenders is psychopathy. The "modern" understanding of psychopathy was realized in 1941 with the publication of Hervey Cleckley's *The Mask of Sanity*. Cleckley's work was the most systematic clinical study of psychopathy of its day, and it laid the groundwork for contemporary research. His theory is helpful in discerning subtle but important differences among even the most violent offenders:

> *In repetitive delinquent behavior, the subject often seems to be going a certain distance along the course that a full psychopath follows to the end. In the less severe*

disorder, antisocial or self-defeating activities are frequently more circumscribed and may stand out against a larger background of successful adaptation. The borderlines between chronic delinquency and what we have called the psychopath merge in this area. Although anxiety, remorse, shame, and other consciously painful subjective responses to undesirable consequences are deficient in both as compared with the normal, this callousness or apathy is far deeper in the psychopath (Cleckley, 1941, p. 268).

Focusing on the individual-level characteristics of the most serious criminals has also enabled criminologists to make strides in understanding crime from general and more specific theoretical perspectives. As the examples cited earlier indicate, serious criminals are not a monolithic group. While great commonalities often exist in the lives of serious offenders, there are also important distinctions between them. In some areas of research, these distinctions necessitate typologies with specific offender types. In a way, Wilson and Herrnstein, Gottfredson and Hirschi, and Moffitt established somewhat of a theoretical paradox, in which similarities and appreciable differences among violent offenders beg for parsimonious general theories that are flexible enough to explain the great diversity of violence. What do the violent criminals look like descriptively and why are they meaningful to criminology? We explore these questions next.

Because murder and rape are the most serious forms of violence, much scholarly attention has been devoted toward the offender and offense characteristics of murderers and various types of sexual offenders. Easily the most serious example of violence is sexual homicide, which occurs in less than 1% of homicides annually and is the primary motivation of serial killers. Sexual homicides are classified as two general types: organized and disorganized. According to Robert Ressler, Ann Burgess, and John Douglas (1988), organized sexual homicides are planned, target strangers, and reflect control. The perpetrator demands submission, uses restraints, is aggressive before the murder, does not use a weapon, and transports and hides the body. Disorganized sexual homicides involve spontaneous attacks on a known but depersonalized victim. These attacks are random and sloppy, involve postmortem sexual acts, have weapons at the scene, and leave the victim at the death scene (Ressler et al., 1988).

J. Reid Meloy (2000) developed a clinical typology that builds upon the organized/disorganized distinction. According to Meloy, organized killers are compulsive sexual sadists with antisocial and narcissistic personality disorders. They are chronically emotionally detached, often primary psychopaths, are autonomically hyporeactive (e.g., have a low resting heart rate), and most likely did not experience early trauma or abuse in childhood. In contrast, disorganized killers are described as catathymic, and perpetrators have mood disorders and avoidant traits. They desire attachment, are moderately psychopathic, are autonomically hyperactive, and have histories of physical and/or sexual trauma (Meloy, 2000).

How dangerous are these characteristics? A study by Jill Levenson and John Morin (2006) examined the factors that predicted selection of sexually violent predators for civil commitment in Florida (civil commitment is the involuntary confinement of violent criminals in treatment facilities following their completion of a prison sentence). Persons with a diagnosis for pedophilia were 4656% more likely to be civilly committed than those without the diagnosis. For sexual sadists, there was a staggering 85,562% increased likelihood of civil commitment. For those diagnosed with paraphilias not otherwise specified, the liability was 10,580% (Levenson & Morin, 2006).

In terms of psychopathology and risk for violence, persons who commit sexual homicides are the most dangerous and disturbed offenders in society; however, they are merely one type of sexual offender. Other types of sexual offenders are distinguished primarily by the characteristics of their victims, such as rapists and pedophiles. Importantly, all criminals are prone to be versatile in their offending behavior, and most do not specialize in narrow or discrete criminal behaviors. What is clear is that persons who commit the most serious crimes are at risk to commit the "other" serious crimes. For example, Brian Francis and Keith Soothill (2000) conducted a 21-year retrospective study of more than 7400 sex offenders in the United Kingdom and found that approximately 3% of sex offenders ultimately commit murder. While this risk seems low, the rate of murder among sex offenders was one homicide for every 400 sex offenders compared to a general population ratio of one homicide per 3000 individuals (Francis & Soothill, 2000). Using a large database of offenders in Illinois, Lisa Sample (2006) found that nearly 8% of child molesters ultimately commit murder. Leonore Simon (2000) compared 142 child molesters, 51 rapists, and 290 other violent offenders in terms of their offending characteristics. Regardless of an offender's current conviction status, sex offenders in her study were found to commit a range of other criminal activity, including violent, property, drug, white-collar, nuisance, and noncompliance offenses (Simon, 2000).

Of course, generalized criminal involvement is not limited to sex offenders or murderers. Even groups that are considered to be comparatively innocuous offenders, such as nonviolent drug offenders, include individuals who behave violently. A recent study found that inmates currently incarcerated for nonviolent drug possession or trafficking charges were also highly likely to be arrested at some point for serious violent crimes, such as murder, rape, robbery, aggravated assault, or kidnapping (DeLisi, 2003b).

Students introduced to criminology often gravitate to the most extreme types of offenders, and there is often the misperception that offender types are just that—narrow categories of offenders who commit single offenses. Empirically, this hypothesis does not hold. All criminals dabble in an assortment of antisocial behaviors. The same applies to those who commit the most serious crimes—they just happen to murder, rape, kidnap, and molest in conjunction with committing robbery, burglary, auto theft, fraud, and drug offenses.

CONCLUSION

Why are violent offenders meaningful to criminology? We offer a handful of broad answers to this question.

First, violent offenders pose a great public health threat to societies in terms of mortality, victimization, lost and diminished productivity, fear of crime, lower quality of life, and untold costs required to manage and supervise them in the criminal justice system. With all apologies to corporate crime, which results in far greater financial losses, violent crime is a direct affront to the social and moral order of society. For this reason, violent offenders are those whom the general public entrusts criminal justice practitioners to apprehend, prosecute, and detain.

Second, violent offenders have forced criminologists to sharpen their theoretical ideas. No longer can banal social processes be cited to explain serious interpersonal violence (although the media still make an attempt to do so). It is doubtful that offenders would disarm and kill police officers, or construct a dungeon to facilitate the kidnapping and rape of runaways, or murder 48 people because of the unemployment rate, general strain, or poverty. These issues might matter, but they are small pieces of the puzzle of violent behavior. Also, the need to understand violence is much broader now, borrowing concepts from across the social and behavioral sciences, medical sciences, and beyond.

Third, the landmark works described in this chapter have afforded criminology greater credibility and authenticity. Not too long ago, it was seen as a dubious proposition to suggest that personality traits were related to crime, that family processes were not just risk factors but causes of crime, and that criminality not only exists but is stable, contributes to heinous forms of violence, and is poorly mollified by punishment. Now that the theories advanced by Wilson and Herrnstein, Gottfredson and Hirschi, and Moffitt have come to the fore, criminology confidently acknowledges that individual-level factors are paramount and recognizes that the pathology of the most violent and serious offenders is largely attributable to biological factors and the complex interplay between nature and nurture.

Finally, violent offenders hold the most promise for criminology to achieve its most noteworthy scientific goals, including a specific account of criminality. Increasingly, the sociological processes of criminological theories will be recast as biosocial; in fact, this is already being done. To illustrate, Kevin Beaver and his colleagues (2007) recently extended Gottfredson and Hirschi's work and provided evidence showing that self-control should be viewed as an executive function that is housed in the prefrontal cortex of the brain. Using data from Early Childhood Longitudinal Study—Kindergarten Class of 1998–1999 (the largest nationally representative sample of children), Beaver and his colleagues found that deficits in neuropsychological functioning were related to levels of self-control in kindergarten and first-grade students. These results held for both genders even after partitioning out the effects of parental and neighborhood influences, and even after controlling for prior levels of low self-control. Contrary to self-control theory, they revealed that most of the parenting measures had relatively small and inconsistent effects on self-control. Overall, the neuropsychological measures were among the most consistent predictors of childhood levels of self-control (Beaver et al., 2007). Over time, many similar advances will occur in criminology, and a scientifically curious eye on the most antisocial and violent offenders will lead the way.

REFERENCES

Beaver, K. M., & Wright, J. P. (2005). Evaluating the effects of birth complications on low self-control in a sample of twins. *International Journal of Offender Therapy and Comparative Criminology, 49*, 450–471.

Beaver, K. M., Wright, J. P., & DeLisi, M. (2007). Self-control as an executive function: Reformulating Gottfredson and Hirschi's parental socialization thesis. *Criminal Justice and Behavior, 33*, 367–391.

Caspi, A., & Moffitt, T. E. (2006). Gene–environment interactions in psychiatry: Joining forces with neuroscience. *Nature Reviews Neuroscience*, 7, 583–590.

Cleckley, H. (1941). *The mask of sanity.* St. Louis: C. V. Mosby.

DeLisi, M. (2001a). Designed to fail: Self-control and involvement in the criminal justice system. *American Journal of Criminal Justice, 26*, 131–148.

DeLisi, M. (2001b). It's all in the record: Assessing self-control theory with an offender sample. *Criminal Justice Review, 26*, 1–16.

DeLisi, M. (2003a). Conservatism and common sense: The criminological career of James Q. Wilson. *Justice Quarterly, 20*, 661–674.

DeLisi, M. (2003b). The imprisoned nonviolent drug offender: Specialized martyr or versatile career criminal? *American Journal of Criminal Justice, 27*, 167–182.

DeLisi, M. (2005). *Career criminals in society.* Thousand Oaks, CA: Sage.

DeLisi, M., & Berg, M. T. (2006). Exploring theoretical linkages between self-control theory and criminal justice processing. *Journal of Criminal Justice, 34*, 153–163.

Francis, B., & Soothill, K. (2000). Does sex offending lead to homicide? *Journal of Forensic Psychiatry, 11*, 49–61.

Frick, P. J., & Hare, R. D. (2002). *The antisocial process screening device.* Toronto, Canada: Multi-Health Systems.

Gottfredson, M. R. (2006). The empirical status of control theory in criminology. In F. T. Cullen, J. P. Wright, & K. R. Blevins (Eds.), *Taking stock: The status of criminological theory. Advances in criminological theory* (Vol. 15, pp. 77–100). New Brunswick, NJ: Transaction.

Gottfredson, M. R., & Hirschi, T. (1990). *A general theory of crime.* Stanford, CA: Stanford University Press.

Heide, K. M. (1995). Dangerously antisocial youths who kill their parents. *Journal of Police and Criminal Psychology, 10,* 10–14.

Junger, M., West, R., & Timman, R. (2001). Crime and risky behavior in traffic: An example of cross-situational consistency. *Journal of Research in Crime and Delinquency, 38,* 439–459.

Levenson, J. S., & Morin, J. W. (2006). Factors predicting selection of sexually violent predators for civil commitment. *International Journal of Offender Therapy and Comparative Criminology, 50,* 609–629.

Loeber, R., Farrington, D. P., Stouthamer-Loeber, M., et al. (2002). Male mental health problems, psychopathy, and personality traits: Key findings from the first 14 years of the Pittsburgh Youth Study. *Clinical Child and Family Psychology Review, 4,* 273–297.

Lombroso, C. (1911). *Crime: Its causes and remedies.* Boston: Little, Brown.

Loza, W. (2003). Predicting violent and nonviolent recidivism of incarcerated male offenders. *Aggression and Violent Behavior, 8,* 175–203.

Lynam, D. R., Piquero, A. R., & Moffitt, T. E. (2004). Specialization and the propensity to violence. *Journal of Contemporary Criminal Justice, 20,* 215–228.

Meloy, J. R. (2000). The nature and dynamics of sexual homicide: An integrative review. *Aggression and Violent Behavior, 5,* 1–22.

Moffitt, T. E. (1990). Juvenile delinquency and attention deficit disorder: Boys' development trajectories from age 3 to age 15. *Child Development, 61,* 893–910.

Moffitt, T. E. (1993). Adolescence-limited and life-course persistent antisocial behavior: A developmental taxonomy. *Psychology Review, 100,* 674–701.

Moffitt, T. E. (2005). The new look of behavioral genetics in developmental psychopathology. *Psychology Bulletin, 131,* 533–554.

Moffitt, T. E., & Caspi, A. (2001). Childhood predictors differentiate life-course persistent and adolescence-limited pathways among males and females. *Developmental Psychopathology, 13,* 355–375.

Moffitt, T. E., Caspi, A., & Rutter, M. (2005). Strategy for investigating interactions between measured genes and measured environments. *Archives of General Psychiatry, 62,* 473–481.

Piquero, A. R., Moffitt, T. E., & Wright, B. E. (2007). Self-control and criminal career dimensions. *Journal of Contemporary Criminal Justice, 23,* 1–18.

Ressler, R. K., Burgess, A. W., & Douglas, J. E. (1988). *Sexual homicide: Patterns and motives.* Lexington, MA: D. C. Heath.

Sample, L. L. (2006). An examination of the degree to which sex offenders kill. *Criminal Justice Review, 31,* 230–250.

Simon, L. M. J. (2000). An examination of the assumptions of specialization, mental disorders, and dangerousness in sex offenders. *Behavioral Sciences & the Law, 18,* 275–308.

Viding, E., Blair, R. J. R., Moffitt, T. E., & Plomin, R. (2005). Evidence for substantial genetic risk for psychopathy in 7-year-olds. *Journal of Child Psychology and Psychiatry, 46,* 592–597.

Walsh, A., & Ellis, L. (2007). *Criminology: An interdisciplinary approach.* Thousand Oaks, CA: Sage.

Watt, M. C., Frausin, S., Dixon, J., & Nimmo, S. (2000). Moral intelligence in a sample of incarcerated females. *Criminal Justice Behavior, 27,* 330–355.

Wilson, J. Q., Herrnstein, R. J. (1985). *Crime and human nature: The definitive study of the causes of crime.* New York: Simon and Schuster.

Wright, J. P., & Beaver, K. M. (2005). Do parents matter in creating self-control in their children? A genetically informed test of Gottfredson and Hirschi's theory of low self-control. *Criminology, 43,* 1169–1202.

CHAPTER 2

The Overlap of Violent Offending and Violent Victimization: Assessing the Evidence and Explanations

Mark T. Berg
Indiana University

Individuals who perpetrate violence and those suffer from it share a similar demographic and social profile. In fact, victims and offenders are not always drawn from distinct groups; they are often one in the same. Numerous studies have revealed that one of the most reliable predictors of violent victimization is violent offending. Nearly five decades of research conducted on self-report and official records of lethal and nonlethal violence using cross-sectional as well as longitudinal designs has persistently observed a strong correlation between offending and victimization. This correlation, known as the victim–offender overlap, is one of the most durable empirical findings in the criminological literature, on par with other prominent findings in the field.

While victimization and offending are oftentimes examined as two separate domains, observers suggest that both are so intimately intertwined that perhaps it is not possible to fully understand them apart from one another. Although an increasing amount of empirical attention is directed at unpacking the complex etiology of this phenomenon, there is only an embryonic understanding of why it comes about. In fact, some scholars claim that more is known about the factors that *do not* explain the victim–offender overlap than about those that might be responsible for its nascence (Esbensen & Huizinga, 1991; Lauritsen & Laub, 2007; Lauritsen, Sampson, & Laub, 1991).

This chapter provides a comprehensive overview of past and ongoing research on the victim–offender overlap. It begins with a description of leading explanations for this phenomenon, along with the relevant body of empirical work. This description is followed by a discussion of the implications of the victim–offender overlap for criminological research and violence prevention policy. Finally, the chapter highlights several directions for future research on the topic.

EARLY RESEARCH ON THE VICTIM–OFFENDER OVERLAP

Classic studies conducted in the early 20th century, such as Addams' (1909) *The Spirit of Youth and the City Streets* and Shaw's (1930) *The Jack Roller*, implicitly portray victims as passive participants in interpersonal violence who are subject to the desires of persistent criminals. The Gluecks were perhaps the first scholars to insinuate a connection between adverse life events such as victimization and one's involvement in offending. Many of their research subjects allegedly led an "uncontrolled street life" and had "early contact with undesirable and dangerous companions" (Glueck & Glueck, 1940, p. 12). By the Gluecks' calculations, approximately 13% of their sample suffered from serious physical ailments, and a small percentage died prematurely from illness, accidents, and other causes. For the most part, the Gluecks' research was descriptive in nature and did not attempt to forge a conceptual link between offending and victimization.

Hans von Hentig's textbook was the first major criminological publication to explicitly recognize and attempt to theorize this phenomenon. Von Hentig argued that certain classes of victims are passive recipients of violence, while another class of victims contributes dynamically to their own misfortunes. It was this latter class of victims that many criminologists and legal professionals had overlooked. In von Hentig's words, by the law's criterion, "Perpetrator and victim are distinguished . . . [But] it may happen that the two distinct categories merge . . . and in the long chain of causative forces the victim assumes the role of a determinant" (von Hentig, 1948). Von Hentig speculated that the overlap among victims and offenders was influenced by the fact that would-be perpetrators seek out victims who are also involved in crime due to their reticence to contact the police. He also believed that some victims actively participate in their own misfortune by tormenting offenders to the point that they incite violence by the aggressor. In sum, von Hentig recognized what much research has since demonstrated: that violent victimization and violent offending are closely connected.

To be clear, von Hentig never offered a formal theory of the victim–offender overlap or even a loosely defined model with testable propositions. Even so, he prefigured the significance of victimization research for criminological knowledge by recognizing that victims and offenders are not always distinct groups.

EXPLAINING THE VICTIM–OFFENDER OVERLAP

Routine Activities/Lifestyle Perspectives

An explicit theory of the victim–offender overlap does not currently exist. Since the publication of von Hentig's textbook some 60 years ago, several leading criminological theories have provided the frameworks from which criminologists have developed hypotheses about the mechanisms underlying the connection between offending and victimization. Among these, lifestyles/routine activities theory has served as a basis for more elaborate conceptual discussions about the etiology of the phenomenon. Hindelang, Gottfredson, and Garofalo's (1978) landmark study of victimization patterns in the National Crime Survey (NCS) set in motion the adoption of a novel theoretical orientation toward explaining victimization risk, but it also has important theoretical implications for knowledge of the victim–offender overlap.

Their exhaustive study of the NCS data indicated to Hindelang and colleagues (1978, p. 251) that victimization risk was not uniformly distributed throughout the population. Certain groups were disproportionately victimized, they found. Moreover, the demographic profile of victims mirrored that of offenders: Both groups tended to be disproportionately made up of young, unemployed, unmarried minority males. Based on their findings, these authors developed an inductive theoretical model organized around the notion of "homogamous interaction." A core premise of this framework suggests that demographic variation in victimization risk is attributable to differences in "lifestyles," a concept comprising routine vocational and leisure activities. In essence, certain lifestyles are seen as being more likely to expose people to situations and places that are conducive to victimization. Moreover, a prevailing interpretation of Hindelang and colleagues' model argues that victims and offenders have a similar sociodemographic profile due to their "homogamous" lifestyles; therefore, the relationship between offending and victimization is explained by factors such as age, leisure activities, and proximity to crime (Cohen, Kluegel, & Land, 1981; Sampson & Lauritsen, 1990).

A related stream of research initiated by Gottfredson (1981) fused the logic of social control theory with Hindelang and colleagues' routine activities/lifestyles model to explain the overlap among victims and offenders. Social control and routine activities/lifestyles theories are compatible in that both assume motivation is a constant, whereas what requires explanation are the conditions inhibiting behavioral expressions (Kornhauser, 1978). To reiterate, lifestyles/routine activities theory proposes that the conditions related to offending are also associated with victimization. Interpreting this notion through the framework of control theory, Gottfredson posited that the same social controls that thwart offending also minimize exposure to motivated offenders, thereby reducing an individual's risk for victimization. Specifically, Gottfredson (1984) proposed that weak social controls produced the pools of offenders and victims as well as "the circumstances that they are likely to come into contact with one another" (p. 726). As Mayhew and Elliot (1990) noted, "underlying common denominators such as being young, male, resident in an urban area and having an active social life increase both exposure to victimization and the likelihood of getting into trouble" (p. 92). According to this formulation, offending and victimization are mutual by-products of shared social conditions and, therefore, the relationship between them is likely spurious.

Gottfredson's analysis of the British Crime Survey showed that victims of crime not only offended at a relatively high rate, but were also more than twice as likely as nonvictims to be involved in accidents. According to Gottfredson (1984), a person's involvement in accidents is in no way causally related to whether he or she is victimized or commits crime; rather, victimization, offending, and traffic accidents are symptoms of ineffective social control mechanisms and other common social conditions. His formulation challenged established notions that victimization and offending have a genuine causal influence on each other.

Other theoretical models make a similar set of assumptions about the etiology of the victim–offender overlap. For example, Osgood and colleagues (1996) articulated a situational theory that conceptualized violent offending as an *outcome* of routine activities, rather than as a *form* of routine activity as it was commonly perceived. Some individuals are more frequently exposed to situations with stronger inducements to commit crime, where a deviant act is easier and more rewarding for them to carry out. Moreover, according to Osgood et al. (1996), situations that are especially conducive to deviance are more common during times spent with peers (because peers enhance the symbolic value of certain potentially deviant actions and facilitate crime commission), in the absence of authority figures, and during times of unstructured socializing. Osgood and colleagues' reformulation of routine activities emphasized that victimization and offending were both products of similar lifestyles, including time-use patterns, deviant associations (which enhance exposure to situational inducements), and the presence of social controls. With regard to the etiology of the victim–offender overlap, their model *implicitly* assumes that routine activities both structure offending opportunities and contribute to victimization risk; as a consequence, controlling for time use, peer associates, and unstructured socializing should explain the link between offending and victimization.

Other researchers have criticized routine activities/lifestyles models that posit the victim–offender overlap is the product of homogamous interaction, ineffective social bonds, or situational inducements. For instance, Jensen and Brownfield (1986) suggested that existing research artificially separated victims and offenders as if they are two separate categories. Contrary to prevailing conceptual frameworks, they proposed that offending is not a product of routine activities or social controls, but rather is a *type of routine activity or lifestyle characteristic* that directly influences risk for victimization (Jensen & Brownfield, 1986). After the publication of Jensen and Brownfield's study, the conceptual classification of victims and offenders as distinct groups grew increasingly inappropriate in research on interpersonal violence. No longer was it the case that victimization and offending were viewed as opposite sides of the same crime coin (Esbensen & Huizinga, 1991).

Scholars have increasingly relied on a revised model of routine activities/lifestyles theory as an explanatory framework. For example, Sampson and Lauritsen examined the victimization–offending link based on a multilevel interpretation of routine activities/lifestyle theory. These authors offered a general hypothesis inspired by Jensen and Brownfield's research: Offending should predict risk for victimization, irrespective of demographic indicators; therefore, offending is a risky routine activity/lifestyle characteristic. According to Sampson and Lauritsen's (1990) analysis of the British Crime Survey data, offense activity had the strongest effect on victimization, and the relationship between the two factors was not, as earlier notions suggested, a product of homogamous social interaction. By their account, the relationship appeared to be causal.

Subsequent empirical research has demonstrated that the link between victimization and offending exists over time, despite controls for social control and social learning variables. In combination, the findings from this line of research have contradicted Gottfredson's claim that the link is a by-product of weakened social controls.

Individual Differences Perspective

An alternative conceptual explanation for the victimization–offending link has developed around the scientific notion of individual differences or population heterogeneity. The scientific notion of heterogeneity is based on the idea that characteristics such as genetic makeup, executive functioning, intelligence, and personality factors are distributed within the population at nonconstant levels (DeLisi, 2005). This perspective represents a stark contrast to other frameworks suggesting that victimization and offending are causally related due to subcultural systems, or that both are artifacts of homogenous interactions and routine activities. According to the population heterogeneity perspective, a diverse range of behaviors is governed by unobserved and observable time-stable traits that vary across people and account for within-person and between-person stability in behaviors (Nagin & Farrington, 1992). Any correlation that is observed between patterns of conduct such as drinking and fighting, or drug use and burglary, or victimization and violent offending, is spurious, according to this view, as all are manifestations of an underlying heterogeneous condition.

Sparks introduced the notion of individual differences to the study of victimization patterns after observing that a relatively small number of individuals in the National Crime Panel (NCP) data disproportionately suffered multiple victimizations (Sparks, 1982). Offering an explanation for this pattern, Sparks posited that the distribution of multiple-incident victims in the population is closely correlated with the distribution of "victim proneness"—a term noting the degree to which some people, by virtue of their personal characteristics, facilitate violence, are susceptible to predation, and can be exploited with impunity. More specifically, Spark's conceptualization holds that heterogeneity in proneness or risk heterogeneity is an immutable personal characteristic. Building on these ideas, Garofalo modified the original lifestyle model to incorporate the concept of heterogeneity as an exogenous determinant of victimization risk. He conceptualized heterogeneity as risk "that cannot be accounted for entirely at the sociological level of explanation"; in fact, he argued that "psychological or biological variables may be relevant" (Garofalo, 1987, p. 39).

Early conceptualizations of the risk heterogeneity perspective had close concordance with the logical premises of Gottfredson and Hirschi's general theory of crime, or self-control theory. These authors' model suggests that between-person differences in average levels of orientation toward the future, empathy, tolerance for frustration, preference for physical activity, and risk avoidance cluster together to form a common trait of "low self-control." While their model was not explicitly developed to explain variation in victimization risk, Gottfredson and Hirschi appeared to recognize its utility in this regard. Drawing on insights from Gottfredson's earlier work, the authors noted that "that victims and offenders tend to share all or nearly all social and personal characteristics. Indeed the correlation between self-reported offending and self-reported victimization is, by social science standards, very high" (Gottfredson & Hirschi, 1990, p. 17). In fact, victimization researchers argue that Gottfredson and Hirschi's theory would agree with the following notions: Because self-control predicts offending, and offenders and victims overlap on most characteristics, then self-control would predict victimization and, therefore, self-control confounds the victimization–offending link.

A small number of empirical studies have evaluated ideas about the role played by low self-control in the genesis of the victim–offender overlap. For example, Schreck (1999) discovered that low self-control predicted variation in property and violent victimization among a sample of Arizona students; moreover, low self-control accounted for a large percentage of the relationship between violent offending and victimization. Even so, the relationship remained positive and significant, meaning that low self-control was unsuccessful at fully explaining the victim–offender overlap. Later research by Schreck and colleagues (2006) also showed a direct effect of low self-control on victimization, net of routine activities, peer associations, and offending. Taken together, the existing body of research suggests low self-control is not a source of heterogeneity underpinning the link between victimization and offending. More recently, Berg and Loeber (2011) found, using panel data from the Pittsburgh Youth Study (PYS), that a proxy measure of impulsivity—a strong correlate of low self-control—had a negligible effect on the relationship between offending and violent victimization.

Within this vein of research, additional studies have examined the possibility that other important time-stable traits explain the relationship between offending and victimization. For example, Dobrin (2001) applied a case-control design that matched members of his sample of Maryland arrestees on observable characteristics and discovered that each arrest for serious violent crime was related to a sharp increase in the individual's odds of being murdered. Elsewhere, research indirectly focused on the victim–offender overlap has found that high-rate offenders are more likely than low-rate offenders to die prematurely of natural causes such as heart failure or cancer and of unnatural causes such as suicide and homicide (Laub & Valliant, 2000; Nieuwbeerta & Piquero, 2008). Studies have also found that, across time, high-rate offenders display a relatively elevated risk of accidental death—a finding that is surprisingly consistent with Gottfredson's findings showing the same people tend to be involved in nonlethal accidents, victimization, and violent offending.

The authors of the aforementioned studies have generally interpreted their findings through the lens of a heterogeneity perspective and insist that victimization, offending, accidents, and early death are products of a latent trait that renders some individuals especially risk prone. With few exceptions, these studies lack indicators of individual differences or do not implement methodological controls that might account for the influences of self-selection and omitted variable bias. Therefore, it remains unclear to what extent the victimization–offending link is spurious and attributable to time-stable heterogeneity and whether individuals actually sort themselves into "high-risk situations" as theorists claim. Because it is a fairly recent innovation, the trait-based perspective of the victimization–offending link has garnered only modest attention to date.

Cultural Perspectives

Explanations derived from subcultural models have an extended history in research on the victim–offender overlap, dating back to Wolfgang's (1958) study of patterns of homicide in Philadelphia, Pennsylvania. His analysis of 588 incident files showed that many killings were often preceded by relatively innocuous verbal and physical disagreements between the parties involved. More importantly, these findings were some of the earliest to illuminate the empirical correlation between offending and victimization. For example, Wolfgang reported that in many cases the victim was the first party in the events surrounding the homicide to use physical force against the subsequent killer, a situation he termed "victim precipitation." In fact, nearly 26% of victims precipitated their own death. Moreover, nearly two-thirds of homicide victims had arrest records, and half had committed crimes against persons. Wolfgang claimed to have reaffirmed von Hentig's earlier conclusion that in some incidents of violence, the victim assumes the role of determinant (Sparks, 1982). Based largely on these findings, Wolfgang proposed that the exceptional levels of violence among certain populations could be "symptoms of unconscious, destructive impulses laid bare in a subculture where toleration—if not encouragement—of violence is part of the normative structure." Moreover, he proposed that "a subculture of violence exists among a certain portion of the lower socioeconomic group—especially comprised of males and Negroes" (Wolfgang, 1958, pp. 329-330).

Wolfgang's reasoning implies that the distribution of nonviolent conflicts is similar across social classes, but that *vast* differences exist among groups with regard to the amount of conflicts that result in violence (Luckenbill & Doyle, 1989). By his logic, the disproportionate concentration of serious violence among members of the lower class is due to the way in which honor is differentially perceived and socially controlled by subcultural preferences. In fact, according to Wolfgang, the "quick resort to physical combat as a measure of daring, courage or defense of status appears to be a cultural expression, especially for lower socioeconomic class of males." Moreover, he insisted that upper- and middle-class suburban dwellers codify norms that transcend subcultural mores and "consider stimuli that evoke a combative response in the lower class as trivial" (Wolfgang, 1958, pp. 188-189). The ideas put forth by Wolfgang assume that lower-class, minority victims are prone to offend and that the same group is also likely to be victimized because its members operate in a social context where violence receives greater sanction. Actors who are exposed to this normative context are less likely to back down from challengers, more likely to instigate them, and prone to deploy retaliatory violence. Collectively, these ideas imply that exposure to norms favoring violence may explain why the same people often both commit violence and suffer from it.

Wolfgang's research set in motion systematic consideration of whether oppositional conduct norms contribute to the overlap among victims and offenders. For example, Singer drew upon the cultural aspects of Wolfgang's theorizing to explain the homogeneity of victim and offender populations, especially those involved in serious interpersonal violence. Singer argued that victims and offenders share similar responses to perceived situations of physical and psychological threat. He claimed that the social interaction among both groups suggests certain normative constraints wherein a violent outcome is dependent in part on the victim's reaction" (Singer, 1981, 1986). According to Singer, offender and victim populations are not distinct, but rather rotate positions within a web of subcultural relationships. Based on analysis of data from the Philadelphia Birth Cohort, he found that juvenile victims were nearly three times more likely to commit an assault as an adult than were nonvictims. Approximately 94% of juvenile gang members who were victims also engaged in serious adult offending.

Singer argued that the victimization–offending link was one of the hazards of lower-class life. Although his research did not employ a measure of norms or values, he inferred cultural effects based on his observations of a strong correlation between victimization and offending. Moreover, Singer explicitly stated that much of the victimization–offending link was a product of retaliation, and driven by violent conduct norms. He argued that in communities where moral sanctions are not imposed on those who violate the law, there is a high risk that victims may interpret their misfortunes as justification for vengeance; in contrast, in higher-class areas, people who are shot or stabbed do not deliver a retaliatory response because dominant cultural values dictate calling the police. In short, Singer asserted that offending is a causal predictor of victimization, perhaps due to cultural dynamics that sanction retaliatory violence.

A related body of ethnographic and theoretical research has sought to clarify the dynamic interplay between violent offending, victimization, and conduct norms. Implicit in this work is the notion that contextualized cultural processes constitute a backdrop for aggressive interaction, giving rise to homogenous victim and offender populations. Contemporary research shows that socioeconomic adversity, racial discrimination, and alienation from mainstream institutions foster the emergence of an alternative normative system in poor neighborhoods that redefines expectations about personal conduct in ways that are incompatible with conventional culture. Norms inhered in a modern "honor culture" place a premium on the maintenance of respect, lower the threshold of personal insult, define violations of self in an adversarial manner, and endorse violence as an appropriate means to regulate interpersonal disputes. In this environment, an individual's reputation often hinges on his or her ability to overcome adversaries with brute force (Cooney, 1998; Horowitz, 1983; Jacobs & Wright, 2006;

Sampson & Bean, 2006). Anderson refers to the honor culture he observed in poor neighborhoods as the "street code." By his account, young men will precipitate violent altercations to promote their street credibility, and many "crave respect to the point that they would risk their lives to attain and maintain it" (Anderson, 1999, p. 32).

Cultural systems that are organized around codes of honor often sanction retaliatory aggression as an appropriate response to an affront. An individual who has been disrespected is expected to use force (Felson, 1982; Horowitz, 1983). Status is assigned to those who do not allow others to easily exploit them; hence, victims who opt against retaliation may run the risk of imperiling their own reputations. Accordingly, mixed-method research on lethal violence has shown that disputants from distressed neighborhoods often believe they have little choice except to retaliate, even over relatively trivial transgressions (Kubrin & Weitzer, 2003).

Within these environments, violence serves as more than a method of status attainment: It also has a unique deterrent function. By adopting a tough and aggressive demeanor, individuals signal to others that they are not to be mistreated. According to Anderson, in the streets of poor neighborhoods it is imperative not to yield to challengers, because doing so conveys weakness, which ultimately enhances one's probability of future victimization. Affirming these notions, an informant in a recent study reasoned that even if the smallest affront is overlooked, others will "try to come at me, the same day, the next day because [they'll] think, 'Aw he's a punk . . . he can't handle it'" (Kubrin & Weitzer, 2003). Under these circumstances, occasional displays of aggression are "instrumental for marketing [one's] reputation as a badass" (Katz, 1988, pp. 183–184).

Similarly, studies of armed robbers in St. Louis indicated that offenders often rob other offenders, appraising such actions as justified on the ideological grounds that it is a "dog eat dog world." More than 60% of the participants in Wright and Decker's (1997) study reported being robbed; many intended to violently punish the perpetrator if they ever again crossed paths with him or her. Likewise, Katz found in his study that norms of retribution steered many armed robbers' decisions to seriously injure or kill their victims. As he noted, "a reputation for violence, perhaps sustained by irrational brutality against the robbery victim, could be valuable for offenders who are interested in not becoming their colleagues' secondary victims." The fact that a reputation for being an "immoral entrepreneur" is prized in the streets of urban areas speaks to the extent to which cultural processes fuel the victimization–offending link (Katz, 1988).

A small but growing body of quantitative research has begun to analyze whether violent conduct norms explain the relationship between offending and victimization as implied by cultural models. Two interrelated ideas have been suggested to explain how culture may operate in this regard. First, cultural mechanisms may confound the effect of offending on victimization. Second, these mechanisms may cause the effect to be more or less divergent. Regarding the former notion, a recent investigation found that youth who adhered to the "street code" were more likely to be victimized—especially those from distressed neighborhoods; however, adherence to the street code did not explain the positive relationship between violent offending and victimization (Stewart, Schreck, & Simons, 2006). With regard to the latter idea, Berg and Loeber discovered in multilevel panel data that the offending–victimization relationship was magnified in highly disadvantaged neighborhoods and significantly weaker in low-poverty areas. Insofar as neighborhood compositional characteristics capture cultural processes, results from Berg and Loeber's (2011) study support the notion that the existence of a culture of honor intensifies the victim–offender overlap. Nevertheless, it is plausible that their findings may reflect the causal effects of other aggregate mechanisms that also vary closely with other neighborhood characteristics, making it difficult to determine if cultural processes are, indeed, at work.

RESEARCH SUMMARY

With few exceptions, studies find that the relationship between victimization and offending is persistent, but its meaning is not entirely clear. The leading theoretical explanations for the victim–offender overlap, as described in the previous section, contain two organizing assumptions about the precise nature of the phenomenon. On the one hand, some explanations suggest the relationship is spurious. On the other hand, a competing viewpoint suggests that the relationship is causal, and that offending has a genuine influence on a person's future risk of being victimized (Lauritsen & Laub, 2007). These ideas are discussed next.

Spuriousness

A main theme found throughout the literature suggests that victimization and offending are mutual outcomes of common factors and, therefore, are unrelated. Nested within this idea are two perspectives. One suggests that the factors that produce a spurious link between victimization and offending are *time stable*, while the other perspective holds that these factors are *time varying*.

Regarding the former view, the evidence accumulated to date does not suggest that serious offending and victimization are unrelated outcomes of an unobserved or observed time-stable characteristic. As noted earlier, net of controls for observed sources of heterogeneity, including self-control, pubertal onset, mental health, and genetic makeup, and unobserved propensity, offending has a strong influence on victimization.

The other perspective regarding spuriousness implies that common factors underlying the victimization–offending link are not time stable but rather are reflected in time-varying social processes. In other words, this view indicates that variables thought to confound the victimization-offending correlation may vary within individuals over time. Generally, theories describe such variables as observable conditions. For example, the model articulated by Osgood and colleagues (1996) implies that changes in time use, peer associates, and unstructured socialization (i.e., routine activities) are related to changes in both offending and victimization. According to this view, changes in social bonds influence the probability of becoming a victim or perpetrating an offense. Likewise, strands of subcultural theory can be interpreted to suggest that changes in actors' beliefs regarding the appropriateness of violence are related to changes in risk for both victimization and offending. Other sources suggest that peer associates and substance use represent different types of time-varying etiology (Jensen & Brownfield, 1986; Sampson & Lauritsen, 1990).

Several investigations have examined whether changes in each of the social processes mentioned above explain the victimization and offending link, but all have found it to be resilient to a wide variety of theoretically relevant variables. For instance, Lauritsen and Quinet (1995) discovered that offending predicted risk for victimization across several waves of panel data from the National Youth Survey despite controls for peer associations and family bonds. Wittebrood and Nieuwbeerta (2000) found that among a sample of adults from the Netherlands, the link between victimization and offending was resilient to measures of attachment to children (i.e., having children) and marriage. Berg and Loeber's (2011) investigation using the Pittsburgh Youth Study was among the first to find that, even after purging model estimates of all unobserved time-stable heterogeneity, the effect of offending and victimization was strong and positive. While these researchers did not find evidence to support the notion that the effect is spurious, a supplementary set of analyses indicated that unobserved time-stable heterogeneity played a partial role in causing this effect to emerge. Using a longitudinal design, Schreck and colleagues' (2006) research suggested a positive link persists between offending and risk for victimization despite controls for low self-control and peer processes.

In short, theoretical notions suggesting that the victim–offender overlap is spurious as a result of either time-stable or time-varying conditions have not proven empirically valid so far.

Causal Connection

A second common theme found in the literature reviewed previously assumes that victimization and offending are causally related to each other. According to this view, violent conduct genuinely promotes a person's risk probability of suffering future victimization. Conversely, people who do not frequently perpetrate aggressive behavior will have a lower probability of being targeted for victimization. The fact that the correlation between victimization and offending is persistent in empirical research despite numerous theoretically relevant control measures tends to support notions suggesting the two roles are related in a causal fashion. According to some scholars, if victimization and offending are causally related, then victimization should be considered a genuine predictor of offending and not an unrelated by-product of risk heterogeneity.

The basic logic underlying a state-dependence explanation is that a person's prior behavior—that is, offending—fundamentally alters his or her probability of future victimization. An interpretation of the empirical literature suggests that this causal connection is partially a product of a state-dependent system. With regard to the victim–offender overlap, hypotheses consistent with this perspective maintain that offending has an authentic behavioral influence on subsequently experiencing victimization. Based on this formulation, retaliation would be a manifestation of a state-dependent cycle because it involves an initial incident of offending followed by a violent response from the aggrieved party. Some scholars have argued that the reluctance of many victims to contact the police or enlist neutral third-party support merely fuels the state-dependent cycle (Jacobs & Wright, 2006). For example, offenders are legally vulnerable; recognizing this fact, their victims may capitalize as what they see as a good opportunity to victimize them. Stated simply, offenders may be victimized with impunity.

Incidents of homicide involving respondents in the Pittsburgh Youth Study are illustrative of how retaliatory violence unfolds in a state-dependent process, whereby offenders become victims:

> **Case A:** "[Sample member A] was shot three times in the side while riding in a car in a parking lot of a bar . . . Offender thought [sample member A] was the man who robbed his home three years earlier; he told police he shot immediately because he thought [sample member A] was armed."
>
> **Case B:** "[Sample member B stole] about $100 and a pager from a man. Later in the day, the man and two others found him pumping gas at a . . . station on Blank Street . . . where they beat him and shot him three times in the back as he tried to crawl away."

If these examples are representative of nonlethal incidents, they would suggest that a portion of the offending–victimization correlation involves retributive violence, meaning the linkage is partially produced by a causal process.

Subcultural theory is perhaps most relevant to understanding the nature of the presumed causal connection, given that retaliation is underpinned by violent subcultural norms. While cultural models may help to understand these dynamics, there is reason to assume the causal relationship between offending and victimization is not just a product of vengeance, but also guided by meaningful changes in the nature of an offender's routine activities and lifestyle. For instance, studies have shown that as street criminals become increasingly successful (and unsuccessful) in their underworld pursuits, they tend to engage in riskier crimes involving potentially dangerous targets within threatening environments, which ultimately enhances their likelihood of being victimized (Wright & Decker, 1997).

Another way to understand the nature of the causal connection, in terms of a state-dependence viewpoint, is to adopt a more finite situational focus and understand the ways in which *conflicts* unfold to produce the victim–offender overlap. Some scholars argue that violent conduct norms dictate whether conflicts escalate to the point where participants use serious physical violence. For instance, Wolfgang's and Singer's research assumes subcultural conduct norms shape the unfolding of situational dynamics involving conflicts, and may prompt antagonists to resist with violence so as to save face or deter attacks down the road. Jacobs and Wright's research shows that offenders make quick decisions to kill threatening adversaries for the strategic reason of preventing future attacks. When serious disputes occur within the context of an honor culture, antagonists may evolve from victim to offender within a matter of seconds or minutes.

Once again, actual examples from cases of homicide involving respondents in the Pittsburgh Youth Study are illustrative of how offenders can quickly become victims in conflict situations, causing an overlap between victims and victimizers:

Case C: "[Sample member C] was pursuing an ongoing dispute with a rival . . . [sample member C] verbally provoked the [other disputant] into gunplay . . . [sample member C] fired but missed, and was then killed with a semi-automatic firearm."

Case D: "[Sample member D] and another man, attending a party . . . argued over the cover charge. Police thought [sample member D] shot and wounded the other man, who then shot [sample member D] in the chest and hand and killed him; the other man died of his wound two hours later."

Taken together, the line of research developed around notions of state dependence implies that the causal connection between offending and victimization is forged against a backdrop of subcultural norms and unfolds in a protracted sequence, but also in a more truncated fashion within the context of rapidly evolving violent incidents.

Mixed Model

An alternative perspective was recently formulated that combines premises from both the spurious and state-dependence perspectives. Known as a mixed model, this perspective assumes the victim–offender overlap reflects the summed or total effect of risk heterogeneity and state dependence. In other words, part of the overlap is considered spurious, while another part is actually causal.

According to Ousey and colleagues (2011), if risk heterogeneity is "peeled away" from the victim–offender overlap, this change would reveal the *true* causal effect of offending on victimization because the so-called true effect is masked by person-level traits. According to the mixed model, risk heterogeneity may cause the oft-cited positive relationship between offending and victimization; however, by purging any source of heterogeneity from model estimates, the actual effect becomes observable. Traditionally researchers observe a positive effect, but a negative effect is possible according to the mixed model. For individuals who are not prone to taking risks, who have careful foresight, and who tend not to act impulsively, their involvement in violent behavior may operate as a deterrent, thereby diminishing their likelihood of becoming someone else's victim. Put differently, a mixed perspective would suggest it is important to wash away spuriousness to isolate the precise nature of the victim–offender overlap.

IMPLICATIONS FOR RESEARCH AND PUBLIC POLICY

A small but mounting body of scholarly research suggests that a complete scientific explanation of the victim–offender overlap may contribute to the development of violence theories that are capable of explaining patterns of both victimization and offending. On several occasions scholars have discussed the possibility of developing a unified theoretical model on the grounds that the victim–offender overlap supports such a formulation. Among those who hold this view, offending and victimization are opposite sides of the same coin, meaning that both are derived from a common explanatory framework. Similarly, citing the parallels between predictors of victimization and offending, Reiss (1981) wrote, "any theory that assumes no overlap exists between populations of victims and offenders or that they are distinct types of persons *distorts* the empirical research [italics added]." But Reiss was especially critical of researchers who failed to explicitly recognize the extent to which offending and victimization had a shared etiology. He argued that "theories that divorce properties of victimizer and victim from behaving units . . . will be most consistent in explaining what has up to now been treated as an etiology of offending distinct from an etiology of victimization" (p. 711).

Currently, there is a trend in criminology toward adopting theories of violent offending to study victimization and theories of victimization to understand violent offending. While these theories have generally developed along separate dimensions, the successful cross-application of offending theories to explain patterns of victimization, and vice versa, suggests a comprehensive theory of violent encounters may be warranted on empirical grounds.

The objective of a general theory of violence is to explain violent conduct by specifying its etiology, including under which conditions such conduct is most probable and least probable. Nevertheless, an important hurdle to overcome in the debate over a unified theory of violence is whether offending and victimization are products of still other variable processes, or whether they exert a true generative influence on each other. If the latter assumption is empirically valid—that offending and victimization are causally related—it would mean they actually serve as predictors of each other and are not unrelated behavioral by-products of the same syndrome(s). Many scholars overlook the implications of these competing ideas. Stated plainly, the logical content of a unified violence theory hinges on the precise nature of the relationship between victimization and offending. A general theory of violence cannot assume, based on recent findings, that violent offending and victimization are entirely unrelated outcomes of a common set of social, situational, psychological, or contextual circumstances. Rather, the evidence suggests that scholarly efforts to understand the etiology of victimization must account for offending. As a result, a general theory of violent encounters is likely justifiable.

With regard to policy, some scholars have speculated that prevention programs designed to reduce offending may be effective at lowering the risk for victimization, and vice versa (Lauritsen & Laub, 2007). To reiterate a point made earlier, existing research on the victim–offender overlap suggests that violent victimization and violent offending share a similar etiology, although they may also have a significant influence on each other. Based on these findings, policy makers should not only target the underlying conditions that explain both offending and victimization (spurious conditions), but also design interventions that prevent conflicts from erupting into cycles of retaliatory violence (causal mechanisms).

On the one hand, policy programming should be focused on restricting individuals' interactions with violent peers, improving their psychological morbidity, and tempering their attitudes toward aggressive behavior. Ameliorating those conditions responsible for the overlap would ultimately translate into lower rates of both violent victimization and offending. On the other hand, to address the causal link between offending and victimization, policy makers should make available nonpartisan third parties who could act as intermediaries between disputants involved in ongoing conflicts. Some conflicts are prone to escalate into serious violence and become more protracted in the absence of legitimate third-party mediators. Police officers and respected members of the community may serve such a role. Unfortunately, in poor minority communities, police–citizen relations are often tenuous and characterized by a deep level of distrust (Rosenfeld, Jacobs, & Wright, 2003). Residents in disadvantaged neighborhoods may not question the authority of the law, but they remain highly suspicious of the means police officers use to enforce it. As a result, what is needed in these environments is a form of policing that does not rely on aggressive policies of arrest and interdiction. Thus the popular strategy of order maintenance policing adopted by many urban departments may represent a model with problematic repercussions for police–citizen relations.

On a practical level, serious attempts should be made to improve the legitimacy of the police and restore public confidence in the ability of law enforcement officials to control crime, especially in historically disenfranchised impoverished neighborhoods. A specific goal of such an effort should be to encourage disputants to enlist police services in mediating their disagreements. The fact that victimization and offending do not influence each other in the least disadvantaged neighborhoods, however, suggests that violence reduction policies should be context specific. For instance, efforts to interfere in conflicts for the purpose of reducing retributive violence may be necessary only in highly disadvantaged environments.

To thwart incidents of retaliatory violence, policy makers must also encourage local well-known and respected citizens to assume the task of serving as intermediaries between disputing parties. Several nonprofit organizations and city leadership groups have developed such policies in various U.S. cities. For instance, a violence prevention program implemented in the Bronx, which relies on indigenous members of the community to negotiate nonviolent resolutions to interpersonal conflicts, has reported modest success (Kotlowitz, 2008). When a similar project was implemented in poor Chicago neighborhoods, ex-offenders proved successful at preventing conflicts from erupting into lethal violence (Loeber, Farrington, Stouthamer-Loeber, & Raskin-White, 2008). Ultimately, policy makers must develop programs that make violence a less appealing form of "social control" for the actors involved in interpersonal disputes. If successfully implemented, such strategies may lead to a reduction in the rate of retaliatory violence, meaning that the strong overlap among victims and offenders will become weaker, especially in disadvantaged urban areas.

DIRECTIONS FOR FUTURE RESEARCH: WAYS FORWARD

Since the publication of Wolfgang's foundational study several decades ago, scientific research on the victim–offender overlap has developed along several lines. As a result, far more is currently known about the etiology of the overlap compared to the body of knowledge that existed in the past. Even so, as noted previously, much remains to be learned about the nature of this phenomenon. There are several possible ways to move this line of research forward. We outline a small number of these possibilities in this section.

First, relatively few truly longitudinal studies of the victimization–offending link exist. This is a critical shortcoming given that longitudinal research designs hold many advantages over traditional cross-sectional designs; for instance, longitudinal designs permit researchers to examine intra-individual change and interpersonal differences in intra-individual change. They also allow for researchers to control for unobserved heterogeneity. The latter advantage is worth emphasizing. Unobserved time-stable mechanisms may confound the link between victimization and offending. A failure to model unobserved or observed heterogeneity may contribute to omitted-variable bias. If not sufficiently modeled, these mechanisms may increase the likelihood of incorrectly rejecting null hypotheses and overestimating the actual strength of the relationship between offending and victimization. Given that a key theoretical explanation for this phenomenon is based on the role of risk heterogeneity, it is important that researchers quantify its contribution to the victim–offender overlap. Longitudinal designs are perhaps best suited for this task.

Because victimization research is only rarely conducted via longitudinal framework, there is a paucity of information regarding the nature of the overlap across developmental time periods. A number of questions have not been evaluated: Is the overlap specific only to adolescence? Is it evident throughout adulthood? Is the onset of offending related to the onset of victimization, and vice versa? As noted, most research conducted on this topic to date has either applied a cross-sectional design or used only a small number of waves of data. These shortcomings may be due to the fact that few data sources are capable of analyzing the victim–offender overlap across multiple time periods, using contiguous waves of data. From a substantive perspective, it is plausible to assume that offending and victimization are causally related only within a specific time period of the life-course—namely, the "peak" violence years. An emerging viewpoint maintains that victimization may be related to patterns of desistance, in that a reduction in the rate of victimization may prompt a reduction in the rate offending, or vice versa (Berg & Loeber, 2011; Lauritsen & Laub, 2007).

Second, with few exceptions, most research conducted on the victim–offender overlap has focused on individual-level mechanisms. Only a small number of studies have integrated neighborhood-level factors, despite good theoretical reasons for doing so. For instance, as noted earlier, strands of subcultural models suggest that the nature of the victim–offender overlap will vary according to neighborhood-level conduct norms. Likewise, interpretations of multilevel versions of routine activities/lifestyles theory imply that offending will have stronger effects on the probability of victimization in some places. Based on this perspective, "victims and offenders are inextricably linked in an ecology of crime" and offenders represent attractive targets by virtue of their legal vulnerability; however, those who reside in impoverished areas may have an especially heightened risk of victimization because the ambient "criminal opportunity structure" is conducive to predatory behavior (Meier & Miethe, 1993, p. 495).

Moreover, developmental criminology makes implicit assumptions about the implications of contextual conditions for the nature of the victim–offender overlap. According to the "person-by-environment" perspective, the importance of individual risk factors for involvement in antisocial behavior depends on the context in which youth are embedded (Piquero, MacDonald, Dobrin, Daigle, & Cullen, 2005). A trajectory of life-course persistent offending is more likely to develop when risk factors, such as cognitive deficits, arise within adverse family or neighborhood contexts. As a consequence, structural adversity may exacerbate the degree to which "high-risk" youth become involved in violent encounters. Ultimately, this chain of events may culminate into a strong, positive correlation between offending and victimization within disadvantaged environments. Given their potential to shed new light on the origins of the victim–offender overlap, these ideas warrant additional empirical attention.

Third, much remains unknown about the degree to which *individual characteristics moderate* the effect of offending on victimization over time. For example, all else equal, two individuals with similar levels of involvement in serious offending may have a disparate probability of victimization due to differences in their constitutional makeup or other person-level characteristics. Recent research on victimization offers some empirical support for this idea, showing that low self-control influences both offending and victimization and moderates the strength of the relationship between the two factors. Developmental criminology has identified a number of social and biological mechanisms that may account for between-person variation in patterns of violent offending. Future research should attempt to evaluate whether any of the conditions identified in this literature give rise to between-person differences in the magnitude and algebraic direction of the victim–offender overlap.

CONCLUSION

In short, a number of questions remain unresolved about the linkage between offending and victimization—the answers to which may help to elucidate the puzzling etiology of this phenomenon. Additional research on the social processes that generate the victim–offender overlap will enrich our understanding of the phenomenon and ultimately advance the development of violence theories as well as violence prevention policies.

REFERENCES

Addams, J. (1909). *The spirit of youth and the city streets.* New York: Macmillan.

Anderson, E. (1999). *Code of the street.* New York: W. W. Norton.

Berg, M. T., & Loeber, R. (2011). The neighborhood context of the relationship between offending and victimization: A prospective investigation. *Journal of Quantitative Criminology*, in press.

Cohen, L., Kluegel J., & Land, K. (1981). Social inequality and predatory criminal victimization: An exposition and test of a formal theory. *American Sociological Review, 46*, 505– 524.

Cooney, M. (1998). *Warriors and peacemakers: How third parties shape violence.* New York: New York University Press.

DeLisi, M. (2005). *Career criminals in society.* Thousand Oaks, CA: Sage.

Dobrin, A. (2001). The risk of offending on homicide victimization. *Journal of Research in Crime and Delinquency, 38*, 154–173.

Esbensen, F.A., & Huizinga, D. (1991). Juvenile victimization and delinquency. *Youth & Society, 23*, 202–228.

Felson, R. B. (1982). Impression management and the escalation of aggression and violence. *Social Psychology Quarterly, 45*, 245–254.

Garofalo, J. A. (1987). Reassessing the lifestyle model of criminal victimization. In M. Gottfredson & T. Hirschi (Eds.), *Positive criminology* (pp. 23-42). Beverly Hills, CA: Sage.

Glueck, S., & Glueck, E. T. (1940). *Juvenile delinquents grown up.* New York: Commonwealth Fund.

Gottfredson, M. (1981). On the etiology of criminal victimization. *Journal of Criminal Law and Criminology*, 72, 714–726.

Gottfredson, M. (1984). *Victims of crime: Dimensions of risk. Home Office research study*. London: Her Majesty's Stationary Office.

Gottfredson, M., & Hirschi, T. (1990). *A general theory of crime.* Stanford, CA: Stanford University Press.

Hindelang, M., Gottfredson, M., & Garofalo, J. 1978. *Victims of personal crime: An empirical foundation for a theory of personal victimization*. Cambridge, MA: Ballinger.

Horowitz, R. (1983). *Honor and the American dream.* Chicago: University of Chicago Press.

Jacobs, B. A., & Wright, R. A. (2006). *Street justice: Retaliation in the criminal underworld.* New York: Cambridge University Press.

Jensen, G., & Brownfield, D. (1986). Gender, lifestyles, and victimization: Beyond routine activity theory. *Violence and Victims, 1*, 85–99.

Katz, J. (1988). *Seductions of crime.* New York: Basic Books.

Kornhauser, R. (1978). *Social sources of delinquency.* Chicago: University of Chicago Press.

Kotlowitz, A.N. (2008, May 4). Blocking the transmission of violence. *New York Times Magazine.* http://www.nytimes.com/2008/05/04/magazine/04health-t.html

Kubrin, C., & Weitzer, R. (2003). Retaliatory homicide: Concentrated disadvantage and neighborhood culture. *Social Problems, 50*, 157–180.

Laub, J. H., & Vaillant, G. (2000). Delinquency and mortality: A 50 year follow-up study of 1000 delinquent and non-delinquent boys. *American Journal of Psychology, 157*, 96–102.

Lauritsen, J. L., & Laub, J. H. (2007). Understanding the link between victimization and offending: New reflections on an old idea. In M. Hough & M. Maxfield (Eds.), *Surveying crime in the 21st century* (Vol. 22, pp. 55-75). Monsey, NY: Criminal Justice Press.

Lauritsen, J. L., & Quinet, K. (1995). Patterns of repeat victimization among adolescents and young adults. *Journal of Quantitative Criminology, 11*, 143–166.

Lauritsen, J. L., Sampson, R. J., & Laub, J. H. (1991). The link between offending and victimization among adolescents. *Criminology, 29*, 265–291.

Loeber, R., Farrington, D. P., Stouthamer-Loeber, M., & Raskin-White, H. (2008). *Violence and serious theft: Development and prediction from childhood to adulthood.* New York: Routledge.

Luckenbill, D., & Doyle, D. P. (1989). Structural position and violence: Developing a cultural explanation. *Criminology, 27*, 419–436.

Mayhew, P., & Elliott, D. (1990). Self-reported offending, victimization, and the British Crime Survey. *Violence and Victims, 5*, 83–96.

Meier, R. F., & Miethe, T.D. (1993). Understanding theories of victimization. *Crime and Justice, 17*, 459–499.

Nagin, D. S., & Farrington, D. P. (1992). The stability of criminal potential from childhood to adulthood. *Criminology, 30*, 236–260.

Nieuwbeerta, P., & Piquero, A. R. (2008). Mortality rates and causes of death of convicted Dutch criminals 25 years later. *Journal of Research in Crime and Delinquency, 45*, 256–286.

Osgood, D. W., Wilson, J. K., O'Malley, P., Bachman, J. G., & Johnston, L. (1996). Routine activities and individual deviant behavior. *American Sociological Review, 61*, 635–655.

Ousey, G. C., Wilcox, P., & Fisher, B. (2011). Something old, something new: Revisiting competing hypotheses of the victimization–offending relationship among adolescents. *Journal of Quantitative Criminology, 27*, 53-84.

Piquero, A. R., MacDonald, J., Dobrin, A., Daigle, L., & Cullen, F. (2005). Self-control, violent offending, and homicide victimization: Assessing the general theory of crime. *Journal of Quantitative Criminology, 21*, 55–71.

Reiss, A. (1981). Towards a revitalization of theory and research on victimization by crime. *Journal of Criminal Law and Criminology*, 72, 704–713.

Rosenfeld, R., Jacobs, B., Wright, R. (2003). Snitching and the code of the street. *British Journal of Criminology, 43*, 291–309.

Sampson R. J., & Bean, L. (2006). Cultural mechanisms and killing fields: A revised theory of community-level racial inequality. In R. Peterson, L. Krivo, & J. Hagan (Eds.), *The many colors of crime: Inequalities of race, ethnicity, and crime in America* (pp. 8-36). New York: New York University Press.

Sampson, R. J., & Lauritsen, J. L. (1990). Deviant lifestyles, proximity to crime, and the offender–victim link in personal violence. *Journal of Research in Crime and Delinquency, 27*, 110–139.

Schreck, C. J. (1999). Criminal victimization and low self-control: An extension and test of a general theory of crime. *Justice Quarterly, 16*, 633–654.

Schreck, C. J., Stewart, E. A., & Fisher, B. S. (2006). Self-control, victimization, and their influence on risky activities and delinquent friends: A longitudinal analysis using panel data. *Journal of Quantitative Criminology, 22*, 319–340.

Shaw, C. (1930). *The Jack Roller: A delinquent boy's own story.* Chicago: University of Chicago Press.

Singer, S. A. (1981). Homogeneous victim–offender populations: A review and some research implications. *Journal of Criminal Law and Criminology, 72*, 779–788.

Singer, S. A. (1986). Victims of serious violence and their criminal behavior: Subcultural theory and beyond. *Violence and Victims, 1*, 61–70.

Sparks, R. (1982). *Research on victims of crime*. Washington, DC: U.S. Government Printing Office.

Stewart, E. A., Schreck, C. J., & Simons, R. L. (2006). I ain't gonna let no one disrespect me: Does the code of the street reduce or increase violent victimization among African American adolescents? *Journal of Research in Crime and Delinquency, 43*, 427–458.

von Hentig, H. (1948). *The criminal and his victim: Studies in the sociobiology of crime*. New York: Shocken Books.

Wittebrood, K., & Nieuwbeerta, P. (2000). Criminal victimization during one's life course: The effects of previous victimization and patterns of routine activities. *Journal of Research in Crime and Delinquency, 37*, 91–122.

Wolfgang, M. A. (1958). *Patterns in criminal homicide*. Philadelphia: University of Pennsylvania Press.

Wright, R., & Decker, S.A. (1997). *Armed robbers in action: Stickups and street culture*. Boston, MA: Northeastern University Press.

CHAPTER 3

Community-Level Correlates of Crime

Matthew R. Lee
Louisiana State University

Graham C. Ousey
College of William and Mary

When most people try to explain criminal behavior, their natural tendency is to think about what it is about individuals that causes them to commit their crimes. Are they mean? Do they simply not care about right and wrong? Are they calculating manipulators who are willing to do anything to get what they want? Or perhaps are they people of decent character who simply learned to do wrong by hanging around with the wrong crowd? All of these individualistic explanations may contain some element of truth, but from a sociological perspective they are incomplete. The reason is that it is also necessary to consider how behavior is shaped by the broader social context in which individuals carry out their daily lives. Indeed, some would suggest that the central argument of the sociological discipline is that "context matters."

In simple terms, this supposition means that human behavior does not take place in a vacuum, but rather is situated in a broader social environment of opportunities and constraints. This broader environment or "social context" consists of structural and cultural components. Structural components include social institutions, organizations, friendship and kinship networks, and demographic aspects of the population such as age structure, race and ethnicity, and social class characteristics. Cultural components refer to symbolic aspects of communities, such as language and communication styles, values, norms, and behavioral expectations.

This chapter focuses on a particular context that criminologists, dating back nearly 100 years, have viewed as a critical ingredient in the cause of crime: the local community (Shaw, McKay, Zorbaugh, & Cottrell, 1929). The term "local community" is used here instead of the more generic word "community" because a community can take on many different forms, including variations such as an online community or a virtual community. In contrast, our attention here is focused exclusively on the importance of structural and cultural factors that are spatially bounded, referring specifically to a neighborhood or neighborhood cluster. Use of this spatially bounded definition enables us to maintain conceptual and practical clarity in the ensuing discussion, which addresses state-of-the-art research findings on the correlates and causes of crime in local communities. The purpose of the discussion is to demonstrate that individual criminal behavior cannot be fully understood without considering how contextual conditions motivate, constrain, or modify it.

CONCEPTUAL ISSUES

Local communities can be viewed as consisting of both structural and cultural features. Structural features can be further subdivided into *compositional* and *social* characteristics. The former include things such as age structure (Is it mostly younger people, a good mix of age groups, or a mostly older retirement community?), racial/ethnic diversity (e.g., predominantly African American, white, Hispanic, or Asian), and the extent of socioeconomic inequality among community residents (lots of poor families, or a big difference between those at the high end and those at the low end of the distribution—what is usually called income or resource inequality). In terms of the last consideration, socioeconomic factors encompass social institutions, organizations, and the networks of association that serve to connect friends and neighbors within a community, and the community to extra-community resources (e.g., government agencies, other neighborhoods).

The culture of local communities also can be divided along two dimensions. The *normative* dimension contains the content of the values and behavioral expectations (norms) that are shared by members of the community. The *interactional* dimension refers the ways that norms and values are actively used and modified by social actors as they navigate their way through daily social life. Although the normative and interactional dimensions of culture are often congruent, in some instances they may diverge.

For example, in local communities where economic success is highly valued but good jobs are scarce, individual residents may find it necessary to find alternative means for achieving economic success, such as participating in markets for drugs or stolen goods. While these alternative means may not be condoned by the broader norms and values of the local community, the individual actor is able to rationalize or find a justification for them because of the severely limited legal economic opportunities. Moreover, if many individuals reach this same conclusion, the local community may begin to develop cultural norms and values that are more tolerant of deviant behavior than would be the case in other local communities where economic opportunities are more plentiful. Similarly, in places where the police are perceived as being biased or ineffective, a set of norms and values that condone the use of interpersonal violence to resolve disputes may supplant the mainstream cultural ethos that the police are trustworthy and should be used to resolve disputes.

Having described the key distinguishing social characteristics of local communities, the remainder of this chapter explores the state-of-the-art empirical research on how the essential structural and cultural features may be related to crime rates. The goal of this discussion is to elucidate answers to a key question: "Why do some communities have high rates of crime while in others crime rates are low?"

RESEARCH FINDINGS

Local Community Structural Factors and Their Effects on Crime Rates

Criminologists have explored a variety of local community factors that are related to crime. An overview of some of the major correlates of crime are examined next.

Age

When considering individual participation in serious crime, one of the strongest correlates is age. Participation in criminal activity typically starts relatively early in life, with the rate of violent crime peaking when individuals are in their late teens and early twenties. It then tapers off very quickly, so that the involvement in violent crime after the mid-twenties is quite rare. Although a small group of violent offenders may persist in their criminal activities until later in adulthood, recent research indicates that this behavior almost never persists across the entire life span (Laub & Sampson, 2003; Sampson & Laub, 1993).

An important conclusion that may be drawn from knowledge about the age patterning of violent offending is that the age structure of a local community may affect its crime rate. In particular, we might expect that in communities with an older age structure—such as a retirement community—crime rates will be quite low; in contrast, in communities dominated by adolescents and young adults, we would expect the opposite—much higher violent crime rates. A community with a balanced mixture of younger and older populations would fall somewhere between the extremes just articulated. In sum, because violent crime behavior is much more common among people in their late teens and early twenties and is rare among older adults, knowing the age structure of the local community tells us something potentially important about the amount of violent crime that may occur in that place. When younger age ranges constitute a larger share of the population, we would expect violent crime rates to be higher; when the elderly population increases as a share of the total population, we would expect that crime rates would be deflated.

Despite these seemingly straightforward expectations, empirical research that has examined the connection between variations in community age structure and rates of violent crime has been hard-pressed to find support for the idea that local communities with a younger age structure will have more violent crime (Browning et al., 2010). Consequently, some recent studies of neighborhood crime rates do not even account for cross-community differences in age structure (Boggess & Hipp, 2010; Graif & Sampson, 2009; Stults, 2010).

Racial and Ethnic Composition

A second compositional variable that is often considered to be a potentially important correlate of community crime rates is racial or ethnic composition. Why might this be an important variable? For instance, African Americans accounted for slightly more than 50% of people arrested for murder and non-negligent manslaughter, the most serious violent crime in the United States. Similarly, they accounted for 56.7% of arrestees for robbery, and 34.2% of arrestees for aggravated assault (Sourcebook of Criminal Justice Statistics, 2011). Yet, according to U.S. Census data, African Americans represent only 12% of the total U.S. population. Thus, compared to their overall representation in the U.S. population, African Americans appear to be disproportionately involved in violent crime.

The reasons for this racial discrepancy in violent crime rates are not extremely well understood, although the economic dislocation and social marginalization of racial minorities—especially blacks—may have the most credibility among crime scholars. Nevertheless, from our standpoint, the key insight to be derived here is that community racial/ethnic composition is correlated with crime rates. In general, local communities with proportionally larger shares of minority group members are expected to have more violent crime than communities with very few minority residents.

Empirical research on the link between racial/ethnic composition and crime has generally supported this expectation, and several studies have illustrated this correlation fairly clearly. For example, scholars have routinely found that communities with substantial African American populations or higher racial heterogeneity have higher rates of violence, even when other crime influences are taken into account (Hipp, 2007, 2010; Warner & Rountree, 1997). Indeed, the connection between racial composition and violent crime is one of the most well-established findings in the research literature on community crime rates.

The long-standing interest in the impact of the African American population share on violent crime rates has recently been joined by a surge of research focused on how the relative concentration of Hispanics may affect neighborhood violent crime rates. Hispanics are an interesting case because their representation in violent crime tends to fall between that of whites, who are on the low end of the crime distribution, and African Americans, who are on the high end of that distribution. Attempts to elucidate this relationship between Hispanic population composition and violent crime rates, however, are often complicated by the fact that local communities with relatively large Hispanic populations also may be immigrant-receiving communities. Hence, to understand the impact of ethnic composition on crime, we must be able to separate out its impact from that of immigration, which also may be correlated with violent crime.

On the question of whether the presence of large Hispanic populations is associated with higher rates of violent crime, evidence suggests that it may be (Hipp, 2007). In regard to the immigration–crime relationship, there has been much recent research activity in this area, much of it spurred on by current controversies about illegal immigration and the necessity of immigration reform. Several studies have found that Hispanic immigration actually may work to lower violent crime rates (Aikins, Rumbaut, & Stansfield, 2009; Lee, Martinez, & Rosenfeld, 2001; Velez, 2009). Other scholars contend that the picture is somewhat more complex. Indeed, in a series of recent papers, two sociologists have argued that the majority of research on the link between Hispanic migration and crime is looking in the wrong places. Specifically, Shihadeh and Barranco (2010a, 2010b, 2011) argue that in "new" immigrant-receiving communities, Hispanic immigration is associated with higher crime rates. In contrast, in places where Hispanic immigrants are already well established, immigration does not appear to put upward pressure on violent crime rates. Moreover, this recent line of investigation indicates that an influx of Hispanic migrants may not cause Hispanic crime to go up, but suggests instead that its crime-producing impact may be diverted onto other social groups. According to Shihadeh and Barranco, Hispanic migration to urban areas increases African American crime rates, due in part to the displacement of the latter from the low-skills job market. In rural areas, the same process occurs for whites; Hispanic migrants displace whites from low-skills jobs, thereby contributing to higher rates of violence among whites.

Socioeconomic Composition

Any discussion of community variations in violent crime is incomplete without a consideration of the influence of socioeconomic composition. Indeed, a long tradition in criminology links intercommunity variations in rates of serious crime to economic deprivation (Blau & Blau, 1982; Messner, 1982, 1983). Two lines of argumentation are evident in this tradition. The first stresses that community variations in rates of *absolute deprivation* (*i.e., poverty*) are predictive of violent crime. Communities with a greater share of their populations living in poverty are expected to have higher violence rates for both instrumental (e.g., committing robbery to get money) and expressive (e.g., reacting to the overwhelming stress and anxiety of severe deprivation) reasons (Agnew, 1992). The second argument suggests that violent crime may be influenced by *relative deprivation* (*i.e., economic inequality*). This thesis argues that a wide gap between the rich and the poor creates motivation for crime, as poorer residents seek the means to attain the living standards exhibited by their neighbors. Under this scenario, the normative constraints that typically restrain criminal behavior become less powerful, and community residents may be pressured into criminal innovations designed to help them make their own economic standing more similar to that of the neighbors to whom they compare themselves (Merton, 1938).

Residential Segregation

The discussion of how racial/ethnic composition and socioeconomic status are related to community crime rates is complicated by a substantial body of research suggesting that residential segregation also plays a key role in determining the incidence of crime. Residential segregation refers to the residential settlement patterns of groups of people within an ecological unit, such as a city. For example, in a city with high levels of racial residential segregation, African Americans and whites generally live in separate neighborhoods. Likewise, where socioeconomic residential segregation is high, poor and nonpoor social classes tend to be found in separate parts of a city, rather than being neighbors in the same community. The relative proximity of blacks to whites and of the poor to the nonpoor may be important structural features of communities because, as crime scholars have argued, high levels of segregation systematically cut off minority and poor communities from access to mainstream middle-class institutions of social mobility such as good jobs, good schools, and stable family structures, all of which are conducive to prosocial behavior patterns (Lee, 2000; Shihadeh & Flynn, 1996).

Empirical research on this hypothesized link between segregation and crime is fairly straightforward. Whether focusing on race, social class, or the intersection of these two factors, most studies have reported that more segregated environments have higher rates of violent crime (Lee, 2000; Lee & Ousey, 2007; Peterson & Krivo, 1993; Shihadeh & Flynn, 1996). Although we discuss residential segregation as a compositional feature of communities, it is also arguably a social feature of communities as well, because it involves the degree of contact that different groups of people have with one another. In a later section of this chapter, we return to this issue, exploring in more depth why this relationship exists.

Social Networks and Collective Efficacy

In addition to the compositional factors that may influence a community's rate of crime, researchers have identified various social factors as affecting levels of crime. Social factors are attributes of communities that measure the degree of interconnectedness between individuals. One important social factor is the nature and extent of social networks in local communities. Social networks refer to associational ties or relationships between people in a community.

To see how such networks function, let's consider an example. Imagine a field in which 10 people are randomly placed, none of whom has any preexisting relationship with the other people. If an observer looking down on the field drew a descriptive image of the community, it would appear essentially as 10 unaffiliated dots (one for each person). However, if those people have relationships with one another, the descriptive image would get more interesting because the observer would represent those relationships by drawing a line between each pair of acquainted individuals. Rather than a drawing of polka dots, the resulting image would look more like a spider's web. Most, if not all, of the dots in the picture would be linked to other dots via lines, which represent social networks or the ties binding people together. These social ties are fundamental aspects of human social life and have important implications for the ability of local communities to act as a collective unit.

Social networks can differ in many ways across communities. Some networks connect two people who have very close or intimate relationships; they are referred to as *strong* social ties. By comparison, many connections between people are less intimate; they are merely acquaintances, or friends of friends. Sociologists call these latter connections *weak* social ties. In addition to varying in terms of the "strength" of the social ties that connect pairs of people, communities may differ in terms of the *network density*.

When residents of a community on average have many ties with other residents, the social network has greater density; when residents have relatively few relationships with others in the community, the network has less density.

Social networks, and the properties of those networks, may significantly affect the crime rate within communities. According to some criminological theories, a strong social fabric—that is, dense ties with people who know one another fairly well—may be a good antidote to serious crime. When lots of ties and a high level of familiarity are present, much of the serious criminal behavior that could take place in public may be averted because criminally inclined neighborhood residents know there is great risk that they will be known and identified by other residents of the community—a phenomenon called a guardianship effect (Bursik & Grasmick, 1993). In addition to increasing guardianship, social networks may help develop the informal social control capacity of communities. Indeed, a well-connected local community is more likely to (1) develop a consensus on approved and disapproved behaviors; (2) effectively teach community youths about the aforementioned approved and disapproved behaviors; and (3) more uniformly reward prescribed behaviors and sanction proscribed behaviors.

Until recently, empirical research on the specific effects of community social network characteristics on serious crime was relatively rare. Primarily, this paucity existed because the types of data needed to properly assess such relationships are fairly difficult and expensive to collect. However, a few studies have overcome these difficulties, at least to some extent, and their results are highly instructive. In one important piece of research, Sampson and Groves (1989) used measures of local social ties from the British Crime Survey to assess the influence of social networks on crime. Specifically, they gauged how many people whom survey respondents said they knew within a 15-minute walk from their house. The researchers found that communities with more dense social ties have lower rates of crime, on average.

Similarly, Warner and Rountree (1997) developed a measure of local social ties using data for communities in Seattle, Washington. Using this measure, they examined whether the influence of social ties on crime depends on the racial characteristics of neighborhoods. Their results indicate that local social ties are associated with lower assault rates in predominantly white neighborhoods, but not in racially mixed or predominantly minority neighborhoods.

In another important study of the influence of social networks on crime, Bellair (1997) asked whether it is more important for social ties to be the intimate connections that are typical between close friends and family members (i.e., strong ties) or the less intimate relationships that link together acquaintances (i.e., weak ties).

Although it might seem logical to expect that strong social ties would be more effective in neighborhood crime control than weak social ties, the results of his analysis of 60 neighborhoods generally suggest the opposite. Measures of infrequent interaction, which Bellair interpreted as measuring weak social ties, were found to provide a stronger protective effect against crime. The conclusion that emerged from this analysis is that communities characterized by high levels of social interaction (strong ties) typically have many small and disconnected "circles of friends," whereas communities featuring more infrequent interactions between neighbors have a more expansive and inclusive social network, which can better facilitate collective action (Bellair, 1997).

Another social factor that has been identified as a strong correlate of cross-community variation in crime rates is *collective efficacy*. This term was coined by Robert Sampson and his colleagues (1997) to describe a community with cohesion (dense social networks and the trust associated with them) and shared expectations for collective action. Part of the value of this concept is that it calls attention to the fact that the mere presence of dense social networks (strong or weak) is not enough to facilitate collective action and achieve specific goals, such as crime control. Rather, communities also need residents who are willing to intercede in public deviant behavior and to work for the common good of the community. In other words, while networks can facilitate a collective response, residents must share a willingness to be engaged in problem solving if this effort is to succeed. From a practical crime control standpoint, this means that residents do not tolerate groups of young people intimidating residents in public spaces, do not allow graffiti or garbage to proliferate on buildings or in vacant lots, and are willing to work to prevent drug dealers and other criminals from plying their trade in community spaces. Some have hailed this concept as one of the most important in recent decades in the communities and crime research literature. The empirical evidence supporting the importance of collective efficacy appears to be strong, and no major challenges to the concept have surfaced.

Cultural Features of Communities and Their Effects on Crime

Whereas structural features of communities are related to the composition of people who live there and the nature and extent of their relationships, the cultural characteristics of communities refer to the norms, values, and mores guiding interactional patterns of community residents. Just as local communities within a given city can have different compositional or social structural characteristics, so too can they exhibit variation in terms of their cultural tapestry. This section describes a few themes commonly found in research that links differential community crime rates to cultural factors.

Street Codes

Some scholars attribute between-community differences in crime rates to differences in the extent to which communities exhibit and adhere to cultural codes that specify tolerance of violent behavior. In his book *Code of the Street* (1999), urban ethnographer Elijah Anderson reports observations from Philadelphia neighborhoods mired down in poverty, broken families, drugs, and violence. Anderson argues that in these contexts where good jobs are not readily available and socioeconomic deprivation is severe, the mainstream routes to social status are effectively removed. Hence, people develop other ways to establish their place in the status hierarchy of the community.

One of these methods, which Anderson discusses in depth, is the "street code." Such codes emphasize the salience of maintaining one's respect, reputation, or social status in the eyes of others, and they outline behavioral norms designed to accomplish that purpose. Anderson suggests that when the chances of achieving mainstream socioeconomic success are slim, individuals learn that presenting themselves through a tough, violent demeanor can earn them recognition and even admiration in the streets. Compelling deference and supplication from others in the community becomes a means by which status and respect are accumulated. Over time, as the notion that "might makes right" gains a foothold, others in the community assume the same tactics, and the cultural tolerance for the use of violence to resolve disputes or to gain the respect of others in the community becomes widely engrained (Anderson, 1999).

Not all members of a community will outright endorse the street code. Indeed, Anderson makes the case that there is an important distinction between "street" and "decent" families. "Street" families are those for whom the commitment to value middle-class standards or expectations has been heavily eroded and the street codes have become a much more prominent guiding force in their lives. They are frequently involved in the criminal underworld, they live on the edge, and their interactions with others, including their own children, is typically harsh and threatening. For them, violence, or the threat of it, is part and parcel of how they navigate social terrain on a daily basis. Children of street families are socialized into this mode of interaction from early on; as adults, they then transmit this cultural orientation to younger generations.

In contrast, "decent" families provide a more sheltered environment for their children, emphasize the importance of school and staying out of trouble, and try to instill mainstream middle-class ambitions in their children. Anderson argues that in environments where "street" people and "decent" families cross paths every day, the decent kids must learn street codes to navigate their way through public aspects of their life. His theory provides an explanation for the diffusion of a cultural orientation tolerating the use of violence in local community contexts where more traditional routes to status such as socioeconomic mobility are not readily available.

Police Distrust

Another cultural mechanism that has been discussed in the research literature on differential community violence patterns involves trust or *distrust in the police.* In some communities, there is very substantial belief in and reliance on the police to mediate disputes. In others, the police operate under a constant cloud of distrust, suspicion, and even contempt (Black, 1984).

In a very interesting discussion of this problem, Kubrin and Weitzer (2003) argue that bad policing practices can influence community cultural patterns in different ways. For example, when the police are not doing an adequate job, residents may feel that they cannot rely on the authorities to solve problems. Alternatively, if the police are engaging in extremely aggressive policing, being disrespectful to community residents, demonstrating police brutality, or being perceived as being discriminatory or random in their treatment of residents, a high level of hostility and resentment of the police can emerge, to go along with feelings of abandonment by the formal authorities. Ultimately, when either an absence of police assistance or an environment of police misconduct erodes collective levels of support for or appreciation of the police, a cultural ethos of "self-help" dispute resolution may emerge, leading to higher rates of violence (Black, 1983).

Recent research has provided some interesting empirical support for these types of cultural effects. For example, Stewart and Simons (2010), have shown that community-level street codes endorsing violence are associated with higher levels of participation in violence, while Kane (2005) reported empirical evidence endorsing the idea that compromised police legitimacy is associated with higher rates of violent crime in disadvantaged neighborhoods.

Culture in Action

Much of the literature on community culture as a source of violence assumes that the essential feature of culture is a set of *values* that guide individual behavior. In this values paradigm, people may commit homicide because they value status and respect, and the use of violence is justified as a means of defending their values. However, recent contributions in this area of scholarship have begun to view culture in more active terms, formulating a "relational" theory of culture as a cause of violence (Sampson & Bean, 2006). This theory draws heavily from innovations in cultural sociology, particularly Ann Swidler's (1986) theory of culture in action.

Although too complex to fully recapitulate here, the essence of relational theory is that culture is created in interactions between people. In essence, it emerges from the skills, habits, social scripts, values, and other aspects that individuals bring into specific interactions. Any given behavior depends on its social context because it is performative. In other words, it emerges from a particular interactional situation in which multiple options are likely to be available; it does not occur because individual actors are slavishly bound to a particular set of values. According to relational theory, violence occurs in some communities with greater regularity not so much because people value violence per se, but because in the interactional situations that they encounter, violence is understood to be the socially preferable option to other courses of action. This is especially true in the case of affronts to personal status that occur in public space.

The most challenging aspect of studying cultural forces in relation to violent crime rates is determining how to effectively measure culture. While a number of studies have conducted surveys of the "values" held by individuals as a means of studying the behavioral effects of culture, the culture in action model does not appear—on its face, at least—to be readily amenable to this approach. In fact, the level of abstraction of this model makes it extremely difficult to measure with conventional approaches, and qualitative research strategies seem better suited for this purpose. At present, not enough work has been done on the relational culture theory of community violence to effectively evaluate its validity.

BRINGING IT ALL TOGETHER: THE INTERACTION OF STRUCTURE AND CULTURE IN THE PRODUCTION OF COMMUNITY CRIME RATES

We conclude by bringing our discussion full circle and revisiting the question asked earlier this chapter: Why is it that some communities have high rates of crime and some have low rates of crime? As described in the preceding sections, research shows that both structural and cultural factors are related to violent crime. One interesting fact not yet discussed is that the criminological research literature has for quite some time entertained a debate about whether crime rates are better explained by structural factors or by cultural factors. Although no widely accepted resolution of this debate has been established, we contend that structural and cultural elements interact to produce communities with either higher or lower rates of violent crime. This idea can be examined in more detail by using a fabricated illustration of how some of the actual mechanisms work.

Imagine a neighborhood in a major U.S. city that is mostly inhabited by African American and other minority group residents. The community itself is characterized by high rates of unemployment, and poverty is widespread. The community is also residentially segregated by both race and class, meaning that even if you travel 20, 30, or 40 blocks in any direction, you are still surrounded by poor and primarily minority residents. Because of the endemic poverty and a lack of investment in the community by the municipal government, local institutions that normally serve as the backbone of communities, such as social and civic organizations and community clubs, are weak or nonexistent. While social networks might be fairly dense, the lack of prosocial community institutions that might otherwise help foster a collective sense of identity and purpose translates into community members not being able to mobilize their networks to achieve goals regarding the well-being of the community (Lee, 2006).

In this climate, alternative methods of making money often emerge. Some are relatively harmless, such as small engine repair or salvaging and selling other people's "trash." Other ways to make a living are not so harmless. When they are in desperate straits, many people in this community turn to illegal markets for drugs, sex, and stolen goods as a means of meeting their financial needs, and others begin to tacitly tolerate this behavior, given the short supply of legal economic opportunities available to their neighbors and friends (Venkatesh, 2006, 2008). Thus we see how both the compositional and social structure of communities can provide the setting in which cultural tolerance for some forms of deviance emerges.

Within some of the markets themselves, particularly drug markets, competition is intense, demand is high, and there is lots of money to be made. Under the conditions of a regular cash-for-goods market dealing in legitimate goods, some regulatory agencies or laws can typically be relied upon to ensure that fair business practices are followed and that business disputes are resolved fairly and efficiently. In the context of illegal markets, however, all of this is moot. In street-level drug markets, the illicit nature of the product means that all participants have to rely on themselves to enforce whatever rules of the market they see fit. The police cannot be relied upon to intervene, and the use of violence, especially in public places, often becomes widespread in an effort to shore up turf and protect the criminal enterprise (Goldstein, 1985). As Anderson notes, children from "decent" families who live in the neighborhood are exposed to this activity on a daily basis, and become familiar with the cultural ethos of violence as a survival mechanism.

When the structural conditions set the stage for emergence of cultural tolerance of violence, widespread violence and illegal activity may then start to affect social structural conditions (South & Messner, 2000). For example, local merchants find it hard to do business in communities where illegal activities are widespread, their clientele is at risk of getting robbed in the parking lot, or their employees are at risk of being the victims of robbery. Hence, businesses are difficult to retain in such communities. Moreover, when communities become so socioeconomically disadvantaged that traditional family structures are decimated, schools fail to provide an education or even a safe haven for local children, and community organizations are unable to thrive, the major institutions of socialization and social control are not operative. In this context, other types of violence start to emerge, particularly predatory and instrumental forms of violence that have very little to do with dispute resolution, but rather are geared toward socioeconomic gain or are just outright predatory, such as criminal rape (Liska, Logan, & Bellair, 1998). Some evidence suggests that these kinds of violence can cause outmigration, with more stable elements of the community leaving for safer environments. As Liska et al. (1998) detail, robbery in particular tends to cause white flight, leading to increased African American population concentration, and thus affecting the overall structural composition of the community.

Based on this illustration, it should be clear that structure and culture are rarely alternative forces that "compete" to wield more influence over community crime rates. Rather, they are intricately linked dimensions of the social organization of local communities that exert joint impacts on the nature and magnitude of illegal behavior.

REFERENCES

Agnew, R. (1992). Foundation for a general strain theory of delinquency. *Criminology, 30*, 47–87.

Aikins, S., Rumbaut, R. G., & Stansfield, R. (2009). Immigration, economic disadvantage, and homicide: A community-level analysis of Austin, Texas. *Homicide Studies, 13*, 307–314.

Anderson, E. (1999). *Code of the street: Decency, violence, and the moral life of the inner city*. New York: W. W. Norton.

Bellair, P. E. (1997). Social interaction and community crime: Examining the importance of neighbor networks. *Criminology, 35*, 677–703.

Black, D. (1983). Crime as social control. *American Sociological Review, 48*, 34–44.

Black, D. (1984). *Toward a general theory of social control.* Orlando, Fl: Academic Press.

Blau, J., & Blau, P. (1982). The cost of inequality: Metropolitan structure and violent crime. *American Sociological Review, 47*, 114–29.

Boggess, L., & Hipp, J. (2010). Violent crime, residential instability, and mobility: Does the relationship differ in minority neighborhoods? *Journal of Quantitative Criminology, 26*, 351–370.

Browning, C., Byron, R., Calder, C., Krivo, L., Kwen, M., Lee, J., & Peterson, R. (2010). Commercial density, residential concentration, and crime: Land use patterns and violence in neighborhood context. *Journal of Research in Crime and Delinquency, 47*, 329–357.

Bursik, R. J., & Grasmick, H. G. (1993). *Neighborhoods and crime.* New York: Lexington Books.

Goldstein, P. (1985). The drugs/violence nexus: A tripartite conceptual framework. *Journal of Drug Issues, 39*, 143–174.

Graif, C., & Sampson, R. (2009). Spatial heterogeneity in the effects of immigration and diversity on neighborhood homicide rates. *Homicide Studies, 13*, 242–260.

Hipp, J. (2007). Income inequality, race, and place: Does the distribution of race and class within neighborhoods affect crime rates? *Criminology, 45*, 665–697.

Hipp, J. (2010). A dynamic view of neighborhoods: The reciprocal relationship between crime and neighborhood structural characteristics. *Social Problems, 57*, 205–230.

Kane, R. (2005). Compromised police legitimacy as a predictor of violent crime in structurally disadvantaged communities. *Criminology, 43*, 469–498.

Kubrin, C., & Weitzer, R. (2003). Retaliatory homicide: Concentrated disadvantage and neighborhood culture. *Social Problems, 50*, 157–180.

Laub, J. H., & Sampson, R. J. (2003). *Shared beginnings divergent lives: Delinquent boys to age 70.* Cambridge, MA: Harvard University Press.

Lee, M. (2000). Concentrated poverty, race, and homicide. *Sociology Quarterly, 41*, 189–206.

Lee, M. (2006). The religious institutional base and violent crime in rural areas. *Journal for the Scientific Study of Religion, 45*, 309–324.

Lee, M. T., Martinez, R., & Rosenfeld, R. (2001). Does immigration increase homicide: Negative evidence from three border cities. *Sociology Quarterly, 42*, 559–580.

Lee, M., & Ousey, G. (2007). Counterbalancing disadvantage? Residential integration and urban black homicide. *Social Problems, 54*, 240–262.

Liska, A., Logan, J., & Bellair, P. (1998). Race and violent crime in the suburbs. *American Sociological Review, 63*, 27–38.

Merton, R. (1938). Social structure and anomie. *American Sociological Review, 3*, 672–682.

Messner, S. F. (1982). Poverty, inequality, and the urban homicide rate: Some unexpected findings. *Criminology, 20*, 103–114.

Messner, S. F. (1983). Regional differences in the economic correlates of the urban homicide rate: Some evidence on the importance of the cultural context. *Criminology, 21*, 477–488.

Peterson, R., & Krivo, L. (1993). Racial segregation and black urban homicide. *Social Forces, 71*, 1001–1026.

Sampson, R., & Bean, L. (2006). Cultural mechanisms and killing fields: A revised theory of community-level racial inequality. In R. Peterson, L. Krivo, & J. Hagan (Eds.), *The many colors of crime: Inequalities in race, ethnicity, and crime in America* (pp. 8–36). New York: New York University Press.

Sampson, R. J., & Groves, W. B. (1989). Community structure and crime: Testing social disorganization theory. *American Journal of Sociology, 94*, 774–802.

Sampson, R. J., & Laub, J. H. (1993). *Crime in the making: Pathways and turning points through life.* Cambridge, MA: Harvard University Press.

Sampson, R. J., Raudenbush, S. W., & Earls, F. (1997). Neighborhoods and violent crime: A multilevel study of collective efficacy. *Science, 277*, 918–924.

Shaw, C., McKay, H., Zorbaugh, F. M., & Cottrell, L. S. (1929). *Delinquency areas.* Chicago: University of Chicago Press.

Shihadeh, E. S., & Barranco, R. (2010a). Latino employment and black crime: The unintended consequence of U.S. immigration policy. *Social Forces, 88*, 1393–1420.

Shihadeh, E. S., & Barranco, R. (2010b). Latino immigration, economic deprivation, and violence: Regional differences in the effect of linguistic isolation. *Homicide Studies, 14*, 336–355.

Shihadeh, E. S., & Barranco, R. (2011) Low skill jobs and rural violence: the link between Hispanic employment and non-Hispanic Homicide. *Deviant Behavior*, in press.

Shihadeh, E., & Flynn, N. (1996). Segregation and crime: The effect of black social isolation on the rates of urban black violence. *Social Forces, 74*, 1325–1352.

Sourcebook of criminal justice statistics. (2011). Retrieved January 14, 2011, from http://www.albany.edu/sourcebook/

South, S., & Messner, S. (2000). Crime and demography: Multiple linkages, reciprocal relations. *Annual Review of Sociology, 26*, 83–106.

Stewart, E., & Simons, R. (2010). Race, code of the street, and violent delinquency: A multilevel investigation of neighborhood street culture and individual norms of violence. *Criminology, 48*, 569–605.

Stults, B. (2010). Determinants of Chicago neighborhood homicide trajectories, 1965–1995. *Homicide Studies, 14*, 244–267.

Swidler, A. (1986). Culture in action: Symbols and strategies. *American Sociological Review, 51*, 273–286.

Velez, M. B. (2009). Contextualizing the immigration and crime effect: An analysis of homicide in Chicago neighborhoods. *Homicide Studies, 13*, 325–335.

Venkatesh, S. (2006). *Off the books: The underground economy of the urban poor.* Cambridge, MA: Harvard University Press.

Venkatesh, S. (2008). *Gang leader for a day.* New York: Penguin Press.

Warner, B., & Rountree, P. (1997). Local social ties in a community and crime model: questioning the systemic nature of informal social control. *Social Problems, 44*, 520–536.

CHAPTER 4

Where I'm From: Criminal Predators and Their Environments

Andy Hochstetler
Iowa State University

Heith Copes
University of Alabama at Birmingham

> *"In the favellas [of Rio De Janeiro], buildings are pock-marked with bullet holes, and youths with military-style small arms patrol the streets at night. Incursions by police or rival factions can happen at any time. Jefferson is an 18-year-old former drug trafficker. Where he lives, gangs have dragged concrete pillars across the streets to stop police, and murals of Osama Bin Laden are painted on the walls. Many people there have lost several relatives to the violence. Jefferson says the gun battles affect everyone in the favellas, including the children. 'They can't play in the street. Half their childhood is spoilt. Every time the fireworks go off [let off by child lookouts to warn of incursions] their mothers yell at them to get inside,' he says. 'People can't stand at the bar and have a drink normally—fireworks go off and they have to find shelter' . . . Jefferson says he took part in the execution of a 15-year-old friend who passed information to the police."*
>
> —*BBC News*, 2005

Although it talks about only one dimension of ghetto life, even the most casual observers of conditions in impoverished nations and neighborhoods of the world will find nothing surprising in the preceding quote. Jefferson's story is familiar, if only from the cinema and occasional forays of the news media into the rows of ramshackle cardboard and corrugated tin residences in South American and African cities. It elicits the same sympathies as stories of the AK-47–wielding children of the Congo, the child soldiers of Sierra Leone, and similarly situated persons who appear, if infrequently, on the Western world's television screens.

Typically, we frame the lives of such young persons living in the worst ghettos of the world as devoid of attractive options. Social and geographic distance, as well as the prominent features of abject poverty, allows us to interpret their actions as products of environments. We do not begrudge or condemn Jefferson or others like him for their unusually hardened criminal acts. Abject poverty in some places focuses our critical aim at community or national leaders, and we are likely to look for the economic roots of the trouble. We are not inclined to wonder precisely when a criminal resident's trajectory slipped in an unfortunate direction or how variables interacted and sequenced between the ages of 5 and 15 to culminate in criminal choices. Such questions seem academic, akin to focusing on whether seized pistons or cracked heads are the mediating causes of engine failures in motors without oil.

Conversely, we often do not view crime committed by hardened, youthful offenders here in the United States with the same understanding eyes. Instead, we are apt to point to personal failure, faulty decision making, or moral ineptitude when explaining why those closer to us choose crime. But if we are so willing to sympathize with Jefferson and his kin, why do we have such a hard time doing so for those predators in our own ghettos? Is it not possible that the forces that push and pull people toward crime in *favellas* are also at work elsewhere?

This chapter explores how people incorporate neighborhood characteristics into identity and rely on these areas to provide opportunities and help construct excuses for their own brutal actions. That is, it examines how sane offenders in criminogenic environments interpret their world and how their locations contribute to people choosing crime.

Drawing on the perspectives of persistent and serious street offenders from the United States, predominantly in pre–Hurricane Katrina New Orleans, the discussion here raises questions about the degree to which contemporary criminology and its complex concerns obscure some links between the environment and violent offending. The psychology and conditions linking place to crime are both real and more likely to occur in the most impoverished communities. Environmental and social conditions are shown to affect thinking, as street offenders make sense of emergent opportunities that fit patterns they have seen before in the larger context of their lifestyles.

We will not slow the reader with a thorough description of the methodology. Simply stated, we arrived at our conclusions through intuition gained from previously published interview-based research, offender biographies, and our own research involving in-depth interviews with dozens of violent offenders. We thought about interviews that we had done before, but drew mainly on interviews with 30 people who had committed at least one carjacking in Louisiana. Carjacking is a crime that is almost always associated with violent career criminals who spend much of their leisure time in impoverished neighborhoods with open-air drug markets (Jacobs, Topalli, & Wright, 2003). Here we interpret what these offenders said about the environmental contexts of their crimes. At times, common sense leads us to be skeptical of some details in offenders' accounts, and we expect the reader may be as well—but this is not court, and these are admitted thieves and robbers. Some degree of dishonesty is expected and assumed (Jacobs & Wright, 2006). Distorted facts and tall stories impart meaning. Indeed, exaggerations and fictions may reveal more about a person than fact, especially when individuals are discussing how places shape their identity and behavior.

The thesis presented here is that the outlooks of congregations of the disaffected, violent young persons, mainly men, hustling on street corners, usually in drug neighborhoods (and sometimes engaged in ongoing street warfare), lead to a great many violent criminal incidents. Their criminal decisions are made in an instant, yet they are congruous with a path through life marked by few legitimate successes and fit easily into what the offenders think they are meant to do based on their social surroundings. We are not arguing that structural and economic arrangements are insignificant factors in explaining the distribution of crime. To the contrary, we suggest that characteristics of place have real consequences for how offenders think about crime, evaluate opportunity, identify with a place, and make sense of environments to justify their criminal choices.

PEOPLED STREETS

Most analysts of the spatial distribution of crime accept the premise that criminals choose to offend based on their evaluation of the costs and benefits of the proposed crime (Brantingham & Brantingham, 1984). For example, this belief underlies routine activity theory, which examines patterned convergences of motivated offenders, suitable targets, and an absence of capable guardianship as explanation of crime rates. In their introductory discussion from the most famous statement of this theory, and probably of predatory crime's distribution, Lawrence Cohen and Marcus Felson (1979) state:

> *[I]n the context of [direct predatory violations], people, gaining and losing sustenance, struggle among themselves for property, safety, territory, hegemony, sexual*

> *outlet, physical control and sometimes for survival itself. The interdependence between offenders and victims can be viewed as a relationship between functionally dissimilar groups . . . violations can only be sustained by feeding upon other activities. As offenders cooperatively increase their efficiency at predatory violations and as potential victims organize their resistance to such violations, both groups apply the symbiotic principle to improve their sustenance position.*

Those criminologists interested in the spatial and temporal distribution of crime typically set aside the implications of agency and the mental mechanisms (or motivations) whereby arrangements of opportunities and objects lead to greater or lesser rates of offending. Little is said, therefore, about *how* offenders make crime more attractive by cooperatively increasing their efficiency at predatory violations. All investigators know, however, that efforts by some offenders to reshape and frame their world make crime more likely for others. Through more or less intentional, incremental actions and interactions, offenders cultivate suitable local and cultural environments for monetary success by criminals and improve their chances of success in interpersonal evaluations.

There are innumerable studies of the effects of limited legitimate opportunities and the availability of illicit ones on rates of offending (Cloward & Ohlin, 1960). There also has been considerable speculation (and some study) of how signs of incivility or disorder and the widespread perception that a place is disorderly figure into the decision to commit crime (Sampson & Raudenbush, 1999). What is missing from most of this discussion is the likely offenders' perspectives on their surroundings, including how criminal opportunities are interpreted, created, and viewed. Exceptions to the neglect of the environmental psychology of crime include works that attempt to look to places where crime is common and the outlooks associated with it in these places. These typically appear in ethnographies such as *The Social Order of the Slum* (Suttles, 1968), *In Search of Respect* (Bourgois, 2002), *Ain't No Making It* (MacLeod, 1995), and *A Place on the Corner* (Anderson, 2003). All of these publications are exceptional in that they give due theoretical attention to cultural/geographic space. All show that inhabitants understand intuitively more about place than what the costs and benefits of various actions in immediate circumstances are.

Places have deeper meanings and serve as locales for passing these meanings along. Sometimes places make things seem rational that would not seem rational elsewhere and bend preferences and expectations accordingly—think about aberrant behavior occurring at spring break on Mexico's beaches, during Mardi Gras in New Orleans, or on the Strip in Las Vegas, where advertising confirms that shameful behavior is expected. These expectations may be readily available and can have salient associations with what a place means, such as a crack-dealing set. All of us have some idea of what the expectations would be there, although we might have doubts about whether we would fit in; most of us have no intention of finding out. Criminologists describe the things that predatory street offenders say about the environmental context of their behavioral expectancies and argue that these individuals' understandings of opportunities in the impoverished community occur through a cognitive lens provided partially by an intuitive understanding of places and personal relations to them.

OPEN DRUG DEALING, RELATED ACTIVITIES, AND OPPORTUNITIES FOR CRIME

Crime is extremely concentrated by place (Sherman, Gartin, & Buerger, 1989). New students of the spatial distribution of crime almost always open their eyes when they realize that crime hot spots are characterized by tight concentrations of dots on a map that signify impoverished, urban geographies. Sometimes criminal events that are exceedingly rare in general pile one atop the other in certain cul-de-sacs and street corners.

When looking at crime maps of impoverished areas in large cities, it is preposterous to pretend that the spatial concentration of crime is exclusively the result of economic variables when there are other obvious and direct reasons for this distribution—namely, crime repeatedly occurs in places where many known offenders live and tends to occur disproportionately among people who trade in drugs and frequent drug corners. To ignore this fact leaves unspoken the volatile and violent scenes that drug dealers, their customers, prostitutes, and pimps create; this oversight could lead to faulty assumptions about connections between neighborhood characteristics and crime. By comparison, how important is it that in poor neighborhoods people living in cramped apartments would rather be outside? This idea has been postulated as a reason for concentrated crime (Stark, 1987). Does this explanation justify why persons would go to the crack corner at unusual hours or why some might smack strangers in the head with a pistol? Statistical analysis of spatial antecedents of open-air drug markets and crime without mention of or attention to the markets' presence obscures the impact of dangerous activities on offending opportunities, criminal opportunism, and, most importantly, the outlaw spirit that develops in places where drugs are distributed in this way.

Neighborhoods infested with illicit drug markets create opportunities not only for those few offenders who make money exclusively by dealing, but also for criminals such as fences, small con artists, and street robbers. By the street criminal's standards, the opportunities arising from the constant cash transactions in drug markets are tremendous, even if the risks associated with them are sometimes high. For example, drug markets are one place where there is an almost certain way to get otherwise scarce cash for those who would hazard it. A robber explains his consideration of options:

> *You go to a [convenience store] and they got a little drop box in it and the [clerk] can't get in there. When you hit the register, they got $40. You would be pretty mad. But, you know a big dope dealer, he got a lot of dope and usually money goes along with dope. He might have some nice jewelry. So, usually it works out better when you hit dope dealers.*

Drug corners and drug neighborhoods, especially those that appear to be beyond authorities' control, lead to cavalier attitudes toward violence that may seem incredible to all but those who have stayed there long enough to know that violence erupts frequently and can be sparked by any number of incidents. In these tinderbox conditions, violence may occur simply because someone nearby is perceived by one or more to have wronged or insulted someone else.

Inhabitants of drug corners are prepared for violence to occur at any time. In the worst neighborhoods, in addition to incidental violence, battles may ensue between those who possess drugs, drug money, or convenient access to customers and those who want their assets. A robber of drug dealers recounts the brutality among his cadre when it came to taking their piece of the local drug economy:

> *I don't want to kill. I don't want to go in nobody's house. If it comes down to it, I'll do it, but I'm not gonna go in there and shoot babies and children and women and shit. [My partner] will go in there and [if] a woman flinch[es], he gonna down her, you dig. Whoever move gonna get killed in an instant. [If someone's] like, "Man, I ain't got no dope in here. Ain't nothing in here, bro," then [he'll] threaten him with a gun. Hitting him with pistol. It took an O.G. [Original Gangster] from the hood to tell me how to do it right, bro. Take a stick pin and press it right up under their fingernails and they will show you where the dope is at. A stick pin get their mind all the way right. They gonna tell you everything. Stick pins, that's some painful shit. . . I've had cases where I have to pull out the stick pin.*

Such men know well the surest and easiest way to make a robbery worthwhile, as well as the extreme risks that such worthwhile crimes pose:

> *Might have $10,000 in the house, some guns, and shit, and we don't like him. He can be somebody we fuck with, you dig, but he got it and he don't break bread so we gonna hit him, you dig. Go kicking in doors, doing what we got to take it. But, there's a lot you gotta worry about. You don't know who's upstairs. [That person] maybe going to shoot at you, you know.*

A dealer confirms that this thinking is prevalent in recognizing himself as a high-profile target and pointing to the importance of mental preparation and readiness for violence:

> *See, I am going to put it in the aspects of the game I was in. I'm gonna put it in those aspects. You know what I worried about more than anything? Robbers! Dudes that would come kidnap you because you were doing your thing and had money. And if you have a name . . . If you are really doing it, then you are going to have a name. See a name comes with the game. If you are really doing it, then your name is going to have a rank and the police are going to get to know your name and try to bust you. They gonna harass you, but that is nothing. When the people come, the robbers and the jackers . . . I mean the consequences is different. They gonna kill you to get what you got.*

Only a few have sufficient "heart," as reckless brutality can be called, to rob reputable dealers; even so, such events are not rare (Jacobs, 2000). There are, of course, easier and more convenient targets in neighborhoods where drug trafficking is part of daily life, compared to attempting to extort or rob dope dealers. Where there are drug dealers, there are also addicts and other hangers-on. These peripheral players represent low-hanging fruit. Still, many of these crimes require putting familiarity with violence to use. A man who targeted the drug-trafficking crowd remembers his hunting grounds, "They had this certain place where a lot of drug activity . . . you know, drug dealers would hang out." Another explains, "People would ride through there all through the day wanting to buy drugs." Lines of drug users flowing in and out of places where they may or may not be welcomed as insiders are perfect marks for the many awaiting opportunists known to them only by street monikers, if at all.

> *I go to the car and as he stopped . . . It was like, it was the thing where if you [are] coming through in your car and you come to buy drugs, you have to put your car in park or cut your keys off because dealers knew [that otherwise customers] would take drugs and pull off [without paying]. So, I say, "Yeah I got something. Put your car in park." They were used to that. I grab the door [as if] I am about to show him something, and [then] I grabbed him and I throw the gun at him.*
>
> *I didn't have to like go around and sit on a corner at wee hours of the night or nothing [for victims]. They come through. The neighborhood was that kind of neighborhood. Hey man, that's what happens—that's all.*
>
> *[The victims] were three or four places back [in a row of cars]. Already had four cars in front of it. You got people selling drugs out there . . . And it just seemed like the opportunity when the light didn't turn. It was like it had already turned; it was*

red and it didn't turn back green. And I just [thought], "Pow!" Just something went off. Boom! And the guy just went straight down and we jerked him out. It was on from there.

The flow of drug customers is a significant source of opportunity for the prepared, but other parts of the informal economy also open doors. Heavy traffic and anonymity of impoverished city life allow offenders convenience and mechanisms for getting close to victims (Wright & Decker, 1997).

In New Orleans we got a lot of people, you know, asking, begging for money and stuff. What I'm saying is that you more or less act like you selling something. You see the people are out there with food stamps; people selling flowers; people with big ole cans of change and stuff. [I] got a nice little can and be acting like I'm begging for change or something like that, and walk up to them. Some people ain't gonna let they window down, but some people will let they window down to put change. [That's when I] put a gun at they throat.

These robberies and other crimes are enabled by a neighborhood context in which offenders assume that they can operate with impunity based on both their past ability to do so and their reading of what neighborhood residents will tolerate or be able to stop. The fact that drug markets exist in the open or are only slightly hidden imparts a message clearer than that communicated by other signs of neighborhood deterioration. Several offenders explained that normal activities in their neighborhood created a situation where any people who knew about crime were unlikely to report it. Some thought that it should be understood that to be on certain streets represented tacit entry into the world of the street hustler and acceptance of the risks of street life; their view was confirmed when victims rarely called police or when nothing changed as a result of these calls.

Nah, they wouldn't call the police because they knew they was doing dirty anyway. Some of them have warrants. They don't want to turn themselves in. Some of them be the ones that be like ducking their own people [their own dealers or fellow gang members]. They looking for them. I know I got a chance in jacking them because I know they ain't gonna call people.

We moved mainly in the neighborhoods where . . . See, in our neighborhood, ain't no such thing as a police patrolling. Everybody is black, and everybody is young and thugging. Ain't no cops over there.

The practiced behaviors that some offenders continually engage in firmly establishes in their minds that they exist in places where crime is safe and where everyone understands this to be the case. Cut off socially, geographically, and economically from the conventional world, high on powerful drugs, and familiar with the nighttime streets, they may mistakenly assume that all others recognize dangers and are part of the game. Witnessed crimes where no one is arrested or prosecuted also lead to the immediate perception that the police and law are distant, even if officers are always on patrol nearby. Offenders know well how dangerous and crime ridden their locales are. Several interviewed here described their friends dying or being shot themselves very near places where they committed their own crimes; others used more mundane accounts to illustrate the unusually dangerous conditions in the places where they offended.

> *David and Danny always carried [9-mm guns]. They bragged about how they have it and pointed it toward me and my mother when we got out of the car. They always carried that. Danny always had that big 12-gauge shotgun. It's nighttime in the little area we live in. There's not too much traffic. And there's not cops—I'm talking about they don't come in that area. You know the school, the church, everything is closed down. It's not the first time something like this happened on the street. They've had other killings and everything on this street. So, you don't travel it much at night.*

Places perceived to be crime hot spots provide those living in these places with excuses and anonymity.

OUR PLACE

People interpret their spaces. This concept lies at the heart of the "defensible space" approach to crime (Newman, 1972). Behaviors, movements, and mood are shaped by a shared understanding not only of how places look, but also of what places mean—ask any landscape or structural architect. Abstract understandings of place include degrees of ownership, attachment, and allegiance to them. Needless to say, most street offenders are not metropolitan, as many are tethered to the spots they know best. They often have a special affinity for their neighborhoods and attribute strong meanings to them. They are familiar and relatively comfortable with their places and the rules of conduct there.

Binding oneself to and living in an area brings a familiarity with the people and routines of the territory, which affords opportunities and protections and contributes to a sense of ownership and place. As previously mentioned, targets are easy to find for local offenders, and unusual incidents and opportunities are noticed. As in all communities, access to the grapevine is important and adds to confidence, security, and opportunity. As one man put it, "If you live in a town, you hear everything, [like] that this person has got money. So, I thought, 'Well, look, if he's got money, then I am going to get him for everything he has got.'" Several interviewees mentioned that specific information about targets could be gained by listening to storied events of wealthy street characters and tales of recent big scores.

In addition to information they provide, friends can diminish some of the risks of street life and crime. Ideally, this is true of retaliation from stranger victims:

> *You not really gonna come back there and try to find somebody that done you wrong. Even though I was on drugs, a lot of people would still be there for me, you know. As far as if you come shooting at me, they gonna shoot at you. If you try to kill me . . . you know, I'm saying it was just a raggedy rough life. That's kind of how I could carjack people. You know, that was my setting in the drug area.*

The distinction between outsiders and insiders sometimes is maintained stringently and severely affirming that place is salient in some neighborhoods.

> *There have been times when we'll go jump on somebody or something. See somebody we don't like or from another city—we go jump on them and everything. You know some are ready to kill and everything.*

A recurring theme was to point out that strangers who have the markings of successful but locally unknown street players are in particular danger.

> *If he's not cool with us or is someone we can trust, then we gonna get the fellow. But, that's the only thing that makes them stand out. You know, people flashing jewel or people flashing money—that's why we would get an individual person. But, if they just puttering along on about their business, we don't usually even pay them no attention.*

One of the most obvious ways that people lay claim to territory is with the assertion that people like them or close to them literally control its economy and events. An offender neatly articulates an obvious reason for feeling safe as a criminal in his neighborhood (i.e., murderous retaliation):

> *You would find out if somebody . . . see, them people don't say nothing. They see people get their heads busted in broad daylight on main corners and . . . People come out and they don't know nothing. If they call the police when we doing something, then we gonna. . . see, we don't have to worry about no police. We knew nobody gonna tell the police.*

Because localities have rules of conduct, they can become linked intimately with the identities of inhabitants. To be from a place is understood by some to mean that the individual was part of its character and could get along under its code. Moreover, it is in the interest of those who claim to be at the center of constructing the norms and business of a place to ensure that their attributes are continually appreciated. Many of the challenges, playful and serious confrontations, and instances of criminal victimization designed to show street-corner dominance may be interpreted not only as claims based on maintaining the integrity of one's own personal criminal identity, but also as claims to place and belonging. In enforcing the rules of criminal places through action, offenders maintain worlds and places where others appreciate their attributes. Many come to believe that because of qualities that make them suitable for street life, they prevail—or at least do acceptably well—in criminal areas. Those who share their approaches to life and spend considerable time with them "on the streets" struggle for place by keeping local codes true and real.

GEOGRAPHIC EXCUSES

One of the most important elements in the decision to commit crime is the psychological process of sanitizing the conscience so that the criminal activity can be accomplished. For this reason, a large swathe of the literature in criminal social psychology is devoted to the ways that offenders make sense of or account for their criminal acts and related behaviors (Maruna & Copes, 2005). Offenders' enabling accounts are designed to explain or make excuses for behavior that the culturally distant or otherwise ignorant outsider might assess as aberrant or inexplicable and distasteful. The best possible pitch is made, although it is usually unconvincing. Criminologists have long thought that offenders mentally justify their acts before deciding to commit crime, yet also acknowledge that these explanations might be rough, ill conceived, and only partially articulated in the offenders' conscience before the act takes place (Sykes & Matza, 1957).

Before proceeding with a discussion of space and rationalization, it is important to recognize that for an excuse to work for the offender and for others, it must be viable to the offender's imagined and somewhat conventional audience. The offender explains why he does it in a way that normal people, as he imagines them, might accept. Therefore, rationalizations often have at least a chime of truth, however muted and dull. For example, in the abstract most people understand that oppressive workplaces and poverty might tempt some employees to pilfer products from work. While considering the act, pilferers know that social audiences will generally understand their decisions, which helps them carry out the task. For many of the offenders in the sample present here, their environments (i.e., the drug markets in their neighborhoods and the troublesome activities associated with these markets) serve to provide nearly perfect excuses.

Ordinary Business

Offenders are often keen to acknowledge that crime is seen as ordinary business in their neighborhoods and hangouts. Recognition of this fact led to two significant recurrent themes in their accounts regarding how they view crime and its appeal.

First, violent acts and callous disregard for others are viewed with a morally neutral stance because others are portrayed as having consented to play within the brutal rules of the street game. Crime can be approached with surprising moral neutrality, as some of the quotes presented earlier in this chapter demonstrate, because recognition that crime is a business signifies that offenders bear neither ill will toward their victims nor guilt for acting senselessly or against innocents when it is done.

> *Everything revolves around money and business. It's business. When we go to war with people over there, if we go to war with them, that shit's just business. They got to do it. They gonna kill beaucoup people, but it's just business.*

Being in criminal places or involved in criminal transactions means that one has recognized the risks and decided to take part in the activity, or at least wandered too near to an extremely dangerous place where risks are known. The assumption is that all who occupy this space do so knowing the risks, and spoils, of doing so. The expectation is that, at some point, even the most virulent aggressors will be victimized. Offenders embedded in these areas recognize that they are not alone in their mindset. Many have a surprisingly fatalistic and cavalier attitude about what they see as the inevitable consequences of life and competition in their neighborhoods and on the streets. As they see it, one of the unwritten rules or ethics of their places is that the strong survive and the weak perish. Of the possibility of others robbing or killing him, an offender contends:

> *If they ever catch me down bad, I got to respect it. Because all of this dirt that I have done, and that is the type of life I've lived. I got to respect it, because it comes with it.*

The second form of rationalization that occurs when crime is perceived as ordinary business in a neighborhood is the belief that crime is appropriate behavior there. This belief appears in offender statements that point to the unlikely contention that everyone in the place is doing it. Four of the 30 offenders whom we interviewed stated that everyone in their neighborhood was directly involved in the drug world, and this belief was implied by several more. A more specific and temporally circumscribed example of the same rationalization follows in this carjacker's description of the immediate period preceding a city street party:

> *See, everybody in the 'hood was hitting cars on the regular during the Essence Festival and Bayou Classic to go to the French Quarter, [which is] packed with females. Everybody just jumping shop to go to that park.*

Sources of rationalizations also are found in potentially exaggerated references to the fact that signs indicating a certain place is criminal are so apparent that no one could ignore them. In unquestionably criminal places, what else could happen? An example follows:

> *You know that's where they dump. It's by the Mississippi River; it is right by the Mississippi River. They dump the bodies there. I mean they will dump kills and bodies, mostly drug dealers. So, we go right back there.*

Another offender remembers:

> *On the backway, you got a street only go down so far, and it got a dead end. That's where everybody go that steals cars. Takes them to get what they want out and burns them. So, like that whole street full of cars [is] like a junkyard.*

Local Knowledge

Many individuals interviewed in our study had long histories of crime and had begun offending early in life. Some had worked from an early age in drug distribution networks or gangs. More often, however, as youths they had spent considerable time in proximity to offenders and observing their activities from nearby. Of course, early exposure to criminal influences brings with it not only the objective conditions and opportunities that make crime more likely, but also the mindset that allows one to quickly and conveniently rationalize participation when confronted with potential sources of criminal opportunity. One young man put it succinctly: "See, when you are young, you watch a lot of drugs done and see how to get money and cars and stuff like that." Another offender claimed that the choice of engaging in crime is contingent on what the older men in the neighborhood do: "If they robbers, then the younger ones is going to be robbers; if they dealers, then they going to be dealers; and if they jackers, then they going to be jackers."

In the following passage, an offender recognizes explicitly how his place in the world, locally and writ large, figured in his preparation for crime and recognition of a criminal opportunity that led to a "spontaneous" carjacking.

> *It was a spur of the moment thing, because that morning I went to the store. I went to Winn-Dixie to be exact, and I was like, "Man, how am I going to get home?" When I walked out of the store, I had a little pocket knife and a Glock on me, so I was like, all these cars, and all I got to do is get me a brand-new one. I'll be straight with you: I grew up doing that and watching the older kids in New Orleans. So, I already knew it was easy. I'm in the parking lot, and I just did it.*

Being from certain places brings with it expectations of how one should respond in specific situations. These environmental expectations make choosing crime more palatable. After all, this is what people like them in places like these do.

Passing Through

As important as their acknowledgment that their place led them to be socialized into understanding how to get by in the streets—which can be shown on maps but are also metaphorical avenues—is the ready recognition by many offenders that they are simply bit players in a long-running drama or game. They may make their temporary mark and hoodlum reputation on street scenes, but the game will inevitably outlast them. In fact, characters almost indistinguishable from themselves will emerge to occupy their neighborhood roles when they abandon them. This understanding closely approaches one of the most frequently heard rationalizations among all who bend or break rules: "If I don't do it, someone else will."

In environments where crime has proliferated for years, it is easy to sustain this view. Actors and actions created by the continuing environment are perceived as easily replaced cogs within the larger system. Structural conditions and local histories have, in these portrayals, so constrained personal choices as to make them almost irrelevant for the occurrence of discrete events. Most street actors are quick to point out that they chose their life, but also remind the listener that they simply filled a niche and stepped into ongoing social circles of neighborhood criminals. When they opt out or go away, nothing will change, they say. One offender makes an obvious, and useful, comparison about conditions in his neighborhood and the prospect for continued violence in a more formal sort of behavior system:

> *Any person, you back them against a wall and you never know what they are going to do. You know, people you never believe [would enlist] go and join the Army or the*

Air Force or the Marines, and next thing you know, they're in Afghanistan killing Arabs and Muslims. People you'd never think of doing something like that—they are out there just doing what they are trained to do.

Another man refers to the inevitability of crime in a drug neighborhood. He notes that criminal activity has persisted while he has been in prison, and accepts that it will persist and be "popular" no matter what one says or does.

The drugs just bringing them down. Killing them slowly now, you know what I'm saying. It still goes on. Like in the middle of New Orleans, that is where this goes on. Carjacking is still going to be popular there—I don't care what you say.

No Way Out

Rationalization based on neighborhood conditions or familiarity with space would not be very convincing if offenders thought that they could be physically or socially mobile. But many believe that that their early mastery of the street environment precludes, or at least makes more difficult to obtain, outlooks that would lead to success in other places. The streets provide them with the self-perpetuating logic that plays a small part in their long histories of criminal choices. Although on many levels they are not, the lived conditions of the streets seem inescapable (MacLeod, 1995). Curiously, admitting to simply not paying heed to conventional morality or claiming to have never been taught any better can rationalize the most abhorrent acts, even if one claims to be an inherently valuable and good person. Purported or real ignorance of what law-abiding life is like is a perfect excuse.

If the streets raise you, or if you raised up in the streets, this is going to seem alright to you. It doesn't matter what the point is behind it or if the cops is behind. Everything just seems right. Most people do it for the fame, the cars, the big rides, or whatever. But, as we start getting independent, we start saying, "Man, now you got to make it on your own." Sleeping on the streets ain't no fun thing. Eating out of garbage cans ain't no fun thing. So, we like, "Man, this is just our life." You know it seems right for us to do.

Another offender points to the difficulty of overcoming the pull of the streets in answering a question about his post-release plans. His response reveals a dilemma for the released offender, who will in all likelihood return to an environment very like that known before prison. More frightening is the notion that while this territory and its rules are familiar, in some cases most other ways of life are somewhat foreign. The offender begins the journey knowing not only that returning to crime is possible, but also that it can seem like a legitimate course of action given his background and likely place. He explains:

> *I'm trying very hard—very, very hard. It would be easy for me to go back to doing what I was doing. That's easy there, you know what I'm saying? The hard part is trying not to do it.*

CONCLUSION

Admittedly, the places that produce high rates of violent criminal offenders in the United States are far better off in comparison to their counterparts in economically peripheral nations. The most apparent explanations of place and crime intrude nearly as obviously in parts of the United States as they do in Brazil's *favellas*, however. Concentrated poverty and all the conditions associated with it make it difficult—nearly impossible—to precisely pinpoint the factors in these places that cause crime. While identifying discrete criminogenic factors and placing them in neat succession so as to discover mediators and moderators is a lovely armchair dream, it is important to remember that professors' penchant for precise, neat, and sophisticated explanation sometimes obscures reality. Where many serious and persistent offenders grow up and live, it requires little imagination and only modest inferences to make a few connections between places and criminal mindsets, but this intricate web may be difficult to model formally.

Young men and women on the streets and in drug-ravaged, impoverished communities think differently about the prospects of crime and individual criminal acts than outsiders. Places provide opportunities and transform thoughts. Inhabitants may not give it a great deal of mental attention, but they have an intimate understanding of place based on history and experience; they have also estimated their prospects in the larger world, and found them wanting. Offenders predict what they can hope to accomplish according to where they are and how they live, and those predictions are often bleak. In many places in the United States—the "internal peripheries," as they are fashionably called—the causes of crime are not mild incivilities found in environments or intricacies of childrearing and development. A little trash on the streets, scratched-up school lockers, a few discarded needles, or cantankerous neighbors are not harbingers of the idea that shooting someone is a feasible and reasonable solution in some circumstances. These environmental conditions in combination may play a small role in supporting the latter thinking by making despair seem suitable, but the more important fact is that gunfire in the area is common. The violent view comes from an understanding of brutal environments and its expectations, often gained in the lifelong experiences of street children.

When discussing place and cultures of place, it is tempting to discount offenders' ability to construct meanings and identity. Theorists should remember that offenders seldom shirk from danger or street success. Although most learn the ways of the streets before the age of legal emancipation, they also actively place themselves in situations where they can show what they have learned and further expand their "street" identity. At some deeper psychological level, all may be in a defensive mode, hoping to ward off future attacks and insults by trying to look hard and bad, until ultimately this stance becomes an ingrained habit.

It is true that these offenders' views on how to behave in the streets might result from past confrontations and victimization at the hands of other hard men. But they are predators nonetheless, and may be unsympathetic to those who do not stand up well under the code of the streets; many look constantly for opportunities to assert their criminal selves and are especially dangerous in their familiar environs (Anderson, 1999). Perhaps there are not many options about where one lives, but there are always alternative models for how one lives. It may seem unrealistic to stay in the house, end bad habits, withdraw from drugs, or work all day for low wages, but generally such decisions are possible and are the better choices among unpleasant options.

Neighborhoods do not produce homicides like randomizing shufflers at casinos produce aces; rather, these outcomes occur because such milieus engender twisted logic among a few residents who prey upon others and for whom committing a robbery or pulling a trigger seems to make sense in certain circumstances. These persons pass their logic along to the next generation, thereby perpetuating the cycle of violence. If asked in private, the neighbors in impoverished neighborhoods can point out such hostile, dangerous persons and can identify those who will likely inherit their worldview, if not with perfect predictive accuracy. Given that these environments are perceived as offering few avenues for reaching better places, no small number of persons will find that they can thrive in another way—and in doing so, they strengthen the particular dictates right where they are.

We realize that the depiction of offenders in this chapter may seem harsh to some readers. It is difficult to avoid the potential for condemnation when typifying subcultures that lead to violent street crime. Nevertheless, conclusions need not be as harsh as the depiction of street life. To say that offenders will find opportunities, identities, and excuses linked to places is not to say that nothing can be done. The emphasis on agency says nothing about how we can help potential offenders reach different conclusions. Indeed, the most compelling reason for undertaking efforts to improve conditions and standards of living in deteriorated neighborhoods is that these endeavors remove convenient rationalizations for failure and misbehavior. Even modest improvements may make horrible behaviors seem out of character (Jacobs, 1961). Gangster posses are almost laughable among youth from middle-class and affluent neighborhoods (except in the rare cases where they lead to authentic tragedy) because they are based on different cultural understandings of place and class and the likely outcomes of those understandings. No young person should be able to cast himself or herself as a societal refugee and endanger others easily, but the responsibility to change this behavior is found only in part in offenders' freely formed and subculturally shared perceptions. Those with resources to do something share responsibility for the objective conditions and ease of excuses in chaotic neighborhoods (Anderson, 2003).

Offenders, like the rest of us, know that they are part of a predictable environment. It will outlast them. It will produce successes and failures according to particular environmental dictates. It provides reassurances that personal mistakes and victories are irrelevant to the big picture. It is comforting to think that life's contingencies can be viewed as so structured that individual choices are incidental. Because the fundamental rules and outcomes of the practiced game do not change, it is justifiable to play it reasonably and to conclusion despite the likely potential for unfortunate outcomes. Even if one must walk away from a gambling table busted, it is comforting to know that other players with a chance of winning remain; those familiar with the tables also take comfort in the fact that there is a place where they know their way around and fit in easily, even when the odds are stacked and almost everyone loses eventually. Just as the compulsive gambler is comforted by the casino and knows that walking away from any single event victorious or busted will not change long-run prospects for self or similar others, so the street player finds comfort in the ongoing and familiar game. Pursuits for local acclaim continue because an intrinsic value is placed on established personal patterns of behavior, subcultural dictates, and appropriate thrills. Understandings of place transcend the spatial and occupy thoughts about how to behave. They contain the same metaphorical meaning found in the phrases "living on the streets" or "living on skid row," which can mean spending considerable time on certain streets or, perhaps more importantly, living life according to the dictates of one or another form of street culture. The streets become part of who one is and responsible for decisions.

REFERENCES

Anderson, E. (1999). *Code of the street: Decency, violence and the moral life of the inner city.* New York: W.W. Norton.

Anderson, E. (2003). *A place on the corner* (2nd ed.). Chicago: University of Chicago Press.

BBC News. (2005, October 21). Rio slums blighted by gun crime. Retrieved January 12, 2011, from http://news.bbc.co.uk/1/hi/world/americas/4338652.stm

Bourgois, P. (2002). *In search of respect: Selling crack in el barrio* (2nd ed.). Cambridge, UK: Cambridge University Press.

Brantingham, P., & Brantingham, P. (1984). *Patterns in crime.* New York: Macmillan.

Cloward, R., & Ohlin, L. E. (1960). *Delinquency and opportunity: A theory of delinquent gangs.* New York: Free Press.

Cohen, L. E., & Felson, M. (1979). Social change and crime rate trends. *American Sociological Review, 44,* 588–605.

Jacobs, B. (2000). *Robbing drug dealers: Violence beyond the law.* Somerset, NJ: Aldine.

Jacobs, B., Topalli, V., & Wright, R. (2003). Carjacking, streetlife and offender motivation. *British Journal of Criminology, 43,* 673–688.

Jacobs, B., & Wright, R. (2006). *Street justice: Retaliation in the criminal underworld.* Cambridge, UK: Cambridge University Press.

Jacobs, J. (1961). *The death and life of great American cities.* New York: Random House.

MacLeod, J. (1995). *Ain't no making it: Aspirations and attainment in a low-income neighborhood.* Boulder, CO: Westview.

Maruna, S., & Copes, H. (2005). What have we learned from five decades of neutralization research? *Crime and Justice: A Review of Research, 32,* 221–320.

Newman, O. (1972). *Defensible space: Crime prevention through urban design.* New York: Macmillan.

Sampson, R. J., & Raudenbush, S. (1999). Systematic social observation of public spaces: A new look at disorder in urban neighborhoods. *American Journal of Sociology, 105,* 603–651.

Sherman, L. W., Gartin, P. R., & Buerger, M. E. (1989). Hot spots of predatory crime: Routine activities and the criminology of place. *Criminology, 27,* 27–56.

Stark, R. (1987). Deviant places: A theory of the ecology of crime. *Criminology, 25,* 893–910.

Suttles, G. (1968). *The social order of the slum: Ethnicity and territory in the inner city.* Chicago: University of Chicago Press.

Sykes, G., & Matza, D. (1957). Techniques of neutralization: A theory of delinquency. *American Sociological Review, 22,* 664–670.

Wright, R., & Decker, S. (1997). *Armed robbers in action.* Boston: Northeastern University Press.

CHAPTER 5

Cyber-Related Violence

Brooke N. Miller
The University of Texas at Dallas

Robert G. Morris
The University of Texas at Dallas

Computers and the Internet have changed the way the modern world communicates. Indeed, there are now approximately 2 billion users of the Internet worldwide. The widespread adoption of the Internet, along with the availability of more affordable computing hardware, has turned this technology into a staple for most U.S. households and has changed the way that people can communicate. It is not uncommon for Americans to communicate primarily or perhaps solely online. We shop, chat, meet new friends (and sometimes romantic partners), play, learn, and express ourselves in the cyberworld. Indeed, the World Wide Web had had a profound impact on human behavior.

While exciting and beneficial, the cyberworld also has a dark side. Our transition into the information age has spawned opportunities for new forms of crime and deviance, including some forms of violence, and has allowed new methods for the commission of some traditional crimes to emerge. Only recently have scholars had an opportunity to try and understand the dark side of the Internet. Subsequently, the research is sparse in this regard. The goal of this chapter is to shed some light on the current state of our understanding several online behaviors that relate to violence, including both behaviors that occur solely online and cyberviolence that is ultimately translated into physical violence.

As most readers will be acutely aware, cyberactivities encompass a broad spectrum of behaviors, ranging from online gaming, e-mail, and chatting, to deviant behaviors including cyberbullying, malicious computer hacking, and cyberterrorism. While many computer-related behaviors are neither criminal nor violent, increasingly we hear about cases of violence stemming from some form of online interaction. The dark side of the Internet, by and large, manifests itself under a shroud of user anonymity. Users can spawn a new identity and convey themselves as whatever, or whoever, they want to be at any given time for any given purpose. This reality has both positive and negative attributes. Such anonymity allows for more freedom, which has positive attributes such as enabling an individual to explore and learn beyond what may be normally available. On the downside, there is little or no oversight of a person's online behavior (particularly among adults), which can lead to deviance that might not have occurred otherwise. In short, the Internet has expanded access to friends, family, and information previously difficult to acquire. For individuals with deviant/criminal intentions, however, this expansion has created greater access to potential victims (i.e., opportunities for crime) in a world that is difficult to police.

Currently, there is no formal tracking system for cyber-related activities that have an increased potential to lead to violence, and criminologists are still developing methods by which to measure and assess this problem. While some targeted users can be tracked, apprehended, and prosecuted, others—particularly those who have not attracted the attention of authorities—can typically carry out their online activities with impunity. On a weekly basis, new media stories highlight the victimization of individuals via a variety of cyber-related offenses. As an extreme example, some youth have become so distraught over being bullied online that they have committed suicide. In addition, victims of cyberstalking discuss their suffering and feelings of being violated through such activities; stories tell of children abducted by "friends" whom they met online but who turned out to be cyberpredators; new leaks are discovered in social networking sites exposing personal information for users of certain components of the websites. On a broader scale, we are presented with terroristic messages inciting violence against innocent victims and warning of attacks that could occur at any time in any place. Given that these behaviors pose an increasing threat for the safety of millions of Americans, and others across the globe, our need to better understand cyber-related violence, and its correlates, is essential. At present, there is much to learn about individual offenders and victims of these offenses.

This chapter addresses a variety of deviant cyber (online) behaviors that have the potential to evolve into actual violence (both physical and nonphysical). The topics covered here include cyberbullying, cyberstalking, cyberpredators, online gaming, and social networking environments as platforms of cyberviolence. Beyond explaining these growing concerns, the chapter presents relevant anecdotal and scholarly evidence surrounding these topics, including a discussion of cyberterrorism and the need for further research in this area.

CYBERBULLYING

As new forms of media technology have become readily accessible to adolescents and children, the potential for harassment and threats of physical harm to peers has increased. Cyberbullying has been defined as the situation in which a child, preteen, or teen is tormented, threatened, harassed, humiliated, embarrassed, or targeted by another child, preteen, or teen using the Internet, interactive and digital technologies, or mobile phones (Aftab, 2010). Other definitions specifically identify cyberbullying to include the use of e-mail, instant messaging, text digital imaging messages, and digital messages sent via cell phones, webpages, blogs, chat rooms or discussion groups, or other information communication technologies (Kowalski, Limber, & Agatston, 2008). Hinduja and Patchin (2009) add that cyberbullying generally involves malicious aggressors seeking either implicit or explicit pleasure or profit through the mistreatment of others. One thing is certain: Cyberbullying is not limited to youthful victims. The studies that do exist, while limited, have demonstrated the emotional consequences of cyberbullying, including fear of offline victimization, stress, anxiety, depression, and lower self-esteem (Grills & Ollendick, 2002; Hawker & Boulton, 2000; Kochenderfer-Ladd & Skinner, 2002; Nansel, Overpeck, Pilla, Ruan, Simons-Morton, & Scheidt, 2001; O'Moore & Kirkham, 2001). Much as with the potential consequences of traditional bullying, recent studies have linked cyberbullying victimization with delinquent behaviors such as vandalism, shoplifting, truancy, dropping out of school, fighting, and drug use (Ericson, 2001; Loeber & Disheon, 1984; Mangusson, Stattin, & Dunner, 1983; Olweus, Limber, & Mihalic, 1999; Patchin, 2002; Rigby, 2003; Tattum, 1989).

The possibility that cyberbullying victims may experience equivalent symptoms as victims of traditional bullying victimization is of growing concern. In addition to the possible outcomes noted previously, several studies have linked traditional bullying to serious forms of school violence such as school shootings (Patchin, 2002; Vossekuil, Fein, Reddy, Borum, & Modzelski, 2002). As cyberbullying victimization increases, the ability to address the potential threats of offline violence becomes more important. As youths become increasingly exposed to the cyberworld, new opportunities for victimization emerge. Thus it takes no stretch of the imagination to consider cyberbullying as having the potential to transcend into physical victimization—in the school environment, for example—including violence and physical aggression. In this sense, it is important to consider the various methods of victimization used by bullies in the cyberworld.

Method of Victimization

In recent years, the media have reported on a variety of cases of cyberbullying, profiling the victims of such bullying as well as their attackers. Simultaneously, identifying the characteristics of both cyberbullies and their victims has become a popular topic among scholars from varying disciplines (e.g., criminology and psychology). A 2005 survey by the National Children's Home Charity and Tesco Mobile of 770 youths between the ages of 11 and 19 reported that 20% of respondents reported cyberbullying victimization (National Children's Home, 2005). Similarly, Patchin and Hinduja (2006) found that almost 30% of the adolescents in their study reported having been victims of online bullying, including behaviors such as being ignored, disrespected, called names, threatened, picked on, or made fun of, or having had rumors spread about them by others. For these individuals, the prevalence of bullying in a variety of forums presents a new challenge in figuring out a way to avoid being bullied.

In a recent study, Wang and colleagues (2009) suggested that adolescent "prevalence rates of having bullied others or having been bullied at school at least once in the last 2 months were 20.8% physically, 53.6% verbally, 51.4% socially, or 13.6% electronically" (p. 372). While social and verbal forms of bullying were found to be the most prevalent behaviors in this study, the occurrence of physical and electronic bullying as a continual threat could influence the behavior and emotional well-being of victims. Ybarra and Mitchell (2004) discovered that 19% of young, regular Internet users were involved in online harassment over a specific period of time. Similarly, Wolak et al. (2006) found a 50% increase in youth victimization among those suffering online harassment in the form of threats or other offensive behavior, either sent online to the victims or posted online about the victims for others to see. As technologies such as cell phones (e.g., via text messaging) and laptop computers become more accessible for youth, the potential for bullying victimization is likewise expected to increase.

Evidence suggests that the development of new media technologies has provided a variety of new forums in which bullying can occur. While bullying victimization was previously restricted to environments in which the victim and the offender were face to face, the Internet has added a new dimension to the bullied/bully relationship. Further, such technology offers a previously unavailable means for bullies to target victims whom they have already targeted at school. For example, when Ybarra and Mitchell (2004) examined groups of adolescents who had been either victimized or engaged in online aggression or online sexual solicitation against a peer, they found that only 23% of youths victimized by electronic aggression also experienced harassment at school. Li's (2007) examination of 177 seventh graders in an urban city discovered that 15%

of respondents had victimized others using electronic communication tools. These findings suggest that the use of new forms of media technology may have created a vulnerability for victims who may not have experienced bullying elsewhere. In other words, youth who would traditionally have a low probability of being bullied now may become targets in the cyberworld. Additionally, 24% to 76% of respondents reported experiencing offline physical victimization (Ybarra & Mitchell, 2004).

When Raskauskas and Stoltz (2007) studied 84 youths, they found that the most common form of electronic victimization was via text messaging (32.1%), followed by Internet or website (15.5%), and picture phone (9.5%). Text messaging was also the most likely source of electronic bullying within this sample. An example of text messaging victimization was presented by Tench (2003), who reported on a 15-year-old girl described as experiencing bullying both at school and on the Internet. While the bullying initially consisted of only name calling, this behavior was later followed by the creation of a website that posted her picture and made fun of her appearance. The website was available for several weeks before the girl became aware of it. The findings in older studies corroborate those mentioned previously. For example, a British study conducted in 2002 found that one in four youths aged 11 to 19 had been cyberbullied (National Children's Home, 2002). Thorp (2004) found that approximately 6% of youths had experienced online harassment.

Gender Differences

Several studies have looked at gender differences between cyberbullies and cybervictims. Some have found that boys and girls may experience different types of bullying, whereas others have suggested that girls are more likely to experience victimization (Hinduja & Patchin, 2009; Raskauskas & Stoltz, 2007; Wang et al., 2009). Others, however, have not found any gender differences in cyberbullying victimization or perpetration (Berson, Berson, & Ferron, 2002; Williams & Guerra, 2007). Related research suggests that boys are more likely than girls to be cyberbullies (Li, 2007).

Collectively, these results suggest that males may be more likely to engage in cyberbullying as offenders, whereas females are more likely to be the victims of cyberbullying. These findings are not clear as to the relationship between the victim and the offender, however. For example, it is not clear if girls experiencing cyberbullying are being victimized by boys, and vice versa.

In the end, it is important to acknowledge that the research surrounding cyberbullying is in its infancy. Thus readers should take caution in making generalizations given the current state of the literature on the topic.

Feelings Resulting from Cyberbullying

Individuals experiencing cyberbullying may fear for their safety offline due to intimidation or negative treatment occurring online. Research on the influence of victimization resulting from traditional bullying have found a variety of negative consequences from this behavior, including general psychological distress, poor psychosocial adjustment, heightened anxiety, depressive symptoms, and a lower sense of self-worth (Grills & Ollendick, 2002; Kochenderfer-Ladd & Skinner, 2002; Nansel et al., 2001; O'Moore & Kirkham, 2001). As research continues to develop surrounding the impact of cyberbullying, it is possible that investigators will confirm that this form of victimization has similar consequences. For example, several studies have found that youth experiencing online harassment suffer from stress and embarrassment (Finkelhor, Mitchell, & Wolak, 2000; National Children's Home, 2005; Raskauskas & Stoltz, 2007; Ybarra & Mitchell, 2004).

A respondent from Hinduja and Patchin's study described the feelings resulting from online victimization in the following way:

> *I was talking to someone in a chat room and they started telling me things. Like was I really that stupid and making fun of me. I told them privately to please stop and they wouldn't. They then told me they were going to harm me and I was scared because I don't know how but they knew where I lived. I am scared sometimes. One time someone made me feel so bad that I wanted to kill myself because I believe those things that they said. My friends calmed me down and told me not to do anything dumb. I dislike it when people spread rumors online about you and it has happened to mostly everyone who chats (Hinduja & Patchin, 2008, pp. 142–143).*

Cyberbullying victims have reported responding to these behaviors in a variety of ways, including removing themselves from the online forum in which the bullying occurred or feeling forced to stay offline for a more extended period of time. Although it might seem as though these are reasonable alternatives for victims of cyberbullying to decrease the frequency or occurrence of victimization, youths today spend so much of their time and life online outside of school hours that the feeling of forced avoidance may be considered a form of bullying in and of itself.

While cyberbullying takes place online, it has the potential to evolve into offline violence. Hinduja and Patchin (2007) discuss the possibility that outcomes of online bullying could lead to real-life violence in a manner similar to traditional bullying. For example, previous studies have linked traditional bullying victimization to the most serious form of school violence—that is, school shootings. While research has yet to determine if the same outcomes may result from cyberbullying victimization,

Hinduja and Patchin (2007) argue it is possible for several reasons: "1) the permanence of computer-based messages (compared to verbal statements), 2) the ease and freedom with which statements of hate can be made, 3) the invasive nature of malicious text via personal cellular phones and personal computers at all hours of the day and all days of the week" (p. 93). Once information has been put online and disseminated to a vast number of people, it is difficult to undo any harm suffered by the victim. Additionally, the constant and persistent access to this information may ensure continual discomfort for the victims.

Increasingly, media reports have cited youths who have not only been bullied at school but also experienced some form of cyberbullying, leading to them to hurt themselves or be hurt by other students. Raskauskas and Stoltz (2007) have argued that in extreme cases, electronic bullying has been linked to adolescent suicide. Using a sample of 1963 middle-school-aged students, Hinduja and Patchin found that youths who had experienced cyberbullying or traditional bullying had more suicidal thoughts or were more likely to attempt suicide than those who had not. This finding was true for those who were victims of cyberbullying as well as cyberbullying offenders. Specifically, these researchers' findings indicate that victims of cyberbullying were almost twice as likely to have attempted suicide as compared to youth who had not experienced cyberbullying (Hinduja & Patchin, 2010). A number of media stories have described teens around the country who were so persistently bullied both online and offline that they ended their own lives to get away from the torment of their attackers.

For example, on September 19, 2010, Tyler Clementi, a Rutgers University student, was recorded in an intimate moment with another male student. Clementi's roommate had posted on his Twitter account that his roommate wanted some privacy. The roommate set up a webcam that he then used to view and distribute Clementi's encounter from another student's dorm room. This encounter was broadcast over the Internet by Clementi's roommate, who also included messages stating that the next encounter he recorded of his roommate would be streamed live. On September 22, 2010, Clementi posted the following message to his Facebook account at 8:50 P.M.: "jumping off the GW [George Washington] bridge sorry" (Fenton, Calhoun, & Mangan, 2010). He then actually jumped off of the George Washington Bridge, ending his life. While there had not been a full investigation of the incident at the time of this writing, it has become clear that the recorded student was aware of the dissemination of his encounter online and was distraught by the incident, motivating him to end his life.

There have been many other examples of cyberbullying leading to threats of offline violence. In 1999, a 13-year-old boy was charged with making death threats against a female classmate via a website. Authorities contend that the website he set up featured an icon of a large gun that shot at a picture of the 13-year-old girl. It included a caption: "Hurry! Click on the trigger to kill her." More recently, the ranting of an 11-year-old girl responding to online postings regarding her supposed sexual history resulted in death threats and continued online harassment (Canning, Netter, & Crews, 2010).

Similarly, "swatting" could be considered an extreme threat from cyberbullies. The FBI (2008) describes swatting as individuals "calling 9-1-1 and faking an emergency that draws a response from law enforcement—usually a SWAT team." The practice generally involves a call describing a situation in which hostages are about to be executed or bombs that are about to go off. The call is set up to appear as though it is originating from the victim's residence while the perpetrator may live in another state; the offender will use a "spoofing" technology to change the phone number that is received on the caller ID screen of the 911 operator. The result is that law enforcement responds to a hoax address, where an unsuspecting victim may be surprised by a SWAT team who themselves fear for their own lives or the lives of others. In the worst-case scenario, the victim is hurt or killed by law enforcement officers, who are acting in good faith against an innocent victim.

For example, in 2007 a 19-year-old Washington state man pretended to call from the home of a married California couple, claiming to have just shot and murdered another individual. A local SWAT team responded to the home where the couple lived with their two young children. After hearing a noise outside, the husband grabbed a knife before going outside to find a group of SWAT assault rifles aimed at him (McMillan, 2007). While this case did not escalate to the violent victimization of any of the residents, the potential for violence is clear.

For individuals intent on victimizing unsuspecting others, the relative anonymity of the Internet and other computer technologies has provided a new opportunity to commit their crimes. According to the FBI, swatting has been reported in numerous states and offenders have used varying types of technology to instigate the event, such as telephone chat lines. In fact, between 2002 and 2006, five swatters were responsible for swatting incidents in more than 60 U.S. cities, resulting in more than 100 victims. The financial loss due to these attacks has been estimated to exceed $250,000 (FBI, 2008).

Clearly, cyberbullying encompasses a diverse variety of behaviors. Each of these has the potential to incite negative consequences for the victims of these behaviors. Research addressing the specific nature of the relationship between cyberbullies and their victims continues to develop, but it is also important to consider several other types of cyber-related activities with the potential to spill into offline violence.

CYBERSTALKING

Cyberstalking is similar to cyberbullying, although it is formally defined as involving at least one adult who is engaging in the behavior. Similar to the concerns surrounding cyberbullying, cyberstalking has the potential to evolve into offline violence. Cyberstalking can be just as threatening as stalking in the real world, leaving the victim vulnerable to anxiety, mental anguish, physical harm, and even homicide (Finn, 2004; Kennedy, 2000; Lamberg, 2002).

In 2001, the U.S. Department of Justice issued a report on stalking and domestic violence. This report began with a discussion of the increasing threat of cyberstalking to individuals in the virtual community. Traditional stalking has been described as involving repeated harassing and threatening behavior, such as following a person, appearing at a person's home or place of business, making harassing phone calls, leaving written messages or objects, or vandalizing a person's property. Additionally, stalking laws generally require that the perpetrator make a *credible* threat of violence against the victim or the victim's immediate family, or that the alleged stalker's course of conduct constitute an implied threat (U.S. Department of Justice, 2001). In recent years, definitions of cyberstalking have added to the definition of stalking "the use of electronic communications to stalk another person through repetitive harassing and threatening communications" (Kowalski et al., 2008, p. 50).

Cyberstalking more often involves threats—including violent threats—than pure harassment. Similar to cyberbullying, it can take on a variety of forms: threatening or obscene e-mail; spamming (in which a stalker sends a victim a stream of junk e-mail); live chat harassment or flaming (online verbal abuse); leaving improper messages on message boards or in guest books; sending electronic viruses; sending unsolicited e-mail; tracing another person's computer and Internet activity; and electronic identity theft. The fear in cases of cyberstalking is that it will evolve into offline (i.e., physical) stalking. Victims of cyberstalking may be either children or adults who have had their privacy violated or been the target of a cyberpredator. The vast quantity of personal information that is made available online may contribute to this problem (National Crime Victims Center, 2003).

The Internet has become an integral part of most individuals' personal and professional lives, providing stalkers with ample opportunities to take advantage of the ease of communications and increased access to personal information provided by this medium. According to the National Crime Victims Center, cyberstalkers may target their victims through chat rooms, message boards, discussion forums, and e-mail. A 1999 report by the Department of Justice described stalking motivations, both online and offline, as centered on the stalkers' desire to exert control over their victims. While cyberstalking does not inherently involve physical contact, the potential for violence may distress victims. Additionally, "whereas a potential stalker may be unwilling or unable to confront a victim in person or on the telephone, he or she may have little hesitation sending harassing or threatening electronic communications to a victim . . . as with physical stalking, online harassment and threats may be a prelude to more serious behavior, including physical violence" (U.S. Department of Justice, 1999).

Similar to traditional offline stalking, online stalking can be frightening for victims and place them at risk of both psychological trauma and possible physical harm. Victims of cyberstalking may experience fear or be threatened with offline violence. The availability of personal information on the Internet has resulted in increased opportunities for individuals to retrieve information that was previously difficult to obtain. Riveira (2000) suggests that many cyberstalking situations evolve into offline (i.e., traditional) stalking in which victims experience abusive and excessive phone calls, vandalism, threatening or obscene mail, trespassing, and physical assault. For example, Christina DesMarais, a writer who posted her work on the Internet began receiving hurtful e-mails as well as nasty posts in response to her writing over the course of several weeks. Eventually, she received a letter in the mail: "I'm the middle-aged man in the turquoise junker or the teen in the silver Camero [sic] or maybe even the woman in the Suburban—I wouldn't be caught running down my road again if I were you," the anonymous letter said" (Brandon, 2010). While it is not clear whether the sender of the e-mail had access to personal information for the writer, including where she lived, the vast amount of public information available online increases the potential for offline violence.

A number of states have begun to address the use of computer equipment for stalking purposes by including provisions that prohibit harassment activities as well as anti-stalking legislation. According to the National Crime Victims Center, some states—including Alabama, Arizona, Connecticut, Hawaii, Illinois, New Hampshire and New York—have specifically developed prohibitions against harassing electronic, computer, or e-mail communications in their harassment legislation. Additionally, Alaska, Oklahoma, Wyoming, and California have defined electronically communicated statements as conduct constituting stalking in their anti-stalking laws. Other states have passed stalking and harassment statutes that criminalize threatening and unwanted electronic communications. Still other states have laws prohibiting the misuse of computer communications and e-mail as well as laws that can be interpreted to include cyberstalking, as the language used to describe this behavior is broad in scope. Finally, federal statutes have been developed prohibiting online stalking. For example, the Violence Against Women Act, passed in 2000, made cyberstalking a part of the federal interstate stalking statute (National Crime Victims Center, 2003).

A 1999 report by the Department of Justice provides an illustration of the potential for violence resulting from cyberstalking:

> *In the first successful prosecution under California's new cyberstalking law, prosecutors in the Los Angeles District Attorney's Office obtained a guilty plea from a 50-year-old former security guard who used the Internet to solicit the rape of a woman who rejected his romantic advances. The defendant terrorized his 28-year-old victim by impersonating her in various Internet chat rooms and online bulletin boards, where he posted, along with her telephone number and address, messages that she fantasized of being raped. On at least six occasions, sometimes in the middle of the night, men knocked on the woman's door saying they wanted to rape her. The former security guard pleaded guilty in April 1999 to one count of stalking and three counts of solicitation of sexual assault. He faces up to six years in prison. A local prosecutor's office in Massachusetts charged a man who, utilizing anonymous remailers, allegedly engaged in a systematic pattern of harassment of a co-worker, which culminated in an attempt to extort sexual favors from the victim under threat of disclosing past sexual activities to the victim's new husband. An honors graduate from the University of San Diego terrorized five female university students over the Internet for more than a year. The victims received hundreds of violent and threatening e-mails, sometimes receiving four or five messages a day. The graduate student, who has entered a guilty plea and faces up to six years in prison, told police he committed the crimes because he thought the women were laughing at him and causing others to ridicule him. In fact, the victims had never met him.*

Over the last decade, increasing attention has been paid to cyberstalking and its victims. Additional research is needed to determine the frequency of occurrence and nature of such cases that lead to offline victimization. In many cases, the perpetrators of cyberstalking are considered cyberpredators—a breed of criminal that is examined more closely in the following section.

CYBERPREDATORS

The widespread availability and easy access to social networking sites, chat rooms, and other forums on the Internet in which strangers can interact has provided a potential way in which those interested in victimizing others can remain anonymous, at least while forming the relationship. Recent television shows such as *To Catch a Predator* have drawn attention to the threat of cyber (online) predators. The picture has been painted of the older man sitting behind a computer seeking out the young boy or girl, developing a relationship with the potential victim and filling some void the child may have in his or her life, and then setting up some kind of meeting with the child. While many stories surrounding cyberpredators relate to possessing child pornography or viewing it online, the fear by parents and law enforcement alike is that these individuals will extend their behavior from simply engaging with children online to molesting or hurting children offline.

There are a variety of methods by which cyberpredators can seek out underage victims. Because most children are not users of social networking sites, cyberpredators may visit other types of websites geared toward young children. For example, multiple sites designed for children allow them to play interactive games that lead to children receiving a stuffed animal that comes with a code to log on and care for the pet. Within the online world of caring for the pet, children are given the opportunity to set up a profile, add friends, and chat while playing games with their pets. Other sites do not require the purchase of an animal; instead, the child can simply create a profile to play games online and chat with other users of the same website. While all of these sites ask for parents to log on and confirm that their child can access the site, there is the potential for predators to gain access to children on these websites.

Adults are not immune to the threat of cyberpredators. A recent story regarding a man in Canada provides an illustration of a potential method used by these criminals to locate victims online, followed by actual physical violence against the victims. Terrence Moquin has been described as a predator who gained the trust of his victims by telling a false story about his U.S. military history. McIntyre (2009) describes Moquin's tactic for targeting victims and developing a relationship. Moquin would give women a fake name as well as a fictional account of his service as a U.S. Marine who fought in Iraq before moving to Manitoba, Canada, to work in the oil and trucking industry. Using similar tactics, Moquin met close to a dozen women in Manitoba over the Internet. Most of these relationships ended after Moquin was caught stealing money from the women or as a result of him abusing the women and their children. Moquin was a perpetual offender who spent time in jail several times as a result of his predatory behavior. His most latest reported conviction involved befriending a single mother of two children in an online chat room and then assaulting her when the relationship turned sour.

The acts committed by Moquin tap into most people's fear of cyberpredators—namely, both mothers and children suffered offline violence resulting from a relationship established online. However, it seems that the media have paid more attention to cyberpredators than have academics. As such, little is known about these individuals. An even greater problem is the fact that with cyberbullying and cyberstalking, there is currently no formal tracking of cases involving cyberpredators who follow through with offline violence.

In addition to the strategies mentioned earlier, another potential method for cyberpredators to locate victims may be through the use of social networking sites.

Social Networking and Online Games

Social networking and online game sites provide additional avenues through which cyberstalkers and cyberpredators can seek out victims. While social networking sites have provided a unique opportunity for individuals to interact with friends and family, they also provide a forum in which relationships can be developed in an anonymous fashion. Even as social networking sites have made it easier for families and friends to reconnect and stay in touch, they have also provided a system that can be manipulated and used to cause harm to other people.

Social networking sites, such as Facebook and MySpace, allow members to build a profile and share information with the public, upload pictures, and provide an outlet through which they can keep in touch with friends and family members. Problems develop when relationships are sought out by individuals who are not who they say they are or who have ill intentions against others. Over the past several years, a number of stories have been reported drawing attention to cases in which online predators sought out both minor and adult victims to build relationships, which then led to a harmful or violent act.

Consider the following hypothetical situation with the potential for offline violence. Jane joins a social networking site to keep in contact with friends and family, setting up limited blocks on her account while looking to find new friends with whom she does not have an offline relationship. At the same time, John is looking online for new friends and comes across Jane's profile. The profile includes public information, such as Jane's employer, her e-mail address, and a telephone number. John decides to ask Jane to be his friend. Upon acceptance of the friend request, John and Jane start chatting online and sharing more information. Shortly after they become friends, John starts making inappropriate comments to Jane. Jane becomes uncomfortable, removes John from her friends list, and blocks communication with John on this website. At this point, however, John is already in possession of Jane's e-mail address and telephone number, and he begins to contact her using both of these forms of communication. John makes comments about wanting to be with Jane, stating that he will see her soon. Because John also has her full name and work information, he begins to show up at her workplace, making verbal threats or following her home. Jane now is confronted with the decision of how to proceed with the situation.

As social networking sites have become more mainstream and have drawn users from all age groups, many users have become more aware of practices necessary to help protect their privacy or restrict the information they share on these sites. Hinduja and Patchin (2009) provide several tips for securing personal privacy and security when participating in social networking sites, including these:

- Assume that everyone has access to your profile and personal information, and use the privacy controls available to restrict access as much as possible.
- Use discretion when selecting photos or content to put on your profile.
- Assume that people will use the information on your profile to cause you harm, so do not put anything in your profile that could later be used against you.
- Assume there are predators looking for victims based on information conveyed on your profile and avoid posting personal information that may help them find you.

While most users of social networking or online gaming sites do not fall victim to violence by strangers, the potential threat remains. Indeed, for most users, the fear of violence or harm to children remains one of the biggest concerns with these sites.

While the potential for violence stemming from cyberbullying, cyberstalking, cyberpredators, and social networking sites is of continuing concern, there is also the potential for more widespread acts of violence against larger groups of people incited by cyberterrorism (Hinduja & Patchin, 2009).

CYBERTERRORISM

Since September 11, 2001, the threat of terrorist attacks on the United States has received increased attention. While debate continues to swirl around the definition of cyberterrorism (Matusitz & Breen, 2010), the current discussion considers cyberterrorism as "a premeditated, politically motivated criminal act by sub-national groups or clandestine agents, against information and computer systems, computer programs, and data, that results in physical violence, where the intended purpose is to create fear in noncombatant targets" (Colarik, 2006, p. 47). There is a real possibility that terrorist organizations will use the Internet and related technologies to coordinate and carry out acts of violence against innocent civilians. While previous terrorist attacks on U.S. interests went largely unnoticed by the American population, the attacks on September 11, 2001, on American soil brought the threat of future potential terrorist acts to the forefront of law enforcement and government focus. The widespread dissemination of the Internet has allowed terrorist groups to "operate in a virtual electronic world that provides them with numerous advantages for communication and coordination efforts, as well as assist in their ongoing development and expansion efforts" (Colarik, 2006, p. 7).

In recent years, media stories have reported terrorist organizations using websites to convey their ideologies or promote acts of violence against people who do not share their beliefs. While law enforcement agencies at all levels have devoted additional funds to seeking out and detecting terrorist acts before they occur since the 2001 attacks, their ability to do so may be limited by the prevalence of and ease of use of Internet communication. For developed and networked terrorist organizations, technology and access to global telecommunications networks worldwide allow previously localized terrorist organizations to transmit and receive coordination and specialty communications from individuals higher in the organization. A terrorist communication hub may be something as simple as communication via an online gaming portal, such as *World of Warcraft*. Additionally, access to global communication systems allows organizations to disseminate past activities and "accomplishments" for purposes of recruitment and movement validation purposes (Colarik, 2006, p. 10).

The use of the Internet to facilitate terrorist acts is not limited to terrorist organizations overseas or from one particular point of origin. The method of communication allowed by the Internet allows people from any walk of life anywhere in the world to share ideas and develop relationships. It stands to reason that extremist groups within the United States would be drawn to the Internet to spread their word, just as much as any terrorist organization elsewhere in the world.

For example, an MS 13 gang member recently used the Internet to spread hateful, racist messages and threatened to bomb a school after an Oklahoma City news organization posted a story online about a softball coach at Edmond North High School. The gang member had plenty of hateful things to say after the Oklahoma City NBC station posted a story on its website about the alleged bullying going on at Edmond North's baseball diamond. Once the disturbing comments hit the web (now removed), Edmond and Tulsa police, working with the FBI Joint Terrorism Task Force, were alerted. A search warrant stated that the user, whose computer was traced back to a home in West Tulsa, made racist comments about white people, threatened to kill police, and then stated (spelling and grammar unchanged), "I hearq about your parents crying about your kids I will go and bomb the school Monday as a matter infact im going to bomb the school Monday at 10 am if you . . . don't believe me turn on the news on monday and you will see your kids arms and heads scatter on the baseball field no lie" (Alford, 2010).

Consider several recent examples that illustrate the potential use of the Internet and global communications by terrorist organizations to fulfill their organizational goals. Terrorist organizations have used the Internet to incite violence against individuals and government. They have also used it as a way to transmit and mass distribute images of their ability and willingness to carry out violent acts against citizens of countries they view as enemies or threats to their belief system. In 2002, reporter Daniel Pearl was in Pakistan investigating Richard Reid (the "Shoe Bomber"). While traveling to what he believed was a meeting with a contact related to the story, Pearl was abducted by a terror organization and held captive for 9 days prior to his execution. Several recordings of Pearl were made while he was held captive in which his captors made a list of demands to secure his release. His execution was recorded and posted on the Internet as well as broadcast online for the world to see. Years later, graphic depictions of his final moments continue to be accessible on the Internet. While this organization was not using the Internet to commit an offense against an individual, it demonstrated its ability to use new methods of communication and technology to advance its political agenda. More recently, a sting of Russian spies revealed their use of Internet systems to communicate secrets and objectives between Russian agents abroad and sleeper agents in the United States.

Colarik (2006, p. 10) discusses the importance of developing a comprehensive approach to combating and dealing with potential threats from terrorists' use of the Internet:

> *The use of this infrastructure by traditional terrorists has broad implications. They may use such a resource for intimidation; to facilitate political strife or economic gain, or to undermine an economy; assert demands for the creation of an independent state, and/or assist in the destabilization of regimes; as a distribution channel for media empowerment of their activities; and for the organization of minority solidarity and subgroup confederations. A technosavvy terrorist may use such an infrastructure for target intelligence, electronic attacks, and most likely both in conjunction with physical attacks to increase effectiveness and further the traditional goals of terrorist organizations.*

CONCLUSION

The majority of cyber-related crimes that are prosecuted in the United States relate to nonviolent activities. One reason for this trend may be that it is sometimes difficult to develop the connection between online activity and offline violence. In cases where online activity does lead to offline violence, the most serious offense is the one that would be counted by most national reporting systems, so the computer-related deviance might not be tracked or even recorded. While nongovernment agencies have begun to collect data on a variety of cyber-related activities with the potential for violence, such as cyberbullying and cyberstalking, no official records are currently being kept of these incidents, to the authors' knowledge. Most often, the organizations collecting the data focus solely on the types of victims on whose behalf they advocate, rather than examining the broader context of cyberviolence. Additionally, if a prosecutor is able to obtain a conviction based on actual violence or actual physical threats of violence through some other means, it may not be cost-effective to build a case on the computer-related activities.

Additionally, technology develops at a pace that far exceeds the ability of law makers to pass corresponding legislation. This lack of a timely legislative response is a serious issue. Due to the structure of the Internet, it may sometimes be difficult to establish the components necessary for development of legal statutes, although some laws have been developed in response to extreme cases receiving media coverage or in response to public outcry. Cyberdeviance manifested as subsequent serious physical violence may be rare, but when it does occur it is generally something that is visible to police and provides some form of tangible evidence for police and prosecutors. In contrast, tracking online deviance can be much more difficult, time consuming, and expensive.

The increased prevalence of Internet usage and the necessity of computers to perform daily tasks have resulted in the potential for a variety of new criminal opportunities. While cyberbullying, cyber-talking, and related activities may lead to victimization and personal physical threats or aggression, this is not always the case, and this potential should be assessed in context.

Cyberbullying is an increasing threat to the security of youth, with perpetrators using various forms of technology to torment their victims. As research examining this trend increases, additional discussion of the best approach to dealing with the problem will be determined. For instance, it is important to determine who should be responsible for dealing with cyberbullying. Should schools have any authority in responding to cyber-related deviance between students and, if so, to what extent? What is the role of law enforcement in cyberstalking cases that have not resulted in any offline communication? Is this issue appropriate for law enforcement to address prior to any kind of violent act occurring, or does the police role begin only after a violent act is committed against a juvenile? Informing parents, educators, and the public about the dangers of cyberbullying as well as the methods for effectively educating youth regarding the potential harm to themselves and others will be important steps in moving forward in the reduction and prevention of cyberbullying incidents. To date, a variety of preventive and responsive actions have been available to parents and educators in dealing with acts of cyberbullying. Unfortunately, cyberstalking can and has led to offline stalking, which has a strong potential for violence. This phenomenon will certainly continue into the future.

Growing attention has been devoted to cyberpredators as the media continue to publicize cyberpredation. As with the other types of cyber-related deviance discussed, legislation addressing these behaviors will continue to play "catch-up" with cyber-related violence and cyberpredation. Many of the cases involving cyberpredators involve minors, and detection and punishment of these individuals continue to develop support within communities nationwide. It is highly likely that the coming years will feature marked legislative development in response to this form of criminality.

Finally, the potential for violence through cyberterrorism will continue to pose a challenge for law enforcement and the public alike. Devising a strategy to determine which information poses a threat to the public will improve our chances of being able to prevent violence. Further exploration of the methods used by various terrorist organizations to communicate via this means will provide additional insight into these strategies.

Discussions of cyber-related deviance that may lead to violence pose a problem, in that there is currently no organized method by which this information is collected or reported at the national level. This lack of a database may partially be the result of the variety of definitions used for each of these behaviors. To facilitate the collection of data, legal definitions of cyber-related crimes should be standardized across state and federal jurisdictions. The flexibility of the Internet and diversity in activities may make it difficult to specifically identify the characteristics of any given behavior. However, as our ability to distinguish various attributes of these offenses and offenders is refined, legislation as well as enforcement of these behaviors will inherently follow.

REFERENCES

Aftab, P. (2010). Retrieved January 18, 2011, from http://wiredsafety.net

Alford, A. (2010, April 9). Bomb threats made by Tulsa gang. *Fox23.com.* Retrieved January 18, 2011, from http://www.fox23.com/news/local/story/Bomb-Threats-Made-By-Tulsa-Gang/wNcOm6p3nUeeGbGN7royCQ.cspx?rss=77

Berson, I. R., Berson, M. J., & Ferron, J. M. (2002). Emerging risks of violence in the digital age: Lessons for educators from an online study of adolescent girls in the United States. *Journal of School Violence, 1,* 51–71.

Brandon, J. (2010, June 29). Is cyberstalking becoming a serious threat? *FoxNews.com.* Retrieved January 18, 2011, from http://www.foxnews.com/scitech/2010/06/29/cyberstalking-threat-online-safety/

Canning, A., Netter, S., & Crews, J. (2010, July 22). Death threats for ranting 11-year-old highlight cyberbullying epidemic. *WHAS11.com.* Retrieved from http://www.whas11.com/news/Death-threats-for-ranting-11-year-old-highlight- cyberbullying-epidemic-99018784.html

Colarik, A. M. (2006). *Cyber terrorism political and economic implications.* Hershey, PA: Idea Group Publishing.

Ericson, N. (2001). *Addressing the problem of juvenile bullying.* Washington, DC: U.S. Department of Justice, Office of Juvenile Justice and Delinquency Prevention.

Federal Bureau of Investigation (FBI). (2008). Don't make the call: The new phenomenon of swatting. Retrieved January 18, 2011, from http://www.fbi.gov/news/stories/2008/february/swatting020408

Fenton, R., Calhoun, A., & Mangan, D. (2010, September 29). Rutgers student filmed having sex commits suicide, 2 charged with filming. *New York Post online.* Retrieved from http://www.nypost.com/p/news/local/charged_with_taping_rutgers_student_9G3XCTxdLoZ3VPJNyk8spO

Finkelhor, D., Mitchell, K. J., & Wolak, J. (2000). *Online victimization: A report on the nation's youth.* Alexandria, VA: National Center for Missing and Exploited Children.

Finn, J. (2004). A survey of online harassment at a university campus. *Journal of International Violence, 19,* 468–483.

Grills, A. E., & Ollendick, T. H. (2002). Peer victimization, global self-worth, and anxiety in middle school children. *Journal of Clinical Child and Adolescent Psychology, 31,* 59–68.

Hawker, D. S. J., & Boulton, M. J. (2000). Twenty years' research on peer victimization and psychological maladjustment: A meta-analysis review of cross-sectional studies. *Journal of Child Psychology and Psychiatry, 41,* 441–445.

Hinduja, S., & Patchin, J. W. (2007). Offline consequences of online victimization: School violence and delinquency. *Journal of School Violence, 6,* 89–112.

Hinduja, S., & Patchin, J. W. (2008). Cyberbullying: An exploratory analysis of factors related to offending and victimization. *Deviant Behavior, 29,* 129–156.

Hinduja, S., & Patchin, J. W. (2009). *Bullying beyond the schoolyard: Preventing and responding to cyberbullying.* Thousand Oaks, CA: Sage.

Hinduja, S., & Patchin, J. W. (2010). Bullying, cyberbullying, and suicide. *Archives of Suicide Research, 14,* 206–221.

Kennedy, T. (2000). An exploratory study of feminist experiences in cyberspace. *Cyberpsychology and Behavior, 3,* 707–719.

Kochenderfer-Ladd, B., & Skinner, K. (2002). Children's coping strategies: Moderators of the effects of peer victimization? *Developmental Psychology, 38,* 267–278.

Kowalski, R. M., Limber, S. P., & Agatston, P. W. (2008). *Cyberbullying: Bullying in the digital age.* Malden, MA: Blackwell.

Lamberg, L. (2002). Stalking disrupts lives, leaves emotional scars. *Journal of the American Medical Association, 286,* 519–522.

Li, Q. (2007). New bottle but old wine: A research of cyberbullying in schools. *Computers in Human Behavior, 23,* 1777–1791.

Loeber, R., & Disheon, T. J. (1984). Early predictors of male delinquency: A review. *Psychology Bulletin, 94,* 68–99.

Magnusson, D., Stattin, H., & Dunner, A. (1983) Aggression and criminality in a longitudinal perspective. In K. T. V. Dusen & S. A. Mednick (Eds.), *Prospective studies of crime and delinquency* (pp. 277–301). Netherlands: Kluwer Nijoff.

Matusitz, J., & Breen, G. M. (2010). Cyberterrorism: A description from multiple perspectives. *International Journal of Information Technology, 16,* 44–60.

McIntyre, M. (2009, March 24). Online predator targeted moms: Man acted out death threats against children; Crown seeks more prison time. *Winnipeg Free Press.* Retrieved January 18, 2011, from http://www.winnipegfreepress.com/local/online-predator-targeted-moms-41739702.html?viewAllComments=y

McMillan, R. (2007, October 17). Couple warned by SWAT team after 911 "hack." *PCWorld.com* Retrieved January 18, 2011, from http://www.pcworld.com/article/138591/couple_swarmed_by_swat_team_after_911_hack.html

Nansel, T. R., Overpeck, M., Pilla, R. S., Ruan, W. J., Simons-Morton, B., & Scheidt, P. (2001). Bullying behaviors among U.S. youth: Prevalence and association with psychosocial adjustment. *Journal of the American Medical Association, 285*, 2094–2100.

National Children's Home. (2002). NCH 2002 survey. Retrieved January 18, 2011, from http:www.nch.org.uk/itok/showquestion.asp?faq=9andfldAuto=145

National Children's Home. (2005). Putting U in the picture: Mobile bullying survey. Retrieved January 18, 2011, from http://www.nch.org.uk/uploads/documents/Mobile bullying %20report.pdf

National Crime Victims Center. (2003). Retrieved January 18, 2011, from http://www.ncvc.org/ncvc/Main.aspx

Olweus, D., Limber, S., & Mihalic, S. (1999). *Bullying prevention programs.* Boulder, CO: Center for the Study and Prevention of Violence.

O'Moore, M., & Kirkham, C. (2001). Self-esteem and its relationship to bullying behavior. *Aggressive Behavior, 27*, 269–283.

Patchin, J. W. (2002). Bullied youths last out: Strain as an explanation of extreme school violence. *Caribbean Journal of Criminology and Social Psychology, 7*, 22–43.

Patchin, J. W., & Hinduja, S. (2006). Bullies move beyond the schoolyard: A preliminary look at cyberbullying. *Youth Violence and Juvenile Justice, 4*, 148–169.

Raskauskas, J., & Stoltz, A. D. (2007). Involvement in traditional and electronic bullying among adolescents. *Developmental Psychology, 43*, 564–575.

Rigby, K. (2003) Consequences of bullying in schools. *Canadian Journal of Psychiatry, 48*, 583–590.

Riveira, D. (2000, September/October). Internet crimes against women. *Sexual Assault Report*, 4.

Tattum, D. P. (1989). Violence and aggression in schools. In D. P. Tattum & D. A. Lane (Eds.), *Bullying in schools* (pp. 7–22). Stroke-on-Trent, UK: Trentham.

Tench, M. (2003, January 21). Schools struggling to stop tech-savvy bullies who have taken their taunting to cyberspace. *Boston Globe*, p. B1.

Thorp, D. (2004). Cyberbullies on the prowl in the schoolyard. Retrieved January 18, 2011, from http://australianit.news.com.au/articles/0,7204,9980900^15322^^nbv^15306,00.html

United States Department of Justice, Office of Justice Programs, National Institute of Justice. (1999). 1999 Report on cyberstalking: A new challenge for law enforcement and industry. A report from the Attorney General to the Vice President. Retrieved January 18, 2011, from http://www.justice.gov/criminal/cybercrime/cyberstalking.htm

United States Department of Justice, Office of Justice Programs, Violence Against Women Office. (2001). Stalking and domestic violence report to Congress. Retrieved January 18, 2011, from http://www.ncjrs.gov/pdffiles1/ojp/186157.pdf

Vossekuil, B., Fein, R., Reddy, M., Borum, R., & Modzelski, W. (2002). The final report and findings of the Safe Schools Initiative: Implications for the prevention of school attacks in the United States. Retrieved January 18, 2011, from http://www.secretservice.gov/ntac/ssi_final_report.pdf

Wang, J., Iannotti, R. J., & Nansel, T. R. (2009). School bullying among adolescents in the United States: Physical, verbal, relational, and cyber. *Journal of Adolescent Health, 48*, 368–375.

Williams, K. R., & Guerra, N. G. (2007). Prevalence and predictors of Internet bullying. *Journal of Adolescent Health, 41*, S14–S21.

Wolak, J., Mitchell, K., & Finkelhor, D. (2006). *Online victimization: 5 years later.* Alexandra, VA: National Center for Missing and Exploited Children. Retrieved from http://www.missingkids.com/en_US/publications/NC167.pdf

Ybarra, M. L., & Mitchell, J. K. (2004). Online aggressor/targets, aggressors and targets: A comparison of associated youth characteristics. *Journal of Child Psychology and Psychiatry, 45*, 1308–1316.

CHAPTER 6

Still Psychopathic After All These Years

Matt DeLisi
Iowa State University

Michael G. Vaughn
Saint Louis University

> *"His mouth is full of curses and lies and threats; trouble and evil are under his tongue. He lies in wait near the villages; from ambush he murders the innocent, watching in secret for his victims. He lies in wait like a lion in cover; he lies in wait to catch the helpless; he catches the helpless and drags them off in his net. His victims are crushed, they collapse; they fall under his strength."*
>
> —Psalms 10:7–9 (cited in Meloy & Gacono, 1998)

> *"Humans have long been concerned by or fascinated with the concept of evil and the people thought to personify evil. Say the word* psychopath *and most people can easily conjure up an image of someone they believe to embody the word."*
>
> —James Blair, Derek Mitchell, and Karina Blair (2005, p. 1)

With a clinical and criminological history that spans more than two centuries, psychopathy is among the most popular, controversial, and empirically evaluated constructs in the behavioral sciences. Robert Hare (1998), arguably the most accomplished psychopathy researcher, noted that even those opposed to the very idea of psychopathy cannot ignore its potent explanatory and predictive power—if not as a formal construct, then as a static risk factor. Indeed, some researchers have found evidence indicating that psychopaths constitute a taxon, meaning a natural, discrete class of persons among the criminal population (Harris, Rice, & Quinsey, 1994; Skilling, Quinsey, & Craig, 2001). This chapter argues that psychopathy is the purest, most parsimonious, and, frankly, best explanation of serious antisocial and violent behavior. More than any other theory of crime, the construct of psychopathy brilliantly forges the connection between the individual-level traits of the actor and his or her antisocial behavior.

Psychopathy is a clinical construct usually referred to as a personality disorder defined by a constellation of interpersonal, affective, lifestyle, and behavioral characteristics that are manifested in wide-ranging antisocial behaviors. The characteristics of psychopathy read like a blueprint for violence. Psychopaths are impulsive, grandiose, emotionally cold, manipulative, callous, arrogant, dominant, irresponsible, short-tempered persons who tend to violate social norms and victimize others without guilt or anxiety. In short, they are human predators without conscience.

At the heart of psychopathy is the complete unfeeling for other people, evidenced by callous-unemotional traits, remorselessness, and the absence of empathy. These individuals do not experience the feelings that naturally inhibit the acting out of violent impulses, and their emotional deficiency is closely related to general under-arousal and the need for sensation seeking (Herpetz & Sass, 2000). Because of this vacancy in the moral connection to other people, psychopaths are qualitatively distinct from other offender groups. But psychopaths go beyond that qualitative distinction: They are quantitatively worse than other offenders. A study by David Simourd and Robert Hoge (2000) speaks to the virulence of the personality disorder even among a sample of dangerous criminals. Simourd and Hoge examined the case histories of 321 felons who were incarcerated for violent crimes. Among this sample, 36 inmates were psychopaths and 285 were not. Compared to non-psychopaths, psychopaths had more previous, total, violent, noncompliant, and different types of criminal convictions; more arrests; greater criminal sentiments and pride in antisocial behavior; and, almost without exception, greater needs in terms of supervision.

What is the prevalence of psychopathy in the general population? It is difficult to know because population-based studies have not been carried out; however, studies in correctional facilities have shown that approximately 25% of persons with antisocial personality disorder—a psychiatric disorder closely associated with psychopathy—meet the criteria for psychopathy (Hart & Hare, 1996). Given that psychiatric epidemiological studies of antisocial personality disorder indicate approximately 4% of the adult population possesses this disorder, we can infer an estimate of roughly 1% of the total population having psychopathy (Hare, 1996).

If someone is interested in understanding criminal violence, psychopathy is a good place to start. This chapter briefly highlights empirical issues pertaining to criminological theory, career criminality and recidivism, murder and sexual offending, and institutional violence as they relate to psychopathy. While these concepts are informed by many academic disciplines and encompass an array of topical areas, psychopathy is central to each.

CRIMINOLOGICAL THEORY

One of the most popular and widely studied theories of crime is the general theory of crime advanced by Michael Gottfredson and Travis Hirschi (1990), which asserts that low self-control is the chief variable that predicts crime and analogous behaviors. The profile of persons with low self-control is well known; however, consider the following description of criminal offenders:

> *[O]ver-evaluation of immediate goals as opposed to remote or deferred ones; unconcern over the rights and privileges of others when recognizing them would interfere with personal satisfaction in any way; impulsive behavior, or apparent incongruity between the strength of the stimulus and the magnitude of the behavioral response; inability to form deep or persistent attachment to other persons . . . poor judgment and planning in attaining defined goals . . . almost complete lack of dependability of and willingness to assume responsibility; and, finally, emotional poverty (Gough, 1948).*

While this reads like a description of an offender with low self-control, it is actually a profile of psychopathic offenders published in 1948. Indeed, the core characteristics in Gottfredson and Hirschi's general theory of crime (e.g., hot tempered, impulsive, action oriented, unempathetic, unable to delay gratification) could be construed as a softened abbreviation of psychopathy. In this way, the most talked-about, controversial, and cited theory in mainstream criminology borrows much of its empirical heft from the construct of psychopathy (Wiebe, 2003).

Psychopathy has proved useful in the integration of previously disparate literatures. For instance, Donald Lynam has shown that a small cadre of children with hyperactivity, impulsivity, attention problems (e.g., ADHD), and conduct disorder are afflicted with a virulent strain of psychopathology best described as "fledgling psychopathy." Lynam's work has strengthened developmental psychology, psychopathology, and criminology by illustrating the "worst of the worst" in terms of violent and antisocial behavior and explaining how it unfolds over the life span (Lynam, 1996; Lynam & Gudonis, 2005). For instance, in one of the earliest studies of adolescent psychopathy, Adelle Forth and her colleagues (1990) found that psychopathic youths had criminal histories marked by more previous violent offending and institutional violence. Even as adolescents, psychopathic offenders are more likely than non-psychopathic youths to receive a swift juvenile court referral, commit a violent offense upon release, and engage in both instrumental ("cold-blooded") and reactive ("hot blooded") forms of aggression (Brandt, Kennedy, Patrick, & Curtin, 1997; Loper, Hoffschmidt, & Ash, 2001; Stafford & Cornell, 2003).

Other authors, such as Michael Vaughn and David Farrington, suggest that psychopathy could be a useful construct for organizing the study of serious, violent antisocial behavior among children and adolescents (Farrington, 2005; Vaughn & Howard, 2005). For instance, three notable longitudinal studies—the Denver Youth Survey, Pittsburgh Youth Study, and Rochester Youth Development Study—revealed that between 14% and 17% of the youths in these samples were habitual offenders who accounted for 75% to 82% of the incidence of criminal violence. These adolescents in Denver, Pittsburgh, and Rochester tended to be "multiple-problem youths" who experienced an assortment of antisocial risk factors, such as mental health problems, alcoholism and substance abuse histories, and sustained criminal involvement. Within this violent group, a small minority of youths were the most frequent, severe, aggressive, and temporally stable delinquent offenders. These youths, all of whom were males, were reared in broken homes by parents who themselves had numerous mental health and parenting problems. These boys were also notable for their impulsivity, emotional and moral insouciance, and total lack of guilt regarding their commission of crime. In other words, these studies indicate that the most violent young offenders in the United States display many of the characteristics of psychopathy (Loeber et al., 2002).

Even those who are critical of the notion of using psychopathy as a general theory of crime recognize how parsimoniously and accurately it describes crime and violence (Walters, 2004). For the more extreme forms of crime, psychopathy is an intuitive heuristic for understanding behavior; however, it seems too severe when attempting to explain mundane forms of crime. For instance, it might seem foolish to suggest that behaviors such as shoplifting, forgery, and drunk driving are the expression of psychopathy, because they are more common and often committed by seemingly "normal" persons. However, the very nature of minor crimes such as stealing and drunk driving reveals something about offenders—namely, those who are willing to take from others, satisfy their desires at the possible expense of others, and flagrantly violate law and morality. To borrow from Gottfredson and Hirschi, all crimes are acts of force and fraud against others in the pursuit of self-interest. This perspective is not unlike that advocated by psychopathy.

CAREER CRIMINALITY AND RECIDIVISM

It is well established that a minority of criminals perpetrate the majority of crimes in a population. Career criminals begin their antisocial careers early, commit greater and more varied crimes, and are the most violent (DeLisi, 2005). Various scholars have empirically explored the links between psychopathy and assorted dimensions of career criminality, especially recidivism and noncompliance with criminal justice sanctions. For example, Grant Harris and his colleagues (1991) examined the recidivism rates of 169 male offenders released from a psychiatric facility and followed up one year later. Nearly 80% of psychopathic offenders committed a new violent offense. Moreover, psychopathy was the strongest predictor of recidivism. In fact, its effects were *stronger than the combined effects of 16 background, demographic, and criminal history variables* (Harris, Rice, & Cormier, 1991). In the Pittsburgh Youth Study, boys who presented with psychopathic traits were between 480% and 630% more likely to be multiple-problem offenders. These most frequent, severe, aggressive, and stable delinquents were prone to externalizing disorders, but remained seemingly immune from internalizing disorders, such as anxiety (Loeber et al., 2002).

Recently, David Farrington (2006) explored etiological predictors of psychopathy using data from the Cambridge Study in Delinquent Development, a 40-year prospective longitudinal survey of the criminal careers and social histories of 411 London males. Using the Psychopathy Checklist Revised Screening Version (PCL-R: SV), Farrington compared the offending careers of the top 11% of the sample who scored 10 or higher on the PCL-R: SV (deemed the most psychopathic) to the remaining members of the panel. The most psychopathic group totaled significantly more convictions, had greater involvement in the criminal justice system, and presented with more criteria for antisocial personality disorder diagnosis. Nearly half of these men were chronic offenders. In his analysis of this sample, Farrington (2006) discovered that an assortment of background factors was predictive of psychopathy at age 48. The strongest predictors (with corresponding odds ratios) were uninvolved father (6.5), physical neglect (5.9), convicted father (5.1), low family income (4.6), and convicted mother (4.5).

With respect to persistence, frequency, and severity, male psychopaths are believed to constitute the most violent population of human aggressors known (Harris, Rice, & Lalumiere, 2001, p. 406). When Mary Ann Campbell and her colleagues (2004) studied 226 incarcerated adolescent offenders, they found that approximately 9% of the sample exhibited high levels of psychopathic traits; however, this small selection of youths had the most violent and versatile criminal histories. Richard Rogers and his colleagues' (2000) analysis of 448 prisoners revealed that, as children, psychopathic inmates forced others into sexual activity, were physically cruel to others, used weapons in fights, deliberately destroyed property, committed arson, and were cruel to animals, among other crimes. In the Cambridge Study in Delinquent Development, children with psychopathic personalities were significantly more likely to be chronic offenders, and these traits had predictive power in regard to criminal behavior decades later (Farrington, 2000). Importantly, high scores on psychopathy measures have also been correlated with early-onset for violent offending (Forth, 1995).

Michael Vaughn and Matt DeLisi (2008) explored the relationship between psychopathic personality traits and career criminality within a large sample of more than 700 incarcerated adolescents. Youths who presented with psychopathic characteristics were approximately 300% to 400% more likely than offenders without psychopathic traits to be classified as career criminals. Subsequently, these researchers found that psychopathy measures were moderately able (between 70% and 73% of the time) to correctly classify career criminal membership. When higher threshold specifications were used, the classification accuracy improved to an impressive 88%, with many of the highly relevant characteristics relating to impulsivity, callousness, fearlessness, and narcissism.

Interestingly, some of the interest in psychopathy centers on the notion that psychopaths' behavior is in some way innate. A recent analysis of 626 twin pairs indicated that nearly 50% of the variation in fearless dominance (i.e., resiliency to internalizing disorders) and impulsive antisociality (i.e., liability to externalizing deviance) was attributable to genes (Bloningen et al., 2005). Put another way, the constellation of negative personality traits imbued in psychopathy have been there since the beginning of the psychopath's life.

Which feature of psychopathy among adolescents is particularly worrisome? Recent research suggests that callous-unemotional traits may be the pathway that leads to severe and persistent aggression in youth. For instance, in a sample of 169 adolescents, Dustin Pardini (2006) found that low empathy was mediated by a nonconcern for the consequences of punishment. If social learning and behavior conditioning principles have little effect on restraining behaviors, then these youth are relatively "free" to do as they please. Similarly, in a sample of 376 boys and 344 girls, Darrick Jolliffe and David Farrington (2006) found that these same low empathy traits were associated with increased frequency of bullying behavior overall, and violent bullying among the boys. It appears that the hallmark feature of primary psychopathy—lack of conscience—may be recognizable among children and adolescents.

Several meta-analyses have indicated that psychopathy is the indispensable predictor of violent recidivism among children, adolescents, and especially adults (Dolan & Doyle, 2000; Edens, Skeem, Cruise, & Cauffman, 2001; Howard, Williams, Vaughn, & Edmond, 2004). For instance, among a sample of incarcerated U.S. adolescents, psychopathic youths were found to have higher rates of instrumental violence, violence where the victim required medical attention, assaults with deadly weapons, and both self-reports and criminal records of violence (Murrie et al., 2004). Among Canadian detained adolescents, psychopathic youths were more likely to offend after release from custody, committed more nonviolent and violent crimes, and recidivated more quickly than other offenders (Corrado, Vincent, Hart, & Cohen, 2004). As recidivists, psychopaths are quicker, more productive, and more severe once released back to the community.

MURDER AND SEXUAL OFFENDING

The violence perpetrated by psychopaths is more instrumental, dispassionate, and predatory than that of other offenders. As such, psychopathy is an important risk factor for homicide and sexual offending. Much of this relationship derives from the ease with which psychopaths can inflict violence. A recent study published in *Nature* found that psychopathic murderers have diminished negative reactions to violence compared to non-psychopaths and other violent offenders, almost as if violence is a facile, unexceptional event to these individuals (Gray et al., 2003).

For serial murder and single sexual homicides, psychopathy is a basic personality characteristic of the offender (Geberth & Turco, 1997; Myers & Monaco, 2000). For instance, Theodore Millon and Roger Davis (1998) suggest that many murderers could be characterized as malevolent psychopaths, which is a particularly negative subtype of offender characterized as belligerent, mordant, rancorous, vicious, brutal, callous, and vengeful. What does psychopathic malevolence look like? Park Dietz and his colleagues (1990) conducted a descriptive study of 30 sexually sadistic criminals. All of these men intentionally tortured their victims for purposes of their own sexual arousal. Their crimes often involved careful planning, the selection of strangers as victims, approaching the victim under a pretext, participation of a partner, beating victims, restraining victims and holding them captive, sexual bondage, anal rape, forced fellatio, vaginal rape, foreign object penetration, telling victims to speak particular words in a degrading manner, murder or serial killings (most often by strangulation), concealing victims' corpses, recording offenses, and keeping personal items belonging to victims. These are some call signs of psychopathy, albeit coupled with sexual sadism.

A study of 125 murderers found that more than 93% of homicides committed by psychopaths were "cold-blooded," in that they were instrumental, completely premeditated, and not preceded by an explosive emotional interaction, such as an argument. When Stephen Porter and his colleagues (2003) compared sexual homicides committed by psychopathic and non-psychopathic offenders in Canadian prisons, they found that nearly 85% of psychopathic murderers engaged in some degree of sadistic behavior during the course of their crimes. Moreover, homicides committed by psychopaths involved significantly greater levels of gratuitous and sadistic violence. Citing a study conducted by the Federal Bureau of Investigation (FBI), Robert Hare (1999) noted that more than half of the law enforcement officers killed on duty were murdered by offenders whose personality profile matched that of the psychopath.

Paul Mullen (2004) reviewed data suggesting that mass killers were isolated individuals who had rarely established themselves in effective adult roles. Persons who committed massacres were usually men roughly 40 years old, who had been bullied or isolated as children, demonstrated an affinity or preoccupation with weaponry and violence, and showed psychopathic-like personalities marked by rigid or obsessive beliefs, delusional suspiciousness, narcissism, and grandiose ideas that they had been persecuted.

When Wade Myers (2004) examined the psychiatric history, criminal history, and family background of 16 juvenile sexual homicide offenders, he discovered a laundry list of severe risk factors, many of them related to psychopathy. The most prevalent of these traits were an impaired capacity to feel guilt, neuropsychiatric vulnerabilities, serious school problems, child abuse victimization and family dysfunction, history of interpersonal violence, prior arrests, sadistic fantasy, psychopathic personality, and personality disorder diagnosis. Nearly 90% had elevated psychopathy scores.

Matt DeLisi (2001) interviewed 500 adult offenders with a minimum of 30 prior arrests. The sample included 42 murderers, 80 rapists, and 38 kidnappers. All of these offender groups showed versatility, as evidenced by multiple arrests for assorted violent and property crimes, recurrent imprisonments, and criminal careers that averaged roughly 25 years. During the interviews, the most violent offenders, especially the rapists, demonstrated prototypical psychopathic traits, such as pathological lying, irresponsibility, malevolent egocentricity, pronounced anger, and little regard for their victims.

Psychopathy figures prominently in the personality profile of sexually offending groups. Roy Hazelwood, the renowned FBI profiler, and Janet Warren developed profiles of serial sexual offenders based on actual cases. They described impulsive serial sexual offenders as persons motivated by a sense of entitlement and the perception that anything (or anyone) is there for the taking—in other words, classic psychopathic symptoms (Hazelwood & Warren, 2000). Based on data from 329 Canadian prisoners, Stephen Porter and his colleagues (2000) found that a substantial number of offenders who commit various types of sexual crimes are psychopaths. Specifically, 64% of the inmates with convictions for rape and child molestation, 36% of rapists, 11% of intrafamilial child molesters, and 6% of extrafamilial child molesters were psychopaths.

Psychopathy also escalates the risk posed by adolescent sex offenders. Heather Gretton and her colleagues (2001) studied 220 adolescent males in an outpatient sex offender treatment program to assess linkages between psychopathy and recidivism. Youths with high psychopathy scores on the Psychopathy Checklist Revised Youth Version (PCL-R: YV) posed multiple threats to public safety. Notably, they were more likely than other offenders to escape from custody, violate probation, and commit violent and nonviolent crimes after release. Moreover, some highly psychopathic youths exhibited deviant sexual arousal as measured by phallometric tests.

Although the relationship between psychopathy and sexual offending is multifaceted, it is undeniable that psychopathy figures prominently in this linkage. When Raymond Knight and Jean-Pierre Guay (2006) summarized 50 years of research on the topic, they arrived at three general conclusions. First, psychopaths are significantly more likely than non-psychopathic criminals to rape and are over-represented in clinical samples of sexual offenders. Additionally, psychopathic traits predict rapacious behavior among noncriminal samples. Second, psychopaths constitute a small subgroup of rapists who are extraordinarily violent and recidivistic. Third, the underlying processes that contribute to psychopathy are similar to those associated with sexually coercive behavior. In other words, psychopathy is inextricably linked to the most heinous forms of violent criminal behavior.

INSTITUTIONAL VIOLENCE

Psychopathy is a strong predictor of whether an inmate will continue to misbehave while incarcerated, especially among those offenders convicted of committing the most physically aggressive types of offenses (Edens, Buffington-Vollum, Colwell, Johnson, & Johnson, 2002). In fact, psychopathic inmates tend to be the most aggressive and difficult-to-manage inmates (Dolan & Blackburn, 2006). Glenn Walters (2003) conducted a meta-analysis of 41 studies and found a moderate correlation between psychopathy and institutional adjustment (r = .27). The studies encompassed adults and juveniles, offenders from four countries, various follow-up periods, and inmates from prisons, forensic hospitals, and psychiatric facilities. Upon release, psychopaths were significantly likely to commit general, violent, and sexual recidivism. Robert Hare and his colleagues (2000) found that psychopathic inmates accumulated more incident reports for violating prison rules, were more likely to assault staff, and were more likely to assault other inmates. Nearly one in two psychopaths (with scores greater than 30 on the PCL-R) had assaulted another inmate.

Similarly, Sarah Spain and her colleagues (2004) found that psychopathic adolescents accumulated more total, violent, verbal, and administrative violations while in custody and also had significantly worse treatment outcomes. In other words, psychopathic youths took much longer to complete or achieve minimal success in treatment. Among a sample of adjudicated adolescents, Daniel Murrie and his colleagues (2004) found that the risk of prison violence increased 10% for each point above the mean PCL-R score. In addition, Mairead Dolan and Charlotte Rennie found that youth psychopathy scores were predictive of assault on others in a secure facility (Dolan & Rennie, 2006).

A link between psychopathy and institutional violence has also been found among mentally disordered offenders. Kirk Heilbrun and his colleagues (1998) administered the PCL-R to 218 clients following their admission to an inpatient forensic hospital. Significant correlations between the PCL-R total scores and both nonphysical and physical aggression during the first two months of hospitalization were observed. The PCL-R total scores were also significantly correlated with postdischarge arrests for violent offenses. Psychopathic inpatients were responsible for significantly more aggressive incidents during the first two months of hospitalization. Moreover, psychopathy was significantly correlated with frequency of seclusion or restraint, suggesting that for mentally disordered offenders, psychopathy may serve as a risk factor for institutional aggression.

That psychopathic offenders have poorer adjustment to correctional supervision likely justifies the most punitive forms of criminal sanction. For example, in a sample of 450 sexually violent offenders in Florida, Jill Levenson and John Morin (2006) found that for each point above the mean score on the PCL-R, offenders were 49% more likely to be civilly committed or selected for involuntary confinement after serving a prison sentence. Inmates who met the standard cut-off score of 30 were 490% more likely to be selected for civil commitment. In conjunction with diagnosed paraphilias, psychopathy correctly predicted commitment recommendations in 90% of cases.

In the public mind, psychopaths are deserving of the death penalty perhaps because they bear the label "psychopath." John Edens and his colleagues (2003) presented vignettes of a 16-year-old murderer, described as having classic psychopathic symptoms, to research respondents. These respondents were 130% more likely to recommend that youths should be sentenced to death if they had psychopathic traits. Moreover, respondents indicated that such youths should not receive treatment in prison.

CONCLUSION

Criminal offenders are a heterogeneous group, with diverse characteristics such as gender, race, ethnicity, social class, criminality, criminal history, offense type, risk and protective factors, and personality. Anyone who has worked with criminal offenders in a correctional setting can rather quickly identify recurrent characteristics of serious offenders, however. When considering the most violent types of offenders, for instance—those with convictions for murder, rape, kidnapping, and armed robbery—several thoughts come to mind.

First, virtually all of the most violent offenders are male; predatory violent behavior is simply less prevalent among women. The majority of incarcerated violent offenders have an adverse, often abusive childhood, and most are raised in poverty. These demographic and social correlates are not the only commonalities among violent criminals. Interpersonally, one is immediately struck by their global irresponsibility and basic refusal to handle the important obligations of adult social roles, such as maintaining relationships, maintaining employment, and maintaining sobriety. Incarcerated violent offenders tend to be mean spirited and insensitive, exceedingly manipulative, and utterly narcissistic. On their rap sheets are found multiple arrests for various crimes committed at high rates across their life spans. A synergy between the violent criminals' personality traits, lifestyle, and observed behavior dovetails so exquisitely that it is as if their criminality is wrapped up in a box—and that box is psychopathy.

Psychopathy is an efficient and protean way to understand and explain crime, because the traits that constitute psychopathy correspond to the elemental characteristics of crime itself: a self-serving, uncaring violation of another person. Recent advances in criminological theory, such as the self-control construct in the general theory of crime, are essentially shorthand for psychopathy. For the extremes of criminal behavior, psychopathy is the sine qua non criminological explanation, and one with a long and recurrent history. While other explanatory constructs are also important, it is clear that psychopathic traits are important to understand not only murder, but also serial murder, sadistic murder, and sexually violent murder. In essence, the construct and theory of psychopathy is inescapable (DeLisi, 2009). It is clear, concise, internally consistent, and, perhaps most importantly, plausible. It efficiently conveys how criminal atrocities can flow from people who, because of their lack of empathy, selfish desires, and deficient conscience, impose a heavy toll on society as a whole.

REFERENCES

Blair, J., Mitchell, D., & Blair, K. (2005). *The psychopath: Emotion and the brain.* Malden, MA: Blackwell.

Bloningen, D. M., Hicks, B. M., Krueger, R. F., et al. (2005). Psychopathic personality traits: Heritability and genetic overlap with internalizing and externalizing psychopathology. *Psychological Medicine, 35,* 637–648.

Brandt, J. R., Kennedy, W. A., Patrick, C. J., & Curtin, J. (1997). Assessment of psychopathy in a population of incarcerated adolescent offenders. *Psychological Assessment, 9,* 429–435.

Campbell, M. A., Porter, S., & Santor, D. (2004). Psychopathic traits in adolescent offenders: An evaluation of criminal history, clinical, and psychosocial correlates. *Behavioral Sciences & the Law, 22,* 23–47.

Corrado, R. R., Vincent, G. M., Hart, S. D., & Cohen, I. M. (2004). Predictive validity of the psychopathy checklist: Youth version for general and violent recidivism. *Behavioral Sciences & the Law, 22,* 5–22.

DeLisi, M. (2001). Extreme career criminals. *American Journal of Criminal Justice, 25,* 239–252.

DeLisi, M. (2005). *Career criminals in society.* Thousand Oaks, CA: Sage.

DeLisi, M. (2009). Psychopathy is the unified theory of crime. *Youth Violence and Juvenile Justice, 7,* 256–273.

Dietz, P. E., Hazelwood, R. R., & Warren, J. (1990). The sexually sadistic criminal and his offenses. *Bulletin of the America Academy of Psychiatric Law, 18,* 163–178.

Dolan, M., & Blackburn, R. (2006). Interpersonal factors as predictors of disciplinary infractions in incarcerated personality disordered offenders. *Personality and Individual Differences, 40,* 897–907.

Dolan, M., & Doyle, M. (2000). Violence risk prediction: Clinical and actuarial measures and the role of the psychopathy checklist. *British Journal of Psychiatry, 177,* 303–311.

Dolan, M., & Rennie, C. (2006). Psychopathy checklist: Youth version and youth psychopathic trait inventory: A comparison study. *Personality and Individual Differences, 41,* 779–789.

Edens, J. F., Buffington-Vollum, J. K., Colwell, K. W., Johnson, D. W., & Johnson, J. K. (2002). Psychopathy and institutional misbehavior among incarcerated sex offenders. *International Journal of Forensic Mental Health, 1,* 49–58.

Edens, J. F., Guy, L. S., & Fernandez, K. (2003). Psychopathic traits predict attitudes toward a juvenile capital murderer. *Behavioral Sciences & the Law, 21,* 807–828.

Edens, J. F., Skeem, J. L., Cruise, K. R., & Cauffman, E. (2001). Assessment of "juvenile psychopathy" and its association with violence: A critical review. *Behavioral Sciences & the Law, 19,* 53–80.

Farrington, D. P. (2000). Psychosocial predictors of adult antisocial personality and adult convictions. *Behavioral Sciences & the Law, 18, 605–622.*

Farrington, D. P. (2005). The importance of child and adolescent psychopathy. *Journal of Abnormal Child Psychology, 33,* 489–497.

Farrington, D. P. (2006). Family background and psychopathy. In C. J. Patrick (Ed.), *Handbook of psychopathy* (pp. 229–250). New York: Guilford Press.

Forth, A. E. (1995). *Psychopathy and young offenders: Prevalence, family background, and violence. Program Branch users report.* Ontario, Canada: Minister of the Solicitor General of Canada.

Forth, A. E., Hart, S. D., & Hare, R. D. (1990). Assessment of psychopathy in male young offenders. *Psychological Assessment, 2,* 342–344.

Geberth, V., & Turco, R. (1997). Antisocial personality disorder, sexual sadism, malignant narcissism, and serial murder. *Journal of Forensic Science, 42,* 49–60.

Gottfredson, M. R., & Hirschi, T. (1990). *A general theory of crime.* Stanford, CA: Stanford University Press.

Gough, H. G. (1948). A sociological theory of psychopathy. *American Journal of Sociology, 53,* 359–366.

Gray, N. S., MacCulloch, M. J., Smith, J., et al. (2003). Forensic psychology: Violence viewed by psychopathic murderers. *Nature, 423,* 497–498.

Gretton, H. M., McBride, M., Hare, R. D., et al. (2001). Psychopathy and recidivism in adolescent sex offenders. *Criminal Justice and Behavior, 28*, 427–449.

Hare, R. D. (1996). Psychopathy: A clinical construct whose time has come. *Criminal Justice and Behavior, 23*, 25–54.

Hare, R. D. (1998). Psychopaths and their nature: Implications for the mental health and criminal justice systems. In T. Millon, E. Simonsen, M. Birket-Smith, & R. D. David (Eds.), *Psychopathy: Antisocial, criminal, and violent behavior* (pp. 188–192). New York: Guilford Press.

Hare, R. D. (1999). Psychopathy as a risk factor for violence. *Psychiatric Quarterly, 70*, 181–197.

Hare, R. D., Clark, D., Grann, M., & Thornton, D. (2000). Psychopathy and the predictive validity of the PCL-R: An international perspective. *Behavioral Sciences & the Law, 18*, 623–645.

Harris, G. T., Rice, M. E., & Cormier, C. A. (1991). Psychopathy and violent recidivism. *Law and Human Behavior, 15*, 625–637.

Harris, G. T., Rice, M. E., & Lalumiere, M. (2001). Criminal violence: The roles of psychopathy, neurodevelopmental insults, and antisocial parenting. *Criminal Justice and Behavior, 28*, 402–426.

Harris, G. T., Rice, M. E., & Quinsey, V. L. (1994). Psychopathy as a taxon. *Journal of Consulting and Clinical Psychology, 62*, 387–397.

Hart, S. D., & Hare, R. D. (1996). Psychopathy and antisocial personality disorder. *Current Opinion in Psychiatry, 9*, 129–132.

Hazelwood, R. R., & Warren, J. I. (2000). The sexually violent offender: Impulsive or ritualistic? *Aggression and Violent Behavior, 5*, 267–279.

Heilbrun, K., Hart, S. D., Hare, R. D., Gustafson, D., Nunez, C., & White, A. (1998). Inpatient and post-discharge aggression in mentally disordered offenders: The role of psychopathy. *Journal of Interpersonal Violence, 13*, 514–527.

Herpertz, S. C., & Sass, H. (2000). Emotional deficiency and psychopathy. *Behavioral Sciences & the Law, 18*, 567–580.

Howard, M. O., Williams, J. H., Vaughn, M. G., & Edmond, T. (2004). Promises and perils of a psychopathology of crime: The troubling case of juvenile psychopathy. *Journal of Law and Policy, 14*, 441–483.

Jolliffe, D., & Farrington, D. P. (2006). Examining the relationship between low empathy and bullying. *Aggressive Behavior*, 32, 540–550.

Knight, R. A., & Guay, J. (2006). The role of psychopathy in sexual coercion against women. In C. J. Patrick (Ed.), *Handbook of psychopathy* (pp. 512–532). New York: Guilford Press.

Levenson, J. S., & Morin, J. W. (2006). Factors predicting selection of sexually violent predators for civil commitment. *International Journal of Offender Therapy and Comparative Criminology, 50*, 609–629.

Loeber, R., Farrington, D. P., Stouthamer-Loeber, M., et al. (2002). Male mental health problems, psychopathy, and personality traits: Key findings from the first 14 years of the Pittsburgh Youth Study. *Clinical Child and Family Psychology Review, 4*, 273–297.

Loper, A., Hoffschmidt, S., & Ash, E. (2001). Personality features and characteristics of violent events committed by juvenile offenders. *Behavioral Sciences & the Law, 19*, 81–96.

Lynam, D. R. (1996). Early identification of chronic offenders: Who is the fledgling psychopath? *Psychology Bulletin, 120*, 209–234.

Lynam, D. R., & Gudonis, L. (2005). The development of psychopathy. *Annual Review of Clinical Psychology, 1*, 381–407.

Meloy, J. R., & Gacono, C. B. (1998). The internal world of the psychopath. In T. Millon, E. Simonsen, M. Birket-Smith, & R. D. David (Eds.), *Psychopathy: Antisocial, criminal, and violent behavior* (pp. 95–109). New York: Guilford Press.

Millon, T., & Davis, R. D. (1998). Ten subtypes of psychopathy. In T. Millon, E. Simonsen, M. Birket-Smith, & R. D. David (Eds.), *Psychopathy: Antisocial, criminal, and violent behavior* (pp. 161–170). New York: Guilford Press.

Mullen, P. E. (2004). The autogenic (self-generated) massacre. *Behavioral Sciences & the Law, 22*, 311–323.

Murrie, D. C., Cornell, D. G., Kaplan, S., et al. (2004). Psychopathy scores and violence among juvenile offenders: A multi-measure study. *Behavioral Sciences & the Law, 22*, 49–67.

Myers, W. C. (2004). Serial murder by children and adolescents. *Behavioral Sciences & the Law, 22*, 357–374.

Myers, W. C., & Monaco, L. (2000). Anger experience, styles of anger expression, sadistic personality disorder, and psychopathy in juvenile sexual homicide offenders. *Journal of Forensic Science, 45*, 698–701.

Pardini, D. A. (2006). The callousness pathway to severe violent delinquency. *Aggressive Behavior, 32*, 590–598.

Porter, S., Fairweather, D., Drugge, J., et al. (2000). Profiles of psychopathy in incarcerated sexual offenders. *Criminal Justice and Behavior, 27*, 216–233.

Porter, S., Woodworth, M., Earle, J., et al. (2003). Characteristics of sexual homicides committed by psychopathic and non-psychopathic offenders. *Law and Human Behavior, 27*, 459–470.

Rogers, R., Salekin, R. T., Sewell, K. W., & Cruise, K. R. (2000). Prototypical analysis of antisocial personality disorder: A study of inmate samples. *Criminal Justice and Behavior, 27*, 234–255.

Simourd, D. J., & Hoge, R. D. (2000). Criminal psychopathy: A risk-and-need perspective. *Criminal Justice and Behavior, 27*, 256–272.

Skilling, T. A., Quinsey, V. L., & Craig, W. M. (2001). Evidence of a taxon underlying serious antisocial behavior in boys. *Criminal Justice and Behavior, 28*, 450–470.

Spain, S. E., Douglas, K. S., Poythress, N. G., & Epstein, M. (2004). The relationship between psychopathic features, violence, and treatment outcome: The comparison of three youth measures of psychopathic features. *Behavioral Sciences & the Law, 22*, 85–102.

Stafford, E., & Cornell, D. G. (2003). Psychopathy scores predict adolescent inpatient aggression. *Assessment, 10*, 102–112.

Vaughn, M. G., & DeLisi, M. (2008). Were Wolfgang's chronic offenders psychopaths? On the convergent validity between psychopathy and career criminality. *Journal of Criminal Justice, 36*, 33–42.

Vaughn, M. G., & Howard, M. O. (2005). The construct of psychopathy and its potential contribution to the study of serious, violent, and chronic youth offending. *Youth Violence and Juvenile Justice, 3*, 235–252.

Walters, G. D. (2003). Predicting criminal justice outcomes with the psychopathy checklist and lifestyle criminality screening form: A meta-analytic comparison. *Behavioral Sciences & the Law, 21*, 89–102.

Walters, G. D. (2004). The trouble with psychopathy as a general theory of crime. *International Journal of Offender Therapy and Comparative Criminology, 48*, 1–16.

Wiebe, R. P. (2003). Reconciling psychopathy and low self-control. *Justice Quarterly, 20*, 297–336.

CHAPTER 7

The Behavioral Genetics of Predatory Criminal Behavior

John Paul Wright
University of Cincinnati

Kevin M. Beaver
Florida State University

Even the most casual observer of science must be struck by the rate at which new findings on human development and maladjustment are published. Hardly a day passes in which the media do not report startling results linking brain function to criminal behavior, or report new relationships between specific genes and criminal traits. The individuals who have led the way into this exciting, uncharted territory, however, have not been sociologically trained criminologists. Unfortunately, most criminologists remain on the disciplinary sidelines or—even worse—remain wedded to an ideology that rejects genetic influences in any form. This phenomenon may explain why "the biological sciences have made more progress in our understanding of criminal behavior in the last ten years than sociology has made over the past 50 years" (Robinson, 2004, pp. ix–x).

While some criminologists may bristle at this declaration, there can be little doubt that an explosion of knowledge on the development of criminal conduct has occurred in recent decades, especially knowledge about serious predatory conduct, and that the results have been published in journals that feature studies on genetics and biology. Not a single path-breaking study on criminal behavior has been published in any leading criminology or sociology journal. Even more telling is the fact that the leading criminological theories—that is, social learning theory, strain theory, and self-control theory—all require biological and genetic factors to be valid. Learning, for example, occurs when connections are made between synapses in the brain; stress and strain have been found to impinge on hippocampus–pituitary–adrenal axes of the brain; and self-control reflects a broader set of brain-based abilities known as executive functions.

This chapter seeks to demystify the influence genes have on behavior and serves to introduce the reader to a "biosocial" understanding of predatory offending. In particular, it provides a brief introduction to current knowledge regarding predatory offending, the fundamentals of human genetics, the methods used by behavioral geneticists, and the way in which this information is used to understand predatory human behavior.

THE ROOTS OF PREDATION

Predation involves an intention to do harm to another, or at least a willingness to actively seek out and injure another person. Predation can be seen, for example, when armed robbers make a choice to "hold up" an individual or commercial establishment, or when a rapist takes the time to stalk his victim and then, when the risk of being identified or caught is the lowest, commits his crime. Predation can also be seen when a child molester seeks out and abducts a child (Wright & Decker, 1997).

Even among criminals, predation in criminal conduct is unusual. Only the most serious and habitual offenders are predatory. In contrast, other offenders tend to be opportunistic or influenced by situational contingencies, such as the presence of other criminals or the use of drugs and alcohol. This is not to say that predatory offenders are not also opportunists or that they do not commit crimes when under the influence; rather, the key difference is that predatory offenders do not require or are not driven by these concerns. To be direct, predatory offenders are the truly criminal.

Research on the development of serious criminal conduct has revealed three factors that are important in discerning the emergence of predatory behavior. First, the warning signs for serious predation are visible in infancy and childhood. Infants who are fussy, irritable, and difficult to soothe, and who react negatively to novel situations are significantly more likely to grow into children who have conduct disorders and into adults with antisocial personality disorder (Caspi, Roberts, & Shiner, 2005; Moffitt et al., 1996; Schmitz et al., 1999; Shiner, Masten, & Roberts, 2003).

Second, traits related to later criminal conduct are also visible in infancy and early childhood. While these traits have been labeled in a variety of ways, they generally focus on the ability of the infant and child to increasingly regulate his or her own behavior and to conform to the social expectations found in varied environments. Impulse control, self-regulation, self-control, emotional regulation, and hyperactivity generally fall under this broad umbrella of traits.

Third, and most importantly, studies into the development of aggression have found that its onset occurs around the time when children gain mobility—that is, when they start walking and interacting socially with other children. Richard Tremblay's studies of very young Canadian children, for example, found the peak age for aggression was around 27 months; the same researcher also found that more than 90% of young children had engaged in acts of aggression, such as hitting, kicking, and biting, before 36 months of age (Tremblay, 2006; Tremblay et al., 2004; Vitaro, Brendgen, & Tremblay, 2002).

Physical aggression is a nearly universal human capacity that is "normal" early in life, but becomes more uncommon in children over time (Tremblay et al., 2004). Indeed, children who fail to "age out" of the use of physical aggression by age 4 years are significantly more likely to continue using physical aggression over long swaths of their life-course. Perhaps not surprisingly, an early age of onset is one of the strongest predictors of future adult predatory offending. Reviews by Marvin Krohn and his colleagues (2001) found that early-onset offenders committed 40% to 700% more crimes than individuals who had an onset of problem behaviors later in life. Moreover, virtually every predatory offender had experienced an early age of onset.

In summary, children who exhibit a variety of criminogenic traits, who fail to gain sufficient self-regulatory capacities by age 4 years, and whose behavior remains consistent across time and situation are at substantial risk of developing into predatory adults. That this risk trajectory materializes at such an early point in the life-course necessarily hints at the likelihood that genetic factors are at play.

A BEHAVIORAL GENETIC UNDERSTANDING OF PREDATORY OFFENDING

How do we understand this set of empirical facts? With a few exceptions, such as Moffitt's developmental taxonomy, traditional criminological theories remain silent on this issue, largely because these theories of crime locate the causes of misconduct in adolescence. Even if we broaden their theoretical lens and take a leap of faith, criminologists would likely point to parental rearing environments as the putative source of variation in young children's behaviors. But would they be correct?

Before answering that question, it is helpful to examine how a behavioral geneticist would understand this issue. Behavioral genetics is the field of study that analyzes how much variance in any given trait or behavior is accounted for by genetic *and* environmental influences. Behavioral genetic studies estimate this variance by analyzing samples of monozygotic and dizygotic twins, relying on the laws of genetics, and using sophisticated statistical models. At the heart of the field, however, is the estimation of genetic and environmental influences.

Figure 7.1 shows the hypothetical results of a behavioral genetic analysis of some trait, such as impulsivity. In this pie chart, genetic influences account for most of the variance in impulsivity (65%), while nonshared environmental influences account for 25% and shared environmental influences for only 10%. Estimates of genetic influences are denoted by h^2, which stands for the degree to which a trait, characteristic, or behavior is heritable. The term "heritable" should not be confused with "inherited." Individuals "inherit" DNA that will ultimately code for the creation of the brain, nervous system, and arms and legs. By comparison, heritability reflects the degree to which a complex trait can be influenced by genes. This distinction is critical to understanding behavioral genetics because a person may be endowed with a specific genetic propensity, such as toward alcoholism or drug dependency, yet the propensity may never be realized. Alcoholism in traditional Muslim countries, for example, is very low because of the cultural prohibitions against alcohol use. Of course, this does not mean that certain Muslims do not have a propensity for addiction.

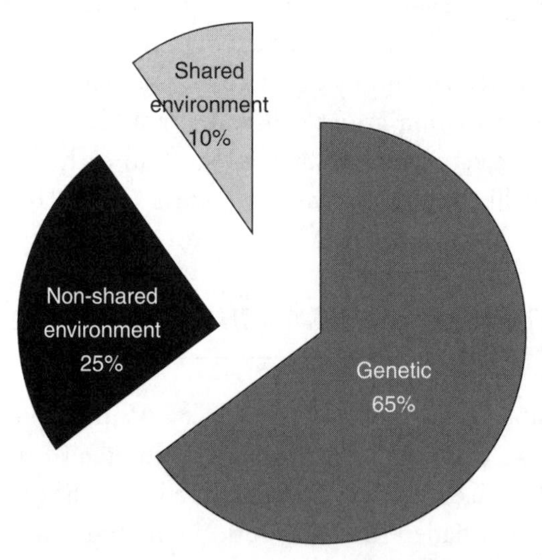

FIGURE 7.1 How behavioral genetics decompose the variance in a trait.

Behavioral geneticists also specify that two types of environmental influences are possible. Nonshared environments are those unique experiences that make individuals more different than alike. Siblings, even twins, often have different peer groups, for example. Thus different peer groups represent a unique (nonshared) environment. Shared environments, in contrast, are thought to make people more alike. Children born to the same parents are exposed to similar broad parental management strategies, for example.

Findings from hundreds of studies now show that virtually every human trait and characteristic is genetically influenced. For certain characteristics, such as vocational interests or religious orientations, the influence of genes is very low. For other characteristics, especially those associated with predatory offending, genetic influences dominate (Plomin, Chipuer, & Neiderhiser, 1994). For example, IQ, impulsivity, and self-control appear to be primarily genetic in origin (Barkley, 1997). The same studies that have considered genetic influences, however, also reveal that unique environmental experiences usually outweigh the influences of shared environmental influences—that is, shared experiences do not appear to make individuals more similar, but instead highlight their differences (Plomin & Daniels, 1987; Plomin, DeFries, McClearn, & Rutter, 1997; Plomin, Owen, & McGuffin, 1994; Rowe & Plomin, 1981).

Would criminologists be correct in predicting that certain parenting practices early in life will produce adult predatory behavior? The behavioral genetic studies typically show that shared environments, or the ways in which parents establish an overall home environment, have little to no effect on their adult offspring (Wright & Beaver, 2005). Does this finding mean that parenting has nothing to do with adult predation? Not exactly. The processes that link parenting practices to human development likely operate through biological mechanisms. When rats lick their pups, for instance, the stimulation releases oxytocin and prolactin, which are hormones that aid in the creation of feelings of safety and love. Studies of Romanian orphans brought up in horrific conditions have shown that these children possess reduced levels of these hormones. Nurturance may thus help build a healthy brain—at least for certain children. Some children may respond well to nurturance; others may have no response at all. This is part of the reason why it may be difficult to detect parental socialization effects. The genetic propensities of the child may interfere with, or cause the child not respond to, parenting efforts.

A Brief Note on Human Genetics

Humans inherit 23 pairs of chromosomes from each parent, one of which is the sex-differentiating chromosome. Males receive a Y chromosome from the father and an X chromosome from the mother (XY); females receive an X chromosome from both parents (XX). Chromosomes are made of deoxyribonucleic acid (DNA). DNA, in turn, is composed of two elongated sections bonded to chemical bases—the now-familiar double helix. The chemical bases are adenine (A), thymine (T), guanine (G), and cytosine (C). Due to their molecular structure, A can bond only with T, and G can bond only with C, thereby forming what are known as "base pairs." Genes, which are embedded in chromosomes, are merely stretches of DNA with a known arrangement of base pairs.

Current estimates place the number of genes in the human genome between 19,000 and 25,000. This number is far less than the number of genes found in other "lower-level" life forms, including some plants. Nonetheless, Mendelian theory tells us that we inherit two copies of each gene, one from the father and one from the mother. At one level, Mendelian theory is correct: We do inherit our genes from our parents. However, research has recently found evidence that for some genes, humans may inherit more than just two copies; that is, one or both parents may pass down more than one copy of particular genes. Three international research projects found that at least 10% of all human genes, or roughly 2900 genes, can vary in their number of copies within an individual. Estimates that used to indicate that humans were 99.9% genetically similar have since been revised to state that we are approximately 99% genetically similar. This restatement translates into a change from a 3 million base-pair difference between humans to at least a 30 million base-pair difference. It also means that it is even more likely that genes play a significant role in serious, predatory behavior.

Even if multiple copies of some genes are present, they are not all turned "on" or "off" at one time. The process whereby genes are made active or inactive is called genetic imprinting. Genes, moreover, come in different varieties. Differences in genes are called alleles. Allelic variation occurs when mutations, genetic drift, cultural selection, evolution, or any combination of factors alters a gene. For example, the dopamine transporter gene DRD4 comes in several allelic varieties, some of which are linked to an increased transmission of impulsivity and attention-deficit/hyperactivity disorder (ADHD) (Arcos-Burgos et al., 2004; Mill et al., 2002). Genes with various alleles are referred to as *polymorphic*.

Individuals can demonstrate significant differences from one another, even at the genetic level. Understanding the role of genes in complex human phenotypes, however, is made even more complicated by the fact that human genetic inheritance does not always follow Mendelian genetic principles. According to these principles, human genetic expression should follow a dominant and recessive form. Under most conditions, dominant genes should be expressed. At one level, some of our genes follow the dominant/recessive framework, such as the genes for eye color. In contrast, for complex traits, human genes do not appear to follow this principle. Instead, functional human genes appear to follow a pattern of incomplete dominance in their relationship to traits and behaviors. *Incomplete dominance* refers to a situation in which the effects of dominant and recessive alleles are blended and then expressed in a phenotype.

How Do Genes Influence Predation?

Complex traits and behaviors are usually not produced by single genes. Rather, multiple genes tend to act in concert to bring about specific genetic potentials (Comings et al., 2001). The term "genetic potentials" is appropriate here because genes create general behavioral tendencies, or propensities, that can sometimes be contingent on the environment for their activation. Recall the example cited earlier in this chapter of the low alcoholism rate in traditional Muslim countries. Clearly, some Muslims will have a genetic potential to become addicted to alcohol, but that propensity will not materialize if the individual never drinks. Single-gene influences are also typically rather small, usually explaining less than 5% of the variance in any complex behavior, such as violence. Research by Comings and his colleagues has shown that genes have an additive influence on ADHD, oppositional defiant disorder (ODD), conduct disorder, and various personality dimensions. It appears that input from many genes working in concert is required to produce traits and behaviors (Comings et al., 2000a, 2000b, 2000c).

How does this knowledge contribute to an understanding of serious, predatory criminal behavior? **Figure 7.2** provides an overview of the respective influences genotypes have on predation. Because the focus here is on genetic influences, the influence of environmental variables is omitted. This omission is done for the sake of brevity and because theory indicates that high-risk genotypes will experience "faulty" development

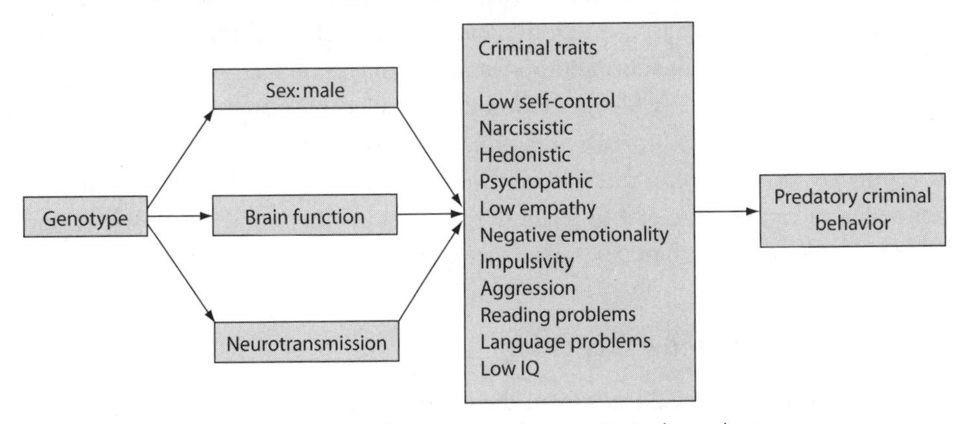

FIGURE 7.2 How the genotype influences predatory criminal conduct.

regardless of their environmental setting—that is, for serious predatory offenders, it is likely that their genotype confers so much risk that environmental mediators and moderators will be rendered ineffectual. Again, we are dealing with the truly criminal.

The first arrow in Figure 7.2 leads from the genotype to sex. In studying predatory offenders, one overriding, consistent fact is obvious: Predatory offenders are almost universally male. At one level, being male is one of the best-known risk factors to criminal involvement overall, but in terms of serious predation, males have the market cornered. Of course, being a male is caused by the father passing along Y chromosome to his offspring. The Y chromosome contains the *SRY* gene, which causes the testes to form and drop in the developing male fetus. This chromosome has only 78 genes, which code for 27 proteins. All other chromosomes contain more genes than the Y chromosome.

Once activated, the *SRY* gene signals the developing testes to flood the brain with androgens, the most commonly known of which is testosterone. Testosterone appears to "masculinize" the developing male brain, leading to several effects. The first effect of this hormone flood is that it creates what has been termed the "male brain." The male brain tends to excel in tests of spatial skills, the ability to focus on an issue or problem, map reading, and mathematics and sciences; it also shows a greater desire to seek risks and has 10% more area dedicated to aggression (Baron-Cohen, 2002). The female brain, in contrast, leads to individuals who are more religious, score higher in tests of empathy and emotional recognition, score better on tests of verbal skills, and test 60% to 70% better in memory for "locations and landmarks" (Craig, Harper, & Loat, 2004).

The second effect of the high testosterone level in the developing brain can be seen in the social behavior of males. Men are generally more status oriented than women and ascribe to status hierarchies more than women. Status hierarchies are also known as dominance hierarchies. Dominance can be achieved through a variety of methods, but the most efficient method—and the one employed more frequently by males than by females—is violence. Dominance fueled by testosterone may explain why overt physical aggression and predatory behavior in men appears to fully materialize during adolescence, a time when testosterone levels are 10 times higher and when males add an average of 1 to 2 feet in growth and 100 pounds of muscle (Booth, Granger, Mazur, & Kivlighan, 2006).

Brain Development and Functioning

The next arrow in Figure 7.2 leads from the genotype to brain structure and functioning. The genotype spends a significant amount of nuclear resources coding for the development of the brain. Given that more than 60% of human genes code for this one organ, it should come as little surprise that the brain itself consumes significant amount of resources. Weighing in at 3 pounds, representing approximately 2% of total body weight, the brain consumes roughly 10 times the amount of glucose as the rest of the organs, and roughly 20% of the body's oxygen intake (Robinson, 2004).

In terms of development, the human brain develops in a linear fashion: from the simplest parts that control life-sustaining nonreflexive activities, such as breathing, to the most complex and most recently evolved systems. The brain stem and lower brain are the first parts to develop. Controlling all necessary functions, the brain stem and lower brain are important for the regulation of sleep and appetite.

Sitting on top of the brain stem is the limbic system. The limbic system is composed of various structures in the brain that regulate hormones, control memory, and create emotions. Fear, anxiety, jealousy, anger, lust, and sadness are biochemical responses to environmental stimuli, the origin of which can be located in the limbic system. Of particular interest to the study of predatory criminal behavior is the amygdala—the "seat of emotions and emotional memory." It provides humans with the ability to match an event to a specific emotion, thereby giving individuals the ability to recall the experiences along with the feelings associated with the experience.

As mentioned earlier, predation involves the willingness, if not the desire, to bring harm to another individual. For most individuals, the thought of hurting an innocent person brings about feelings of guilt, shame, and anxiety. But what if these internal constraints are missing? What if the thought of hurting another person brings instead feelings of joy, excitement, or, even worse, apathy? Several studies show that the lack of empathy—a primary controlling emotion—is absent in psychopaths, and that its absence is due to problems associated with amygdala responses. Moreover, several studies, including studies of active offenders, have found that committing crime can be fun or emotionally rewarding, and that it can build a positive social reputation among other criminals.

The last part of the brain to develop is the neocortex, or the "thinking part" of the brain. The cortex sits atop the limbic system and is responsible for the human abilities of planning, delaying gratification, impulse control, and rational thought. The speech and language parts of the brain, known as Wernicke's and Broca's areas, are also located in the neocortex.

Taking more than 20 years to fully mature, the cortex houses the "executive functions" of the human brain, such as self-control and emotional regulation. Numerous studies have shown that the cortex, especially the orbital-frontal cortex (OFC), is critical to prosocial human behavior. Certain individuals appear to lack self-control because of deficits in the ventrolateral prefrontal cortex, especially deficits in the left hemisphere. This pattern of findings is particularly striking among males who have more problems with impulse control.

All sensory input is channeled first to the limbic system. Under constant monitoring by the cortex, the limbic system may send some of this information to the motor cortex, to the language centers, or to the neocortex—or to all necessary parts of the brain. The cortex has the ability to override the initial limbic impulses, given that the cortex is intimately "wired" into the limbic system. When strong emotions are encountered, the initial limbic impulse may be to act with violence, or to act aggressively. The cortex, however, may intervene to curtail or modify the initial limbic impulse. Obviously, deficits in the cortex will allow those impulses to materialize in the form of violent behavior.

Neurotransmission

The last arrow in Figure 7.2 leads from the genotype to the box labeled "neurotransmission." With 100 billion neurons, and even more glial cells, the human brain is a masterpiece of complexity. Even more astonishing is the fact that each neuron can connect to between 1 and 10,000 other neurons. Neurons communicate with one another by channeling an electric charge down an axon. As the electric charge reaches the end of the axon, it forces the release of neurotransmitters from the synaptic cleft. These neurotransmitters then cross the synaptic space, which is approximately 1/600th the width of a human hair, and bond to receptors located on the dendrite of the receiving neuron.

Neurotransmitters come in two types: excitatory and inhibitory. Excitatory neurotransmitters, such as dopamine, instigate movement; inhibitory neurotransmitters, such as serotonin, reduce motivation. Although neurotransmission occurs in response to environmental input, it is under substantial genetic control (Fishbein, 2001). The genotype, for example, codes for the number of neurotransmitter receptors located on dendrites.

Neurotransmission is also related to brain functioning. Because many genes are expressed in the brain, scientists have recently sought to examine how certain functional genetic variants are related to the overall brain–behavior response. One of the most fascinating studies genotyped subjects to ascertain which form of a gene the subjects inherited. The gene in question codes for monoamine oxidase A (MAOA), an enzyme in the brain that breaks down excess serotonin. Molecular research has shown that there is an "efficient" allele and an "inefficient" allele of the MAOA gene. Individuals with the inefficient allele are thought to have excess serotonin, which theoretically would decrease their impulsivity. Armed with this information, researchers used 3-D magnetic resonance imaging to compare those persons with the more efficient and less efficient alleles (Meyer-Lindenberg et al., 2006).

Compared to individuals with the efficient MAOA allele, individuals with inefficient alleles showed substantial reductions in brain volume (measured by gray matter) in areas of the brain that control attention (anterior cingulated cortex) and emotions (amygdala). For males only, however, researchers also found a 14% reduction in the OFC. Further evaluation revealed that the OFC was less active and less connected to the amygdala in men with the inefficient allele. Recall that the OFC is deeply implicated in self-control and emotional regulation.

To summarize the information presented thus far, human genetics is related to predatory offending through three primary variables: sex, brain structure and functioning, and neurotransmission. How the genotype codes for these biological factors plays a major role in the resulting expression of criminogenic traits. It is worth noting again that the genotype is also influenced by environmental variables, such as the induction of neurotoxins and constant stress and anxiety. However, it is not clear that serious offenders would behave differently even if presented with entirely prosocial environmental stimuli (Caspi & Moffitt, 2006; Rutter, 2006).

CRIMINALITY

As Figure 7.2 illustrates, the genome may not influence behavior directly—that is, genes do not cause behavior. Instead, genes create the conditions that enable various human traits to be expressed in terms of personality, thinking patterns, and ultimately behavior. The concentration of these traits within an individual (typically male) elevates the likelihood that he will engage in predatory conduct for much of his life.

Several dimensions distinguish predatory offenders from non-offenders, and even from minor offenders. Predatory offenders often have problems with self-control, in that they act impulsively and sometimes without accurately assessing the immediate situation. They tend to be narcissistic, thinking primarily of themselves and their needs and wants, and they appear unable to relate to the pain and anguish they cause others. Moreover, they also tend to be below average in measures of IQ, especially on measures of verbal IQ. Finally, they tend to view the world negatively, demonstrating open hostility toward others, and they tend to act aggressively without provocation. Theirs is also a lifestyle characterized by drug use, partying, and general irresponsibility.

Criminality—the propensity to commit crime and other destructive behaviors—manifests itself from the confluence of these traits. A behavioral genetic understanding of these traits highlights three interrelated points.

First, all of these traits have high levels of heritability, especially IQ, impulsiveness, and self-control (Cadoret, Cain, & Crowe, 1983; DiLalla, 2002; Koenen et al., 2006; Moffitt, 2005; Rietveld et al., 2003). This is part of the reason why these traits have been linked to sex, brain functioning, and neurotransmission.

Second, many individuals with relatively low levels of self-control, high levels of impulsivity, and non-empathetic personalities are not criminal. A behavioral genetic understanding would point out that criminality reflects not only the confluence of these multiple heritable traits working in conjunction with one another, but also the degree to which these heritable traits dominate one's personality and behavior.

Third, psychologists use the term "clinical" to indicate that an individual's score on a measured personality characteristic, such as narcissism, is significantly different from the average score in the population. This term also indicates that the characteristic is associated with problems in that person's life or that it has interfered with or delayed the individual's development. Serious, predatory offenders are likely to score in the "clinical" range on all, or at least most, of these criminal traits and characteristics.

CONCLUSION

As noted by Robinson (2004), behavioral geneticists have provided more insights into the origins of serious predatory conduct in the last 10 years than sociology has in the last 50 years. These insights, however, have yet to fully penetrate criminology. While the reasons for this delay are numerous, this reluctance is in part due to many criminologists' fear that recognizing genetic influences will leave them with nothing left to study. Understanding predation from a behavioral genetic viewpoint, however, does not obviate the importance of environmental factors. Indeed, just the opposite is true: A behavioral genetic viewpoint helps to clarify how environmental stimuli operate on the human organism and to specify more precisely which stimuli sponsor criminality and which do not.

Genetic influences on criminality are complex and multifaceted. It is simply not the case that "bad" genes create "bad" people. Any over-simplification of genetic influences does a disservice to the science of human behavior and gives a misleading and incomplete picture of the operation of the human genome on evolved behavioral patterns. While this chapter has endeavored to provide the reader with a basic understanding of the behavioral genetics of serious criminality, we caution against any further simplification.

The astute reader will likely ask about policy consequences associated with a behavioral genetic understanding of pathology. Just as behavioral genetics shed light on the operation of genes and environmental factors, so it also supports various policies. Space limitations preclude the inclusion of an exhaustive listing of potential policies here. Nevertheless, three approaches have emerged at the forefront of this intersection of biology and behavior.

First, good evidence supports the idea that ADHD can be effectively treated and managed through a combination of medications and individual and family counseling. Given the overlap between characteristics of impulsivity and criminality, there is ample reason to believe that serious misbehavior may likewise respond well to this combination of intervention efforts.

Second, the earlier the age of onset of problem behaviors, the more likely those problems are to become resistant to change. Unfortunately, criminal behavioral patterns remain relatively stable for long periods of time (Farrington et al., 1986; Loeber, 1982; Olweus, 1979; Robins, Caspi, & Moffitt, 2002). This evidence points to the need for early intervention with high-risk children and their families. Given the incredible personal and social costs associated with a lifetime of personal pathology, the earlier and more frequent the intervention efforts, the better.

Third, it is unreasonable to expect that even the best efforts to habilitate or rehabilitate a criminal individual will prove successful in all cases. For those individuals who demonstrate an inability to effectively manage and regulate their antisocial tendencies, then nothing else is left except incarceration for lengthy periods of time. As mentioned earlier, even among offenders, serious predation is not common. The crimes these men commit are typically the most serious and the most brutal. Protective social efforts are thus necessary. Moreover, the gravity of their crimes will dictate retribution.

Behavioral genetics is a fascinating, integrative field of study that has much to offer to students of criminal behavior. Perhaps its most important contribution is its focus on consilience. No single field has the capacity to completely unravel the complexities of human behavior. Interdisciplinary efforts are necessary to integrate findings from various disciplines that study the same topic. As an overarching perspective, the field of behavior genetics offers invaluable insights into the origins of predatory criminal conduct.

REFERENCES

Arcos-Burgos, M., Castellanos, F. X., Konecki, D., et al. (2004). Pedigree disequilibrium test (PDT) replicates association and linkage between DRD4 and ADHD in multigenerational and extended pedigrees from a genetic isolate. *Molecular Psychiatry, 9*, 252–259.

Barkley, R. A. (1997). *ADHD and the nature of self-control.* New York: Guilford Press.

Baron-Cohen, S. (2002). The extreme male brain theory of autism. *Trends in Cognitive Science, 6*, 248.

Booth, A., Granger, D. A., Mazur, A., & Kivlighan, K. T. (2006). Testosterone and social behavior. *Social Forces, 85*, 167–191.

Cadoret, R. J., Cain, C. A., & Crowe, R. R. (1983). Evidence for gene–environment interaction in the development of antisocial behavior. *Behavioral Genetics, 13*, 301–310.

Caspi, A., & Moffitt, T. E. (2006). Gene–environment interactions in psychiatry: Joining forces with neuroscience. *Nature Reviews Neuroscience*, 7, 583–590.

Caspi, A., Roberts, B. W., & Shiner, R. (2005). Personality development: Stability and change. *Annual Review of Psychology, 56*, 453–484.

Comings, D. E., Gade-Andavolu, R., Gonzalez, N., et al. (2000a). Comparison of the role of dopamine, serotonin, and noradrenaline genes in ADHD, ODD and conduct disorder: Multivariate regression analysis of 20 genes. *Clinical Genetics, 57*, 178–196.

Comings, D. E., Gade-Andavolu, R., Gonzales, N., et al. (2000b). Multivariate analysis of associations of 42 genes in ADHD, ODD, and conduct disorder. *Clinical Genetics, 58*, 31–40.

Comings, D. E., Gade-Andavolu, R., Gonzalez, N., et al. (2000c). A multivariate analysis of 59 candidate genes in personality traits: The temperament and character inventory. *Clinical Genetics, 58*, 375–385.

Comings, D. E., Gade-Andavolu, R., Gonzalez, N., et al. (2001). The additive effect of neurotransmitter genes in pathological gambling. *Clinical Genetics, 60*, 107–116.

Craig, I. W., Harper, E., & Loat, C. S. (2004). The genetic basis for sex differences in human behavior: Role of the sex chromosomes. *Annals of Human Genetics, 68*, 269–284.

DiLalla, L. F. (2002). Behavior genetics of aggression in children: Review and future directions. *Developmental Review, 22*, 593–622.

Farrington, D., Gallagher, B., Morley, L., et al. (1986). Unemployment, school leaving and crime. *British Journal of Criminology, 26*, 335–356.

Fishbein, D. (2001). *Biobehavioral perspectives in criminology*. Belmont, CA: Wadsworth.

Koenen, K. C., Caspi, A., Moffitt, T. E., et al. (2006). Genetic influences on the overlap between low IQ and antisocial behavior in young children. *Journal of Abnormal Psychology, 115*, 787–797.

Krohn, M., Thornberry, T., Rivera, C., & LeBlanc, M. (2001). Later delinquency careers. In R. Loeber & D. P. Farrington (Eds.), *Child delinquents: Development, intervention, and service needs* (pp. 67–93). Thousand Oaks, CA: Sage.

Loeber, R. (1982). The stability of antisocial and delinquent child behavior: A review. *Child Development, 53*, 1431–1446.

Meyer-Lindenberg, A., Buckhoitz, J. W., Kolachana, B., et al. (2006). Neural mechanisms of genetic risk for impulsivity and violence in humans. *Proceedings of the National Academy of Science, 103*, 6269–6274.

Mill, J. S., Caspi, A., McClay, J., et al. (2002). The dopamine D4 receptor and the hyperactivity phenotype: A developmental-epidemiological study. *Molecular Psychiatry*, 7, 383–391.

Moffitt, T. E. (2005). The new look of behavioral genetics in developmental psychopathology: Gene–environment interplay in antisocial behaviors. *Psychology Bulletin, 131*, 533–554.

Moffitt, T., Caspi, A., Dickson, N., et al. (1996). Childhood-onset versus adolescent-onset antisocial conduct problems in males: Natural history from ages 3 to 18 years. *Developmental Psychopathology, 8*, 399–424.

Olweus, D. (1979). Stability of aggressive reaction patterns in males: A review. *Psychology Bulletin, 86*, 852–875.

Plomin, R., Chipuer, H. M., & Neiderhiser, J. M. (1994). Behavioral genetic evidence for the importance of non-shared environment. In M. Hetherington, D. Reiss, & R. Plomin (Eds.), *Separate social worlds of siblings: The impact of the non-shared environment on development* (pp. 1–21). Hillsdale, NJ: Lawrence Erlbaum.

Plomin, R., & Daniels, D. (1987). Why are children in the same family so different from one another? *Behavioral and Brain Sciences, 10*, 1–60.

Plomin, R., DeFries, J. C., McClearn, G. E., & Rutter, M. (1997). *Behavioral genetics* (3rd ed.). New York: W. H. Freeman.

Plomin, R., Owen, M. J., & McGuffin, P. (1994). The genetic basis of complex human behaviors. *Science, 264*, 1733–1739.

Rietveld, M. J. H., Hudziak, J. J., Bartels, M., et al. (2003). Heritability of attention problems in children: Cross-sectional results from a study of twins, age 3 to 12 years. *Neuropsychiatric Genetics, 1176*, 102–113.

Robins, L. W., Caspi, A., & Moffitt, T. E. (2002). It's not just who you're with, it's who you are: Personality and relationship experiences across multiple relationships. *Journal of Personality, 70*, 925–964.

Robinson, M. B. (2004). *Why crime? An integrated systems theory of antisocial behavior.* Upper Saddle River, NJ: Prentice Hall.

Rowe, D. C., & Plomin, R. (1981). The importance of non-shared (E1) environmental influences in behavioral development. *Developmental Psychology, 17*, 517–531.

Rutter, M. (2006). *Genes and behavior: Nature–nurture interplay explained.* Malden, MA: Blackwell.

Schmitz, S., Fulker, D. W., Plomin, R., et al. (1999). Temperament and problem behavior during early childhood. International *Journal of Behavioral Development, 23*, 333–355.

Shiner, R. L., Masten, A. S., & Roberts, J. M. (2003). Childhood personality foreshadows adult personality and life outcomes two decades later. *Journal of Personality, 71*, 1145–1170.

Tremblay, R. E. (2006). Prevention of youth violence: Why not start at the beginning? *Journal of Abnormal Child Psychology, 34*, 480–486.

Tremblay, R. E., Nagin, D. S., Séguin, J. R., et al. (2004). Physical aggression during early childhood: Trajectories and predictors. *Pediatrics, 114*, e43–e50.

Vitaro, F., Brendgen, M., & Tremblay, R. E. (2002). Reactively and proactively aggressive children: Antecedent and subsequent characteristics. *Journal of Child Psychology and Psychiatry and Allied Disciplines, 43*, 495–505.

Wright, J. P., & Beaver, K. M. (2005). Do parents matter in creating self-control in their children? A genetically informed test of Gottfredson and Hirschi's theory of low self-control. *Criminology, 43*, 1169–1202.

Wright, R. T., & Decker, S. H. (1997). *Armed robbers in action: Stickups and street culture.* Boston, MA: Northeastern University Press.

CHAPTER 8

Vigilance, Violence, and Murder in Mateships

Farnaz Kaighobadi
Florida Atlantic University

Todd K. Shackelford
Oakland University

According to the U.S. Department of Justice, between 2001 and 2005, 22% of reported incidents of nonfatal violence against women aged 12 or older were perpetrated by an intimate partner. This amounts to nearly 600,000 reported incidents of nonfatal violence against women by an intimate partner in just a single year in the United States (Bureau of Justice Statistics, 2007). Partly in response to the tragically high incidence of female-directed violence in intimate relationships and the devastating physical and psychological consequences of these behaviors, a large body of literature has been dedicated to the investigation of risk factors and predictors associated with men's violence against intimate partners. Previous research has identified several proximate predictors of men's partner-directed violence, such as family history of aggression (Busby, Holman, & Walker, 2008; Riggs & O'Leary, 1996), unemployment (Campbell et al., 2003), alcohol use (Klostermann & Fals-Stewart, 2006), and cultural influences (Archer, 2006a; Gage & Hutchinson, 2006), as well as ultimate or evolutionary predictors of such costly behaviors, such as men's perception of partner infidelity risk (Kaighobadi, Starratt, Shackelford, & Popp, 2008), men's life history strategies (Figueredo, Gladden, & Beck, 2010), and male sexual jealousy as a solution to the adaptive problem of paternity uncertainty (Goetz, Shackelford, Romero, Kaighobadi, & Miner, 2008; Shackelford, Goetz, Buss, Euler, & Hoier, 2005).

Over the past several decades, social scientists have increasingly recognized the value of an evolutionary perspective for guiding their research (Archer, 2006b; Daly & Wilson, 1997; Kenrick, Maner, & Li, 2005; Saad, 2007). Evolutionary psychological theories have been applied successfully to the investigation of diverse human behaviors. Evolutionary psychology is concerned with identifying and describing the design and function of psychological adaptations that evolved to solve the specific problems our ancestors faced recurrently over human evolutionary history. These evolved mechanisms include information-processing devices that motivate behavior in response to particular environmental inputs. An evolutionary psychological perspective can guide research on intimate-partner violence—most notably, research on the evolved mechanisms that motivate these behaviors.

This chapter reviews briefly different forms of men's partner-directed violence, including insults, sexual coercion, physical violence, and murder, with a particular focus on the adaptive problem of *paternity uncertainty*. It also considers individual differences in men's perpetration of violence in intimate relationships, including men's stable personality dispositions and life history strategies.

PATERNITY UNCERTAINTY AND MALE SEXUAL JEALOUSY

Over human evolutionary history, men and women have faced the adaptive problems of maintaining relationships and retaining intimate partners. Jealousy is an emotion that motivates behaviors that deter mate-poaching rivals and prevents partner infidelity or outright desertion from the relationship (Buss, Larsen, Westen, & Semmelroth, 1992: Daly, Wilson, & Weghorst, 1982; Symons, 1979). Men and women do not differ in the frequency or intensity with which they experience jealousy (Shackelford, LeBlanc, & Drass, 2000). Nevertheless, they respond differently to two different types of partner infidelity: emotional infidelity and sexual infidelity. Men are more distressed about a partner's sexual infidelity than about her emotional infidelity, whereas the opposite pattern is found among women. This sex difference has been documented in more than a dozen empirical studies using various methods, including forced-choice, self-report assessments (Buss et al., 1999), physiological assessments (Buss et al., 1992), experimental methods (Schützwohl, 2005, 2008; Thomson, Patel, Platek, & Shackelford, 2007), and archival and cross-cultural data (Betzig, 1989).

The sex difference in the experience of jealousy may be attributable to sex-specific adaptive problems humans faced over their evolutionary history (Buss, 2000; Symons, 1979). It is hypothesized that ancestral women faced the recurrent problems of paternal investment and acquisition and retention of resources with which to raise offspring. A partner's emotional infidelity might have predicted his current or future investment of resources in another woman and another woman's children.

Ancestral men, in contrast, faced the adaptive problem of paternity uncertainty. Female sexual infidelity and subsequent cuckoldry—a male's unwitting investment in offspring to whom he is not genetically related—carried substantial reproductive costs for ancestral men. It is theorized that the reproductive costs of cuckoldry in early human life, including the loss of time, energy, resources, and alternative mating opportunities, were potentially so great that men evolved to be sensitive to and to experience more distress about a partner's sexual infidelity. Men also may have evolved mechanisms that assess the risk of partner sexual infidelity and mechanisms that motivate the performance of anti-cuckoldry tactics. Goetz and others hypothesized that to assess the likelihood or risk of partner sexual infidelity, these information-processing mechanisms may use cues such as greater time spent apart from the partner, the presence of potential rivals, and a partner's attractiveness, or her "mate value" as a short-term or long-term partner (Goetz, 2007; Goetz & Shackelford, 2006; Peters, Shackelford, & Buss, 2002; Schmitt & Buss, 1996; Shackelford & Buss, 1997; Shackelford, Goetz, McKibbin, & Starratt, 2007; Trivers, 1972; Wilson & Daly, 1993). The behavioral output of male sexual jealousy varies from subtle nonviolent mate retention behaviors to outright physical violence.

MALE SEXUAL JEALOUSY AND MATE RETENTION BEHAVIORS

One class of behavioral output of sexual jealousy is men's mate retention behaviors, which function to prevent a partner's infidelity or outright relationship defection or to thwart rivals' attempts to encroach on the relationship. Buss (1988a) developed a taxonomy that organized mate retention behaviors into five categories:

- Acts of *direct guarding* function to keep a partner under surveillance.
- Acts of *intersexual negative inducements* include threats to punish a partner's infidelity.
- Acts of *positive inducements* include expressions of affection and care.
- Acts of *public signals of possession* include acts intended to signal possession of a partner to potential rivals.
- Acts of *intrasexual negative inducements* include acts intended to threaten potential rivals and thereby deter them from encroaching on the relationship.

As the risk of a partner's infidelity increases, men perform more frequent mate retention behaviors. For example, Buss and Shackelford (1997) found that men who are mated to younger, more attractive partners (cues to the women's reproductive value or expected future reproduction) and men who perceive greater probability of partner infidelity guard their partners more intensely. Also, men perform more frequent mate retention behaviors when they are mated to women who possess qualities that predict the woman's infidelity, including her personality characteristics such as surgency and openness to experience, and when their partner is near ovulation—a time when a female infidelity would be most costly for the in-pair male (Gangestad, Thornhill, & Garver, 2002; Goetz et al., 2005). Because time spent physically apart from a partner increases the risk of a partner's infidelity, men who have spent a greater proportion of time apart from their partners since the couple last copulated also report engaging in more frequent mate retention behaviors (Starratt, Shackelford, Goetz, & McKibbin, 2007). Finally, men who accuse their partner of infidelity are also more likely to engage in more frequent nonviolent mate retention behaviors (Kaighobadi et al., 2008; McKibbin, Goetz, Shackelford, Schipper, Starratt, & Stewart-Williams, 2007).

A more physically damaging output of male sexual jealousy is partner-directed violence. Male sexual jealousy is one of the most frequently cited causes of men's partner-directed violence, both physical and sexual (Daly & Wilson, 1988; Dobash & Dobash, 1977; Dutton, 1998; Frieze, 1983; Russell, 1982; Walker, 1979). The frequency with which men perform nonviolent mate retention behaviors predicts the frequency with which they inflict physical violence against their partners, because both classes of behavior are hypothesized to be outputs of sexual jealousy.

In three studies, Shackelford et al. (2005) investigated the relationship between men's nonviolent mate retention behaviors and men's partner-directed violence. Based on men's self-reports, women's partner-reports, and cross-partner reports in a sample of married couples, men's use of emotional manipulation as a mate retention tactic—marked by the performance of acts such as "I told my partner I would 'die' if she ever left me"—predicted female-directed violence. Men's monopolization of their partner's time and men's sexual inducements also predicted men's physical violence against their partners.

Because suspicions of partner infidelity explain the variance in men's mate retention behaviors, and because men's mate retention behaviors predict men's partner-directed violence, Kaighobadi, Starratt, Popp, and Shackelford (2008) hypothesized that men's suspicions of their partner's infidelity is linked directly to partner-directed violence. Using men's self-reports and women's partner-reports, these researchers found that men's accusations of their partner's infidelity explain a significant variance in men's partner-directed violence. Moreover, this relationship is mediated by the performance of nonviolent mate retention behaviors. Kaighobadi et al. (2008) hypothesized that men perform nonviolent and violent mate retention behaviors in a temporal hierarchical fashion. Less severe, less costly behaviors might be deployed first, followed by more severe behaviors such that the hierarchy of events leading to partner-directed violence is initiated with men's suspicions of partner infidelity, which are then followed by nonviolent mate retention behaviors, and finally end in acts of physical violence (Kaighobadi, Shackelford, Popp, Moyer, Bates, & Liddle, 2009).

FORCED IN-PAIR COPULATION

Instances of forced in-pair copulation (FIPC) have been documented in avian species that form long-term pair-bonds (Bailey, Seymour, & Stewart, 1978; Barash, 1977; Birkhead & Møller, 1992; Birkhead, Hunter, & Pellatt, 1989; Cheng, Burns, & McKinney, 1983; Goodwin, 1955; McKinney, Cheng, & Bruggers, 1984; Tryjanowski, Antczak, & Hromada, 2007). FIPC is hypothesized to be a form of post-copulatory male–male competition (i.e., a sperm competition tactic), because it often follows a female partner's extra-pair copulation or intrusions by rival males (Lalumière, Harris, Quinsey, & Rice, 2005; McKinney, Derrickson, & Mineau, 1983; Parker, 1970; Seymour & Titman, 1979; Valera, Hoi, & Kristin, 2003). Sperm competition occurs when a female copulates with and is inseminated by more than one male in a sufficiently brief period of time. By forcing the female to copulate shortly after there is the increased risk of insemination by a rival, males place their sperm in competition with any sperm deposited into their partner by a rival male. By engaging in extra-pair copulation in secluded areas, females of some avian species decrease the chance of discovery by their regular partner and, therefore, decrease the likelihood of a retaliatory forced copulation (Tryjanowski et al., 2007). Males also decrease the risk of sperm competition by frequent in-pair copulation and by close mate guarding (Birkhead & Møller, 1992).

Observations of sperm competition in nonhuman species offer a framework with which to consider similar adaptations in humans, who also form long-term socially (but not genetically) monogamous pair-bonds. Recently published evidence suggests that sperm competition has been a recurrent feature of human evolutionary history and that men have physiological and psychological mechanisms that may have evolved to solve related adaptive problems (Baker & Bellis, 1993; Gallup, Burch, Zappieri, Parvez, Stockwell, & Davis, 2003; Kilgallon & Simmons, 2005; Pound, 2002; Shackelford & Goetz, 2007; Shackelford & Pound, 2006; Shackelford, Pound, & Goetz, 2005; Smith, 1984; Wyckoff, Wang, & Wu, 2000). For example, when the risk of a partner's infidelity increases, men display copulatory urgency, perform more semen-displacing behaviors at next copulation, and adjust their ejaculates to include more sperm. Men's perception of their partner's physical and sexual attractiveness—a proxy for risk of sperm competition—also predicts the frequency of in-pair copulations. Men engage in more frequent in-pair copulations when they perceive their partner to be more physically and sexually attractive (Kaighobadi & Shackelford, 2008).

RISK OF SPERM COMPETITION AND SEXUAL COERCION

Male sexual coercion and rape of an intimate partner is another hypothesized manifestation of male sexual jealousy, which may be a response to perceived risk of sperm competition. Between 10% and 26% of women report being raped by their husbands (Finkelhor & Yllo, 1985; Hadi, 2000; Painter & Farrington, 1999; Watts, Keogh, Ndlovu, & Kwaramba, 1998). In a sample of young adults in a committed, sexual relationship, Goetz and Shackelford (2006) found that 7.3% of men admitted to at least one incidence of raping their current partner, and that 9.1% of women reported having experienced at least one incidence of rape by their current partner.

Many studies have investigated men's sexual coercion in an intimate relationship. A number of hypotheses have been formulated to test the proximate or immediate predictors of men's sexual coercion of an intimate partner as well as the ultimate or evolutionary predictors of men's sexual coercion of an intimate partner. Several scholars have argued that men's sexual coercion of their partners is motivated by a desire to dominate and control their partners (Bergen, 1996; Frieze, 1983; Gelles, 1977; Meyer, Vivian, & O'Leary, 1998). For example, several studies have found that men who are physically abusive toward their partners are more likely to be sexually coercive toward their partners than men who are not physically abusive (Apt & Hurlbert, 1993; DeMaris, 1997; Donnelly, 1993; Koziol-McLain, Coates, & Lowenstein, 2001; Shackelford & Goetz, 2004). Shackelford and Goetz (2004) also found a positive relationship between men's nonviolent controlling behaviors and men's sexual coercion of their partners. Gage and Hutchinson (2006) found that women's experience of sexual coercion by their partner is predicted by their partner's jealousy and nonviolent controlling behaviors, but not by differences in social power between the partners.

To address these apparently conflicting results regarding sexual coercion of intimate partners and men's motivation to gain or exert control over their partners, Goetz and Shackelford (2009) investigated several relevant predictors of sexual coercion, securing both men's self-reports and women's partner-reports. The results indicated that men's sexual coercion of their partners is predicted by both suspicions of female infidelity and men's controlling behavior, suggesting that both classes of variables contribute to the explanation of men's sexual coercion in intimate relationships.

Although the desire to dominate and control a partner may explain some portion of the individual differences in men's sexually coercive behaviors, proponents of the domination and control hypothesis argue that men *as a group* are motivated to exert "patriarchal terrorism" or "patriarchal power" over all women through sexual coercion of their own partners (Brownmiller, 1975; Johnson, 1995; Yllo & Straus, 1990). To date, no research has empirically tested these hypotheses. Note, however, that hypotheses proposing coordinated male–male cooperation to dominate and control women are not consistent with substantial theoretical and empirical work highlighting the frequency and intensity of male–male competition (rather than cooperation) for attracting women as intimate partners (Bleske & Shackelford, 2001; Buss, 1988b; Schmitt & Buss, 2001; Trivers, 1972).

Sexual coercion also is hypothesized to function as an anti-cuckoldry tactic (Gallup & Burch, 2006; Thornhill & Thornhill, 1992). According to one hypothesis, by forcing their partners to have sex, men who are suspicious of their partner's infidelity introduce their own sperm into their partner's reproductive tract, thereby decreasing the risk of cuckoldry. As described earlier in this chapter, this sperm competition hypothesis for partner-rape has been applied to nonhumans (notably, several avian species) to account for observations of partner-rape immediately following female extra-pair copulations. Rape of an intimate partner in humans also often follows accusations of female sexual infidelity.

Gallup and Burch (2006) proposed the "intra-pair copulation proclivity model" of female infidelity to predict and explain variance in the likelihood of men's sexual coercion of their partner following female sexual infidelity. These authors argued that, on the one hand, men have a propensity to engage in immediate copulation with their partners when they perceive a high risk of recent female infidelity. On the other hand, women may attempt to avoid copulating with their regular partners immediately after an extra-pair copulation. According to this hypothesis, extra-pair copulations may function to secure "good" genes from men other than their regular partner, whereas copulating immediately with her regular partner may cause displacement of the extra-pair sperm or otherwise interfere with the ability of the extra-pair sperm to fertilize the woman's egg(s). For these reasons, Gallup and Burch (2006) argue that men's copulatory urgency following detection of partner infidelity, and women's concurrent copulatory reluctance, may increase the risk of men's sexual coercion of their partner.

In two studies using data from men's self-reports and women's partner-reports, Goetz and Shackelford found that men's sexual coercion correlated positively with women's past and future likelihood of engaging in sexual infidelity. They also found that men who perform more mate retention behaviors are more likely to perform sexually coercive behaviors against their partners, as reported both by men and by men's partners. A number of studies have documented a positive relationship between men's sexual jealousy and men's sexual coercion of their partners (Goetz & Shackelford, 2006). For example, Frieze (1983) and Gage and Hutchinson (2006) found that men who sexually coerced their wives are more sexually jealous than men who did not. Previous research has found a direct positive relationship between men's suspicions and accusations of partner infidelity and men's sexual coercion of their partners (Starratt, Goetz, Shackelford, & Stewart-Williams, 2008).

According to Goetz and Shackelford (2009), the domination and control hypothesis and the sperm competition hypothesis reflect different levels of analysis. The domination and control hypothesis offers a proximate explanation of partner sexual coercion, including social or cultural causes of behavior. By comparison, the sperm competition hypothesis offers an ultimate explanation of partner sexual coercion, and addresses how adaptations that produce such costly behaviors could have evolved. Goetz and Shackelford do not argue that all sexually coercive behaviors are produced by evolved mechanisms that motivate anti-cuckoldry behaviors. Instead, they attempt to explain the increased likelihood of sexual coercion in the context of risk of female infidelity. Perhaps some instances of sexual coercion are the result of, for example, an antisocial man's motivation to control, dominate, or humiliate his partner (Goetz, Shackelford, Starratt, & McKibbin, 2008). Future research, therefore, might seek to investigate the interaction of individual differences in men's perpetration of sexual coercion and evolutionarily relevant contexts such as the risk of female infidelity and sperm competition to predict men's sexual coercion and rape of their partners.

INTIMATE-PARTNER HOMICIDE

According to the U.S. Department of Justice, between 1976 and 2005, 30% of female homicide ("femicide") victims were killed by an intimate partner; this type of crime accounts for the largest number of victim–offender relationships. In sharp contrast, just 5% of all male homicide victims were killed by an intimate partner. In most categories of intimate-partner homicide, men far outnumber women as the perpetrators and women far outnumber men as the victims (Bureau of Justice Statistics, 2007).

Many studies have investigated the predictors and risk factors associated with female-partner homicide. Campbell and colleagues (2003), for example, used a multisite case-control study to investigate the risk factors associated with intimate-partner femicide. The results indicated that the male partner's unemployment was the most significant sociodemographic risk factor for intimate-partner femicide. These researchers also found that the risk of intimate-partner femicide increased by five times when the victim left the abusive partner for another man or when her partner's sexual jealousy was triggered. Additional empirical evidence suggests that the risk of intimate-partner femicide increases in response to the victim's attempt to leave the relationship (McFarlane, Campbell, Wilt, Sachs, Ulrich, & Xu, 1999; Nicolaidis et al., 2003; U.S. Department of Justice, 1998). Campbell et al. (2003) also documented that having a child from the victim's previous partner living in the home doubled the risk of intimate-partner femicide. This result parallels reports of increased frequency and severity of female-directed partner abuse when children unrelated to the man are living in the household.

Nicolaidis et al. (2003) conducted 30 in-depth interviews with women who survived an attempted homicide by their intimate partner. They found that 20 women (67%) had a history of prior violence with the partner. Twenty-five women (80%) explicitly mentioned their partner's prior stalking, sexual jealousy, accusations of infidelity, social isolation and physical limitation of the woman, or threats of violence against the woman. In 22 (73%) of the cases, the male partner attempted to kill his partner when she threatened to leave the relationship.

Several hypotheses have been advanced to explain the occurrence and frequency of intimate-partner homicide, two of which are summarized here: the "killing as a by-product" hypothesis and the "evolved homicide module" hypothesis. According to the "killing as a by-product" hypothesis, because killing a partner carries substantial and severe costs that might not have consistently produced sufficient benefits over human evolutionary history, it is unlikely to be the product of specialized adaptations. The costs associated with killing a partner include the risk of retaliation by the victim's kin and the local community; the loss of time, energy, and resources a man invested in maintaining the relationship; the damage to the man's reputation; and the loss of maternal investment in any shared offspring. Thus, partner-killing might be a by-product of other male psychological adaptations, including adaptations specialized to motivate nonlethal punishment of a partner's suspected or actual infidelity and to control her sexual interactions (Daly & Wilson, 1988; Wilson & Daly, 1998; Wilson, Daly, & Daniele, 1995).

In keeping with this hypothesis, Wilson and Daly argue that men's "sexual proprietariness" is a key cause of 80% of spousal homicides. Wilson and Daly define sexual proprietariness as a combination of sexual jealousy and men's "presumptions of entitlement" and motivation to control their partner's sexual behavior. According to Wilson and Daly, the variability in men's sexual proprietariness and perpetration of violence is explained by several factors: (1) intensity of intrasexual competition, (2) female-partner attractiveness, (3) suspicions of partner infidelity, (4) degree of female choice, and (5) costs associated with perpetration of violence (Wilson & Daly, 1992).

Proponents of a second hypothesis regarding the origins of intimate-partner homicide argue that the by-product hypothesis cannot explain the large incidence of apparently premeditated partner homicides. Premeditated homicides include hiring someone to kill the partner, aiming at and shooting a partner, and deliberately poisoning a partner. Buss and Duntley (1998, 2003) argue that partner-killing by men might be motivated by specialized adaptation in the context of suspected or actual female infidelity (and, as a consequence, paternity uncertainty). This "evolved homicide module" hypothesis does not imply that discovery of female sexual infidelity will always (or even frequently) lead to partner-killing by men, but instead suggests that the relevant evolved mechanisms are likely to be activated with suspicions of female-partner infidelity and may very occasionally result in partner-killing (Buss, 2005). The results from previous studies may provide evidence that supports the evolved homicide module hypothesis, in that men most frequently kill or attempt to kill their partners when the woman threatens to leave the relationship.

Wilson and Daly's and Buss and Duntley's competing hypotheses have not yet been tested concurrently, so it is not possible to conclude that one hypothesis accounts for the data better than the other. However, given the many costs associated with intimate-partner femicide, the most parsimonious explanation for this crime might be Wilson and Daly's by-product hypothesis. Future research may investigate the adaptation, by-product, and alternative hypotheses concurrently.

INDIVIDUAL DIFFERENCES IN INTIMATE-PARTNER VIOLENCE

Previous research has identified links between men's partner-directed violence and men's personality traits, including antisocial tendencies, self-centeredness, lack of emotional regulation, and impulsivity (Dean & Malamuth, 1997; Dutton, 1994; Dutton & Starzomski, 1993; McNulty & Hellmuth, 2008; Stuart & Holtzworth-Munroe, 2005; White, Darcy, Swartout, Sechrist, & Gollehon, 2008). Hellmuth and McNulty (2008) note that most previous research addressing individual differences in men's perpetration of intimate-partner violence has investigated personality disorders as predictors of partner-directed violence, whereas less research has addressed variations in normal personality traits such as those assessed by the Five Factor Model (FFM). In their work, these authors documented that the husband's and wife's emotional stability interacts with levels of chronic stress to influence the frequency with which violence is inflicted against spouses. Busby et al. (2008) also found a relationship between partner-reports of neuroticism and violence in intimate relationships.

To address this empirical gap in the literature, Kaighobadi et al. (2009) investigated the relationships between men's personality traits as assessed by the FFM and men's partner-directed violence. These researchers hypothesized that men's personality traits may interact with situational contexts that trigger male sexual jealousy, thereby predicting intimate partner-violence. The results of their study indicated that men's emotional stability, agreeableness, and conscientiousness predict men's partner-directed violence. Men who score low on these personality traits are more likely to perpetrate violence against their partners. More importantly, however, the results indicated that men's suspicions of their partner's infidelity moderate the relationship between men's personality traits and men's partner-directed violence. In other words, personality affects partner-directed violence differently depending on men's perceptions of their partner's risk of infidelity (Kaighobadi et al., 2009). Other studies have identified additional sources of individual differences in partner-directed violence, including family history of aggression, childhood sexual and physical abuse, alcohol and drug use, and problem-solving skills (Busby et al., 2008; Klostermann & Fals-Stewart, 2006; Riggs & O'Leary, 1996).

LIFE HISTORY STRATEGIES AND INTIMATE-PARTNER VIOLENCE

Life history theory provides a framework for investigating individual differences in humans as they allocate energy and resources to mating versus parenting efforts (Figueredo et al., 2006). Previous research has linked the *fast* life history strategy (LHS) with high mating effort and low parenting effort. In particular, fast LHS has been linked to risk taking, impulsiveness, sexual promiscuity, and disregard for social rules. *Slow* LHS, in contrast, has been linked to monogamy, high parenting effort, future orientation, and attentive regard for social rules. LHS, therefore, is hypothesized to predict individual differences in mating effort (and, conversely, parenting effort), including, but not limited to, partner-directed sexual coercion and violence.

Figueredo, Gladden, and Beck (2011) argue that because fast-LHS, high-mating-effort men (relative to slow-LHS, low-mating-effort men) are unlikely to commit to a monogamous relationship and concomitant parenting effort, their mating strategies may more often come in conflict with women's mating strategies. This conflict may result in negative attitudes toward the opposite sex and low relationship satisfaction. Furthermore, Gladden, Figueredo, and Snyder (2010) have documented a link between LHS and general perceived mate value. The results of their study indicate that individuals with a slow LHS possess a positive sense of self-worth and perceived mate value, whereas individuals with a fast LHS perceive themselves as ranking relatively low on mate value. Thus, because fast LHS is linked to (1) risk-taking, impulsive, law-breaking behaviors, (2) perceived low mate value, and (3) increase sexual conflict, fast LHS may predict men's partner-directed sexual coercion and violence.

Consistent with this hypothesis, Gladden, Sisco, and Figueredo (2008) have documented a link between a single cluster of slow LHS traits and decreased sexual coercion in a college student sample. Moreover, in a recent study, Figueredo and colleagues (2010) predicted a relationship between LHS and intimate-partner violence mediated by mate value. Their results indicated that (1) a relationship exists between slow LHS and higher perceived mate value, and (2) a relationship exists between higher perceived mate value and decreased partner-directed violence. In other words, slow LHS is indirectly associated in men with decreased intimate-partner violence; this relationship is mediated by men's perceived mate value.

In conclusion, LHS may be a reliable predictor of sexual coercion and violence in intimate relationships, providing an evolutionary framework for explaining some individual differences in the propensity toward these costly behaviors.

CONCLUSION

Evolutionary psychologists are interested in identifying the ultimate or distal explanations of a trait or behavior. Intimate-partner violence and murder are especially costly behaviors, for both the victim and the perpetrator. It is useful to consider an evolutionary perspective to investigate the design features and evolved function of the psychological mechanisms that motivate these behaviors. An evolutionary psychological perspective can guide identification of those contexts that trigger the relevant information-processing mechanisms and motivate the subsequent behaviors.

Many instances of female-directed intimate-partner violence and homicide co-occur with and may be triggered by men's suspicions or knowledge of their partner's sexual infidelity. Nevertheless, individual differences may exist in terms of men's willingness to perpetrate violence against their intimate partners. Previous research has identified several proximate correlates of partner-directed violence, such as a family history of aggression and acceptance of local cultural norms.

Situational factors, including the characteristics of the perpetrator, the victim, and the circumstances in which the violence occurs, also have been considered in research investigating intimate-partner violence and homicide. For example, the perpetrator's age and the victim's age, the perpetrator's mental health, and the availability of weapons have been investigated as risk factors associated with men's partner-directed violence and murder (Dutton & Kerry, 1999; Paulozzi, Saltzman, Thompson, & Holmgreen,

2001; Shackelford, LeBlanc, Weekes-Shackelford, Bleske-Rechek, Euler, & Hoier, 2002; Wilkinson & Hamerschlag, 2005).

In conclusion, it is important to investigate both the proximate and ultimate causes of men's partner-directed violence and murder. The relevant evolved psychological mechanisms interact with stable dispositions and situational factors to produce manifest behavior. Future research might benefit from using an evolutionary perspective to build models of female-directed intimate-partner violence that include both stable dispositions (such as personality traits) and environmental factors (such as a family history of aggression). To achieve a fuller understanding of intimate-partner violence and homicide, researchers must include a careful consideration of the evolved psychological mechanisms that motivate these costly behaviors.

REFERENCES

Apt, C., & Hurlbert, D. F. (1993). The sexuality of women in physically abusive marriages: Comparative study. *Journal of Family Violence, 8*, 57–69.

Archer, J. (2006a). Cross-cultural differences in physical aggression between partners: A social-role analysis. *Personality and Social Psychology Review, 10*, 133–153.

Archer, J. (2006b). Testosterone and human aggression: An evaluation of the challenge hypothesis. *Neuroscience & Biobehavioral Reviews, 30*, 319–345.

Bailey, R. O., Seymour, N. R., & Stewart, G. R. (1978). Rape behavior in blue-winged teal. *Auk, 95*, 188–190.

Baker, R. R., & Bellis, M. A. (1993). Human sperm competition: Ejaculate adjustment by males and the function of masturbation. *Animal Behavior, 46*, 861–885.

Barash, D. P. (1977). Sociobiology of rape in mallards (*Anas platyrhynchos*): Response of the mated male. *Science, 197*, 788–789.

Bergen, R. K. (1996). *Wife rape: Understanding the response of survivors and service providers.* Thousand Oaks, CA: Sage.

Betzig, L. (1989). Causes of conjugal dissolution: A cross-cultural study. *Current Anthropology, 30*, 654–676.

Birkhead, T. R., Hunter, & F. M., Pellatt, J. E. (1989). Sperm competition in the zebra finch, *Taeniopygia guttata. Animal Behavior, 38*, 935–950.

Birkhead, T. R., & Møller, A. P. (1992). *Sperm competition in birds: Evolutionary causes and consequences.* London: Academic Press.

Bleske, A. L., & Shackelford, T. K. (2001). Poaching, promiscuity, and deceit: Combating mating rivalry in same-sex friendships. *Personal Relationships, 8*, 407–424.

Brownmiller, S. (1975). *Against our will: Men, women, and rape.* New York: Simon & Schuster.

Bureau of Justice Statistics. (2007, December). *Intimate partner violence: Victim characteristics, 1976–2005.* Washington, DC: U.S. Department of Justice, Office of Justice Programs.

Busby, D. M., Holman, T. B., & Walker, E. (2008). Pathways to relationship aggression between adult partners. *Family Relations, 57*, 72–83.

Buss, D. M. (1988a). From vigilance to violence: Tactics of mate retention in American undergraduates. *Ethology and Sociobiology, 9*, 91–317.

Buss, D. M. (1988b). The evolution of human intrasexual competition: Tactics of mate attraction. *Journal of Personality and Social Psychology, 54*, 616–628.

Buss, D. M. (2000). *The dangerous passion.* New York: Free Press.

Buss, D. M. (2005). *The murderer next door.* New York: Penguin Press.

Buss, D. M., & Duntley, J. D. (1998, July 10). *Evolved homicide modules.* Paper presented at the Annual Meeting of the Human Behavior and Evolution Society, Davis, CA.

Buss, D. M., & Duntley, J. D. (2003). Homicide: An evolutionary perspective and implications for public policy. In N. Dress (Ed.), *Violence and public policy* (pp. 115–128). Westport, CT: Greenwood.

Buss, D. M., Larsen, R. J., Westen, D., & Semmelroth, J. (1992). Sex differences in jealousy: Evolution, physiology and psychology. *Psychological Science, 3*, 251–255.

Buss, D. M., & Shackelford, T. K. (1997). From vigilance to violence: Mate retention tactics in married couples. *Journal of Personality and Social Psychology, 72*, 346–361.

Buss, D. M., Shackelford, T. K., Kirkpatrick, L. A., Chloe, J., Hasegawa, M., Hasegawa, T., & Bennett, K. (1999). Jealousy and beliefs about infidelity: Tests of competing hypotheses in the United States, Korea, and Japan. *Personal Relationships, 6*, 125–150.

Campbell, J. C., Webster, D., Koziol-McLain, J., Block, C., Campbell, D., Curry, M. A., et al. (2003). Risk factors for femicide in abusive relationships: Results from a multisite case control study. *American Journal of Public Health, 93*, 1089–1097.

Cheng, K. M., Burns, J. T., & McKinney, F. (1983). Forced copulation in captive mallards III. Sperm competition. *Auk, 100*, 302–310.

Daly, M., & Wilson, M. (1988). *Homicide.* Hawthorne, NY: Aldine de Gruyter.

Daly M., & Wilson M. (1997). Crime and conflict: Homicide in evolutionary psychological perspective. *Crime and Justice, 22*, 251–300.

Daly, M., Wilson, M., & Weghorst, J. (1982). Male sexual jealousy. *Ethology and Sociobiology, 3*, 11–27.

Dean, K. E., & Malamuth, N. M. (1997). Characteristics of men who agress sexually and of men who imagine aggressing: Risk and moderating variables. *Journal of Personality and Social Psychology, 72*, 449–455.

DeMaris, A. (1997). Elevated sexual activity in violent marriages: Hypersexuality or sexual extortion? *Journal of Sex Research, 34*, 361–373.

Dobash, R. E., & Dobash, R. P. (1977). Wives: The appropriate victims of marital violence. *Victimology, 2*, 426–442.

Donnelly, D. A. (1993). Sexually inactive marriages. *Journal of Sex Research, 30*, 171–179.

Dutton, D. G. (1994). The origin and structure of the abusive personality. *Journal of Personality Disorders, 8*, 181–191.

Dutton, D. G. (1998). *The abusive personality.* New York: Guilford Press.

Dutton, D. G., & Kerry, G. (1999). Modus operandi and personality disorder in incarcerated spousal killers. *International Journal of Law and Psychiatry, 22*, 287–299.

Dutton, D. G., & Starzomski, A. J. (1993). Borderline personality in perpetrators of psychological and physical abuse. *Violence and Victims, 8*, 327–337.

Figueredo, A. J., Gladden, P. R., & Beck, C. J. A. (2011). Intimate partner violence and life history strategy. In A. Goetz & T. Shackelford, (Eds.), *The Oxford handbook of sexual conflict in humans.* New York: Oxford University Press.

Figueredo, A. J., Vásquez, G., Brumbach, B. H., Schneider, S. M. R., Sefcek, J. A., Tal, I. R., et al. (2006). Consilience and life history theory: From genes to brain to reproductive strategy. *Developmental Review, 26*, 243–275.

Finkelhor, D., & Yllo, K. (1985). *License to rape: Sexual abuse of wives.* New York: Holt, Rinehart, & Winston.

Frieze, I. H. (1983). Investigating the causes and consequences of marital rape. *Signs, 8*, 532–553.

Gage, A. J., & Hutchinson, P. L. (2006). Power, control, and intimate partner sexual violence in Haiti. *Archives of Sexual Behavior, 35*, 11–24.

Gallup, G. G. Jr., & Burch, R. L. (2006). The semen displacement hypothesis: Semen hydraulics and the intra-pair copulation proclivity model of female infidelity. In. S. M. Platek & T. K. Shackelford (Eds.), *Female infidelity and paternal uncertainty: Evolutionary perspectives on male anti-cuckoldry tactics* (pp. 129–140). New York: Cambridge University Press.

Gallup, G. G. Jr., Burch, R. L., Zappieri, M. L., Parvez, R. A., Stockwell, M. L., & Davis, J. A. (2003). The human penis as a semen displacement device. *Evolution and Human Behavior, 24*, 277–289.

Gangestad, S. W., Thornhill, R., & Garver, C. E. (2002). Changes in women's sexual interests and their partner's mate-retention tactics across the menstrual cycle: Evidence for shifting conflicts of interest. *Proceedings of the Royal Society of London, 269*, 975–982.

Gelles, R. (1977). Power, sex and violence: The case of marital rape. *Family Coordinator, 26*, 339–347.

Gladden, P. R., Figueredo, A. J., & Snyder, B. (2010). Life history strategy and evaluative self-assessment. *Personality and Individual Differences, 48*, 731–735.

Gladden, P. R., Sisco, M., & Figueredo, A. J. (2008). Sexual coercion and life history strategy. *Evolution and Human Behavior, 29*, 319–326.

Goetz, A. T. (2007). Violence and abuse in families: The consequences of paternal uncertainty. In C. Salmon & T. K. Shackelford (Eds.), *Family relationships: An evolutionary perspective* (pp. 259–274). New York: Oxford University Press.

Goetz, A. T., & Shackelford, T. K. (2006). Sexual coercion and forced in-pair copulation as sperm competition tactics in humans. *Human Nature, 17*, 265–282.

Goetz, A. T., & Shackelford, T. K. (2009). Sexual coercion in intimate relationships: A comparative analysis of the effects of women's infidelity and men's dominance and control. *Archives of Sexual Behavior, 38*, 226–234.

Goetz, A. T., Shackelford, T. K., Romero, G. A., Kaighobadi, F., & Miner, E. J. (2008). Punishment, proprietariness, and paternity: Men's violence against women from an evolutionary perspective. *Aggression and Violent Behavior, 13*, 481–489.

Goetz, A. T., Shackelford, T. K., Starratt, V. G., & McKibbin, W. F. (2008). Intimate partner violence. In J. D. Duntley & T. K. Shackelford (Eds.), *Evolutionary forensic psychology* (pp. 65–78). New York: Oxford University Press.

Goetz, A. T., Shackelford, T. K., Weekes-Shackelford, V. A., Euler, H. A., Hoier, S., Schmitt, D. P., & LaMunyon, C. W. (2005). Mate retention, semen displacement, and human sperm competition: A preliminary investigation of tactics to prevent and correct female infidelity. *Personality and Individual Differences, 38*, 749–763.

Goodwin, D. (1955). Some observations on the reproductive behavior of rooks. *British Birds, 48*, 97–107.

Hadi, A. (2000). Prevalence and correlates of the risk of marital sexual violence in Bangladesh. *Journal of International Violence, 15*, 787–805.

Hellmuth, J. C., & McNulty, J. K. (2008). Neuroticism, marital violence, and the moderating role of stress and behavioral skills. *Journal of Personality and Social Psychology, 95*, 166– 180.

Johnson, M. P. (1995). Patriarchal terrorism and common couple violence: Two forms of violence against women. *Journal of Marriage and the Family, 57*, 283–294.

Kaighobadi, F., & Shackelford, T. K. (2008). Female attractiveness mediates the relationship between in-pair copulation frequency and men's mate retention behaviors. *Personality and Individual Differences, 45*, 293–295.

Kaighobadi, F., Shackelford, T. K., Popp, D., Moyer, R. M., Bates, V. M., & Liddle, J. R. (2009). Perceived risk of female infidelity moderates the relationship between men's personality and partner-directed violence. *Journal of Research in Personality, 43*, 1033–1039.

Kaighobadi, F., Starratt, V. G., Shackelford, T. K., & Popp, D. (2008). Male mate retention mediates the relationship between female sexual infidelity and female-directed violence. *Personality and Individual Differences, 44*, 1422–1431.

Kenrick, D. T., Maner, J., & Li, N. P. (2005). Evolutionary social psychology. In D. M. Buss (Ed.), *The handbook of evolutionary psychology* (pp. 803–827). Hoboken, NJ: Wiley.

Kilgallon, S. J., & Simmons, L. W. (2005). Image content influences men's semen quality. *Biology Letters, 1*, 253–255.

Klostermann, K. C., & Fals-Stewart, W. (2006). Intimate partner violence and alcohol use: Exploring the role of drinking in partner violence and its implications for intervention. *Aggression and Violent Behavior, 11*, 587–597.

Koziol-McLain, J., Coates, C. J., & Lowenstein, S. R. (2001). Predictive validity of a screen for partner violence against women. *American Journal of Preventive Medicine, 21*, 93–100.

Lalumière, M. L., Harris, G. T., Quinsey, V. L., & Rice, M. E. (2005). *The causes of rape: Understanding individual differences in male propensity for sexual aggression.* Washington, DC: APA Press.

McFarlane, J., Campbell, J. C., Wilt, S., Sachs, C. J., Ulrich, Y., & Xu, X. (1999). Stalking and intimate partner femicide. *Homicide Studies, 3*, 300–316.

McKibbin, W. F., Goetz, A. T., Shackelford, T. K., Schipper, L. D., Starratt, V.G., & Stewart-Williams, S. (2007). Why do men insult their intimate partners? *Personality and Individual Differences, 43*, 231–241.

McKinney, F., Cheng, K. M., & Bruggers, D. J. (1984). Sperm competition in apparently monogamous birds. In R. L. Smith (Ed.), *Sperm competition and evolution of animal mating systems* (pp. 523–545). New York: Academic Press.

McKinney, F., Derrickson, S. R., & Mineau, P. (1983). Forced copulation in waterfowl. *Behavior, 86*, 250–294.

McNulty, J. K., & Hellmuth, J. C. (2008). Emotion regulation and intimate partner violence in newlyweds. *Journal of Family Psychology, 22*, 794–797.

Meyer, S., Vivian, D., & O'Leary, K. D. (1998). Men's sexual aggression in marriage: Couple's reports. *Violence Against Women, 4*, 415–435.

Nicolaidis, C., Curry, M. A., Ulrich, Y., Sharps, P, McFarlane, J., Campbell, D., et al. (2003). Could we have known? A qualitative analysis of data from women who survived an attempted homicide by an intimate partner. *Journal of General Internal Medicine, 18*, 788–794.

Painter, K., & Farrington, D. P. (1999). Wife rape in Great Britain. In R. Muraskin (Ed.), *Women and justice: Development of international policy* (pp. 135–164). New York: Gordon and Breach.

Parker, G. A. (1970). Sperm competition and its evolutionary consequences in the insects. *Biological Reviews, 45*, 525–567.

Paulozzi, L. J., Saltzman, L. E., Thompson, M. P., & Holmgreen, P. (2001). Surveillance for homicide among intimate partners: United States, 1981–1998. In *MMWR: CDC surveillance summaries* (50, SS-3). Atlanta, GA: U.S. Department of Health and Human Services.

Peters, J., Shackelford, T. K., & Buss, D. M. (2002). Understanding domestic violence against women: Using evolutionary psychology to extend the feminist functional analysis. *Violence and Victims, 17*, 255–264.

Pound, N. (2002). Male interest in visual cues of sperm competition risk. *Evolution and Human Behavior, 23*, 443–466.

Riggs, D. S., & O'Leary, K. D. (1996). Aggression between heterosexual dating partners: An examination of a causal model of courtship aggression. *Journal of Interpersonal Violence, 11*, 519–540.

Russell, D. E. H. (1982). *Rape in marriage.* New York: Macmillan.

Saad, G. (2007). *The evolutionary bases of consumption.* Mahwah, NJ: Lawrence Erlbaum.

Schmitt, D. P., & Buss, D. M. (1996). Strategic self-promotion and competitor derogation: Sex and context effects on the perceived effectiveness of mate attraction tactics. *Journal of Personality and Social Psychology, 70*, 1185–1204.

Schmitt, D. P., & Buss, D. M. (2001). Human mate poaching: Tactics and temptations for infiltrating existing mateships. *Journal of Personality and Social Psychology, 80*, 894– 917.

Schützwohl, A. (2005). Sex differences in jealousy: The processing of cues to infidelity. *Evolution and Human Behavior*, 26, 288–299.

Schützwohl, A. (2008). The disengagement of attentive resources from task-irrelevant cues to sexual and emotional infidelity. *Personality and Individual Differences, 44*, 633–644.

Seymour, N. R., & Titman, R. D. (1979). Behaviour of unpaired male black ducks (*Anas rupribes*) during the breeding season in a Nova Scotia tidal marsh. *Canadian Journal of Zoology, 57*, 2412–2428.

Shackelford, T. K., & Buss, D. M. (1997). Cues to infidelity. *Personality and Social Psychology Bulletin, 23*, 1034–1045.

Shackelford, T. K., & Goetz, A. T. (2004). Men's sexual coercion in intimate relationships: Development and initial validation of the Sexual Coercion in Intimate Relationships Scale. *Violence and Victims, 19*, 541–556.

Shackelford, T. K., & Goetz, A. T. (2007). Adaptation to sperm competition in humans. *Current Directions in Psychological Science, 16*, 47–50.

Shackelford, T. K., Goetz, A. T., Buss, D. M., Euler, H. A., & Hoier, S. (2005). When we hurt the ones we love: Predicting violence against women from men's mate retention tactics. *Personal Relationships, 12*, 447–463.

Shackelford, T. K., Goetz, A. T., McKibbin, W. F., & Starratt, V. G. (2007). Absence makes the adaptations grow fonder: Proportion of time apart from partner, male sexual psychology, and sperm competition in humans (*Homo sapiens*). *Journal of Comparative Psychology, 121*, 214–220.

Shackelford, T. K., LeBlanc, G. J., & Drass, E. (2000). Emotional reactions to infidelity. *Cognition and Emotion, 14*, 643–659.

Shackelford, T. K., LeBlanc, G. J., Weekes-Shackelford, V. A., Bleske-Rechek, A. L., Euler, H. A., & Hoier, S. (2002). Psychological adaptation to human sperm competition. *Evolution and Human Behavior, 23*, 123–138.

Shackelford, T. K., & Pound, N. (Eds.). (2006). *Sperm competition in humans.* New York: Springer.

Shackelford, T. K., Pound, N., & Goetz, A. T. (2005). Psychological and physiological adaptations to sperm competition in humans. *Review of General Psychology, 9*, 228–248.

Smith, R. L. (1984). Human sperm competition. In R. L. Smith (Ed.), *Sperm competition and the evolution of animal mating systems* (pp. 601–660). New York: Academic Press.

Starratt, V. G., Goetz, A. T., Shackelford, T. K., & Stewart-Williams, S. (2008). Men's partner-directed insults and sexual coercion in intimate relationships. *Journal of Family Violence, 23*, 315–323.

Starratt, V. G., Shackelford, T. K. Goetz, A. T., & McKibbin, W. F. (2007). Male mate retention behaviors vary with risk of female infidelity and sperm competition. *Acta Psychology Sinica, 39*, 523–527.

Stuart, G. L., & Holtzworth-Munroe, A. (2005). Testing a theoretical model of the relationship between impulsivity, mediating variables, and husband violence. *Journal of Family Violence, 20*, 291–303.

Symons, D. (1979). *The evolution of human sexuality.* New York: Oxford University Press.

Thomson, J. W., Patel, S., Platek, S. M., & Shackelford, T. K. (2007). Sex differences in implicit association and attentional demands for information about infidelity. *Evolutionary Psychology, 5*, 569–583.

Thornhill, R., & Thornhill, N. W. (1992). The evolutionary psychology of men's coercive sexuality. *Behavioral and Brain Sciences, 15*, 363–421.

Trivers, R. L. (1972). Parental investment and sexual selection. In B. Campbell (Ed.), *Sexual selection and the descent of man 1871–1971* (pp. 136–179). Chicago: Aldine.

Tryjanowski, P., Antczak, M., & Hromada, M. (2007). More secluded places for extra-pair copulations in the great shrike *Lanius excubitor. Behavior, 144*, 23–31.

U.S. Department of Justice. (1998, March). *Violence by intimates: Analysis of data on crimes by current or former spouses, boyfriends, and girlfriends.* Washington, DC: Author.

Valera, F., Hoi, H., & Kristin, A. (2003). Male shrikes punish unfaithful females. *Behavioral Ecology, 14*, 403–408.

Walker, L. E. (1979). *The battered woman.* New York: Harper & Row.

Watts, C., Keogh, E., Ndlovu, M., & Kwaramba, R. (1998). Withholding of sex and forced sex: Dimensions of violence against Zimbabwean women. *Reproductive Health Matters, 6*, 57–65.

White, J. W., Darcy, M., Swartout, K., Sechrist, S., & Gollehon, A. (2008). Violence in intimate relationships: A conceptual and empirical examination of sexual and physical aggression. *Child and Youth Services Review, 30*, 338–351.

Wilkinson, D. L., & Hamerschlag, S. J. (2005). Situational determinants in intimate partner violence. *Aggression and Violent Behavior, 10*, 333–361.

Wilson, M., & Daly, M. (1992). The man who mistook his wife for a chattel. In J. H. Barkow, L. Cosmides, & J. Tooby (Eds.), *The adapted mind* (pp. 289–322). New York: Oxford University Press.

Wilson, M., & Daly, M. (1993). An evolutionary psychological perspective on male sexual proprietariness and violence against wives. *Violence and Victims, 8*, 271–294.

Wilson, M., & Daly, M. (1998). Lethal and nonlethal violence against wives and the evolutionary psychology of male sexual proprietariness. In R. E. Dobash & R. P. Dobash (Eds.), *Rethinking violence against women* (pp. 199–230). Thousand Oaks, CA: Sage.

Wilson, M., Daly, M., & Daniele, A. (1995). Familicide: The killing of spouse and children. *Aggressive Behavior, 21*, 275–291.

Wyckoff, G. J., Wang, W., & Wu, C. (2000). Rapid evolution of male reproductive genes in the descent of man. *Nature, 403*, 304–308.

Yllo, K., & Straus, M. A. (1990). Patriarchy and violence against wives. *Crime and Justice, 8*, 383–399.

CHAPTER 9

Early Risk Factors for Young Homicide Offenders and Victims

David P. Farrington
University of Cambridge

Rolf Loeber
University of Pittsburgh

Rebecca Stallings
University of Pittsburgh

Doni Lynn Homish
State University of New York at Buffalo

Homicide is the second leading cause of death for persons aged 15 to 24 in the United States, after accidents and before suicide (Hoyert et al., 2001). According to 1997 data, 1 in 40 African American males is murdered, compared with 1 in 280 Caucasian males. The peak ages at which males died from this cause were between ages 18 and 25, with the rate of death in this group being more than 100 per 100,000 African American males and more than 10 per 100,000 Caucasian males (Federal Bureau of Investigation, 2003). The rates of homicide offenders and victims in ages 13–17 and 18–24 increased from 1985 to a peak in 1993, then declined. For ages 18–24, the rates at the peak in 1993 were almost double the rates in 1985; for ages 13–17, the rates at the peak in 1993 were almost triple the rates in 1985. For individuals between the ages of 18 and 24, the peak homicide offending rate in 1993 was 280 per 100,000 for African Americans and 30 per 100,000 for Caucasians; the peak homicide victimization rate in 1993 was 190 per 100,000 for African Americans and 20 per 100,000 for Caucasians (Cook & Laub, 1998).

In general, the ages of offenders and victims are positively correlated, with the peak risk for 19- and 20-year-olds killing other 19- and 20-year-olds. Most Caucasian victims (84%) are killed by Caucasian offenders; similarly, most African American victims (91%) are killed by African American offenders. Most homicide offenders aged 18–24 killed with a gun; in the peak year of 1993, there were three times as many gun homicides as homicides by other means by this age group (Fox & Zawitz, 1999; Snyder & Sickmund, 1999).

A great deal of research has focused on the characteristics of young homicide offenders. Retrospective case-control studies show that they tend to come from broken homes and violent families, have experienced parental alcoholism and child abuse, have low school achievement, and have run away from home, truanted, and been suspended from school. Not surprisingly, they also tend to have prior arrest histories. Much less is known about the backgrounds of young homicide victims. However, it is known that assault victims disproportionately tend to have committed offenses themselves (Lauritsen, Laub, & Sampson, 1992; Rivara et al., 1995).

As Kathleen Heide (2003) has pointed out, "The available literature on juvenile homicide offenders is retrospective in nature. To say something definitive about etiological factors associated with youth murder requires longitudinal studies of children" (p. 25). Only two previous such studies of youth homicide have been performed, both using data from the Pittsburgh Youth Study. Loeber and his colleagues studied the characteristics of 11 boys killed and 29 wounded by guns (Loeber et al., 1999; Loeber et al., 2005). Compared with control groups, these boys tended to have low academic achievement, depressed mood, poor parental supervision, poor parent–boy communication, behavior problems of the father, the family on welfare, and low socioeconomic status. The researchers also investigated factors that predicted violent offenders out of all boys and factors that predicted homicide offenders out of all violent boys. The factors that predicted homicide offenders were a high risk score, a positive attitude toward substance use, conduct disorder, carrying a weapon, gang fighting, persistent drug use, selling hard drugs, peer delinquency, peer substance use, and being held back in school.

FOCUS OF THIS CHAPTER

The key questions addressed in the research presented in this chapter are as follows:

1. Which risk factors, measured at the beginning of a prospective longitudinal survey, predict homicide offenders and homicide victims out of a community sample 4 to 15 years later? This question has never been addressed before.

2. How accurately can homicide offenders and victims be predicted at ages 7 to 13?
3. To what extent are risk factors similar for homicide offenders and victims?
4. Do early risk factors predict homicide offenders more accurately than homicide victims? This relationship would be expected if homicide offenders were more antisocial and deviant than homicide victims.
5. Do behavioral and attitudinal risk factors predict homicide victims more accurately than explanatory risk factors? This relationship would be expected to the extent that behavioral and attitudinal factors reflect the same underlying construct (e.g., an antisocial personality) as is present in homicide offenders. By definition, explanatory factors measure a different underlying construct. However, it is less clear that behavioral and attitudinal factors would predict homicide victims more accurately, because it is less clear that being a homicide victim reflects an underlying antisocial personality or any other key hypothetical construct (e.g., low self-control, criminal propensity, deviance).
6. Are racial differences in the prevalence of homicide offenders and victims attributable to racial differences in early risk factors?

METHODOLOGY

Samples

The Pittsburgh Youth Study is a prospective longitudinal survey of the development of offending and antisocial behavior in three samples of approximately 500 Pittsburgh boys each, totaling 1517 boys. When they were first contacted to participate in the study in 1987–1988, random samples of first-, fourth-, and seventh-grade boys enrolled in Pittsburgh public schools were selected (Loeber, Farrington, Stouthamer-Loeber, & van Kammen, 1998; Loeber, Farrington, Stouthamer-Loeber, & White, 2008). At that time, 72% of all children residing in Pittsburgh attended public schools. The city of Pittsburgh covers the inner-city population of approximately 370,000 (in 1990), within of the Pittsburgh–Beaver Valley Metropolitan Statistical Area of approximately 2,243,000.

Out of approximately 1000 boys in each grade selected at random for a screening assessment, about 850 boys (85%) were actually assessed. The boys completed a self-report questionnaire about antisocial behavior while their primary caretakers completed an extended Child Behavior Checklist, and their teachers completed an extended Teacher Report Form (Achenbach & Edelbrock, 1983; Edelbrock & Achenbach, 1984; Loeber, Stouthamer-Loeber, van Kammen, & Farrington, 1989). We will refer to the primary caretaker as the mother, because this was true in 94% of cases. Participants did not differ significantly from the comparable male student population in their scores on the California Achievement Test (CAT) or in their racial composition (African American or Caucasian).

From the screening assessment, a risk score was calculated for each boy, indicating how many of 21 antisocial acts he had ever committed. Data from the boy, the mother, and the teacher were then combined. The risk score was used to select the sample for follow-up, consisting of approximately the 250 most antisocial boys in each grade and 250 boys randomly selected from the remaining 600. Hence, the screening sample of about 850 per grade was reduced to a follow-up sample of about 500 per grade. In the first follow-up assessment conducted 6 months after the screening assessment, the three samples of boys were aged (on average) 7, 10, and 13, respectively. This chapter compares child risk factors measured in the screening and first follow-up assessments (in 1987–1988) with later information about homicide offenders and victims.

Risk Factors

Risk factors were classified as explanatory or behavioral/attitudinal. Explanatory variables were those that clearly did not measure antisocial behavior, while behavioral/attitudinal variables could have reflected the boy's antisocial behavior. For example, a young mother and poor parental supervision were classified as explanatory variables, while being truant and being suspended from school were classified as behavioral variables.

The most contentious variable was peer delinquency. We classified it as a behavioral variable because three-fourths of the delinquent acts committed by the boys were committed with their peers. Therefore, boys who had committed delinquent acts usually had delinquent peers. Also, Farrington and colleagues (2002) found that peer delinquency was the strongest correlate of a boy's delinquency in between-individual correlations (both measured at the same time) but did not predict a boy's later delinquency within individuals. These researchers concluded that peer delinquency was not a cause of a boy's delinquency, but rather measured the same underlying construct as delinquency. In contrast, poor parental supervision was correlated with delinquency between individuals and predicted delinquency within individuals.

The 21 explanatory risk factors measured in 1987–1988 and used in the present study were as follows:

- *Child*: Low guilt, hyperactivity–impulsivity–attention deficit (HIA), old for the grade (has been held back), low achievement (both according to the California Achievement Test and according to mother–teacher ratings), depressed mood, callous-unemotional (Frick, O'Brien, Wootton, & McBurnett, 1994).

- *Parental*: Young mother (a teenager when the boy was born), father seeking help for behavior problems (rated by the mother, so available for all fathers whether present or absent), parental substance use.
- *Childrearing*: Poor parental supervision, physical punishment by the mother, poor parent–boy communication.
- *Socioeconomic*: Low socioeconomic status (the Hollingshead measure, based on parental education and occupational prestige), family on welfare, broken family (missing biological parent), large family, small house, unemployed mother, bad neighborhood (both according to 1990 census data and according to mother ratings).

These factors were derived from the 40 distinct explanatory factors, excluding those not known for all three samples, those with a large amount of missing data (e.g., variables referring to the operative father or to the mother–father relationship), and those not strongly related to delinquency in 1987–1988 (odds ratio [OR] > 2.0). Some variables were added because they were related to violence (OR > 2.0) (Farrington, Loeber, & Stouthamer-Loeber, 2003). Most variables were based on interviews with mothers, and in a few cases supplemented by data from boys.

The 19 behavioral/attitudinal risk factors measured in 1987–1988 and used in the present chapter were as follows:

- *Child behavior*: Serious delinquency, covert behavior (concealing, manipulative, untrustworthy), physical aggression, nonphysical aggression, cruel to people, runs away, disruptive behavior disorder according to the Revised Diagnostic Interview Schedule for Children (Costello et al., 1982), high risk score.
- *Child attitude*: Favorable attitude toward delinquency, favorable attitude toward substance use.
- *Parental*: Bad relationship with parents, counter-control (the bad behavior of the boy inhibits parental attempts at socialization).
- *Peer*: Peer delinquency, delinquent or poorly socialized friends, bad relationship with peers.
- *School*: Suspended, truant, negative attitude toward school, low school motivation.

For the present analyses, the explanatory and behavioral risk factors were dichotomized within each sample into, as far as possible, the "worst" quarter (e.g., the one-fourth of the sample with the lowest attainment or poorest supervision) versus the remainder. This dichotomization fostered a "risk factor" approach and made it easy to study the cumulative effects of several risk factors. It also made all variables directly comparable by equating sensitivity of measurement and permitted the use of the OR as a measure of strength of relationship, which has many advantages (Fleiss, 1981). In the Pittsburgh Youth Study, conclusions about the most important explanatory variables for delinquency were not greatly affected by the use of dichotomous as opposed to continuous variables, or by different dichotomization splits (Farrington & Loeber, 2000).

Dichotomizing the variables within each sample made all the samples comparable and made it possible to combine them. Arguably, the quarter of the youngest sample with the poorest parental supervision, for example, was comparable to the quarter of the oldest sample with the poorest parental supervision, even though the absolute levels of parental supervision were different in all three samples.

Homicide Offenders and Victims

Information about homicide offenders and victims was obtained from searches of local, state, and federal criminal records; interviews with the participants; searches of local newspapers; and the local coroner's office. According to information collected up to February 2006, 33 of the study participants had been convicted of homicide. Eight were from the youngest sample, 13 from the middle sample, and 12 from the oldest sample. The first homicide attributable to a study participant was committed in 1992; the offenders committed their homicides between ages 15 and 26 (median age 19). Their victims' ages ranged between 1 and 61 years (median age = 22). The weapon used was a gun in 24 cases; in other cases, the weapons used were hands (2 cases), knives (2), a brick (1), a metal rod (1), an automobile (1), a cord (1), and matches (in a case of homicide by arson). The motives were retaliation (12), robbery (7), a drug deal gone wrong (5), gang related (3), domestic (1), arson (1), and mental problems (1); in 3 cases, the motives were not known by the police.

Prior to February 2006, 30 of the study participants had died because of homicide (and 12 had died from other causes). Two were from the youngest sample, 12 from the middle sample, and 16 from the oldest sample. The first homicide victim was killed in 1993, and all victims died between ages 16 and 27 (median age = 21). The weapon used was a gun in 27 cases. One participant killed someone and then was killed himself. He was counted as an offender, leaving 33 homicide offenders and 29 homicide victims.

In addition to 33 participants convicted of homicide, 30 had been arrested for homicide but not convicted. Some of these individuals may be convicted in the future, but many were not convicted either because they terrorized potential witnesses or because the evidence was too weak. Some, of course, may have been innocent. These 30 individuals were excluded from the analyses. A further four individuals who died between ages 13 and 14 for reasons other than homicide were also excluded, on the grounds that they were not at risk of becoming homicide offenders or victims. Excluding these 34 cases left 33 homicide offenders, 29 homicide victims, and 1421 controls.

RESULTS

Race Differences in Homicide

In light of national statistics, it is not surprising that African American males in this sample were more likely than Caucasian males to be both homicide offenders and victims: 28 of the 33 offenders (85%) were African American, as were 27 of the 29 victims (93%). Both figures were disproportionate to the overall composition of the population (African Americans accounted for 780 of the 1421 controls, or 55%). Prospectively, 3.4% of African American males were convicted of homicide, compared with 0.8% of Caucasian males, and 3.2% of African American males were killed, compared with 0.3% of Caucasian males. Weighting back to the population of Pittsburgh public schools, 2.7% of African American males were convicted of homicide, compared with 0.5% of Caucasian males; and 2.6% of African American males were killed, compared with 0.3% of Caucasian males. Unweighted data are used in this chapter, because the main interest is in studying predictions for individuals rather than making population estimates.

For ease of exposition, this chapter presents retrospective rather than prospective percentages (**Table 9.1**). The OR for race versus homicide offenders was 4.6, while the OR for race versus homicide victims was 11.1. The confidence intervals are large because of the small numbers of homicide offenders and victims.

Later Features

Table 9.1 also shows the relationship between classic homicide features (guns, gangs, and drugs) and homicide offenders and victims. These features were measured in

TABLE 9.1 Features of Homicide Offenders and Victims

Feature	Percentage of Controls (1421)	Percentage of Offenders (33)	Percentage of Victims (29)	Odds Ratio C-O	Odds Ratio C-V
African American	55	85	93	4.6*	11.1*
Carried gun	6	20	33	4.0*	8.0*
Used weapon	8	25	29	3.8*	4.5*
Gang member	14	30	19	2.6*	1.4
Persistent drug user	19	39	25	2.8*	1.5
Sold marijuana	8	28	21	4.2*	2.9*
Sold hard drugs	8	31	30	5.5*	5.1*

Abbreviations: C-O, contrasting controls and offenders; C-V, contrasting controls and victims. Offenders and victims were not contrasted because of small numbers.

*$P < .05$

multiple data waves before any male was a homicide offender or victim. They are not studied as early risk factors in this chapter.

Homicide offenders were not always more antisocial than homicide victims. In particular, homicide victims were more likely to carry guns (33%) than homicide offenders (20%). The OR for victims (8.0) was twice as great as the OR for offenders (4.0). Also, homicide victims were roughly as likely as homicide offenders to have used a weapon (e.g., when attacking someone or in a gang fight or robbery) or to have sold hard drugs. Homicide offenders, in contrast, were more likely to be gang members, to be persistent drug users, and to have sold marijuana.

Explanatory Predictors

Eight of the 21 explanatory factors significantly predicted homicide offenders: broken family, bad neighborhood (according to the census), old for the grade (held back), family on welfare, low guilt, a young mother, an unemployed mother, and low socioeconomic status (**Table 9.2**). The strongest predictor was coming from a broken family: 88% of homicide offenders had a missing biological parent (usually the father) by the first year of the study compared with 62% of controls (OR = 4.4).

Nine of the 21 explanatory factors significantly predicted homicide victims: a broken family, low achievement (according to the California Achievement Test), low guilt, old for the grade, behavior problems of the father, HIA, family on welfare, callous-unemotional, and large family size (Table 9.2). Again, the strongest predictor was a broken family; 90% of homicide victims had a missing biological parent compared with 62% of controls (OR = 5.2).

Based on explanatory risk factors, homicide offenders were not more extreme than homicide victims; if anything, the odds ratios for predicting victims were greater than for predicting offenders. Notably, however, most of the significant predictors of homicide offenders were socioeconomic/demographic factors: a broken family, a bad neighborhood, the family on welfare, a young mother, an unemployed mother, and low socioeconomic status. In contrast, most of the significant predictors of homicide victims were individual factors: low achievement, low guilt, old for the grade (held back), HIA, and callous-unemotional.

Explanatory Risk Scores

A logistic regression analysis was carried out to investigate which of the eight significant predictors of homicide offenders were independent predictors. Three variables were found to be significant or nearly significant in a stepwise analysis: a bad neighborhood

TABLE 9.2 Explanatory Predictors of Homicide Offenders and Victims

Predicting Offenders	**Percentage of Controls (1421)**	**Percentage of Offenders (33)**	**Odds Ratio**
Broken family	62	88	4.4*
Bad neighborhood (C)	32	67	4.2*
Old for grade	25	55	3.6*
Family on welfare	43	71	3.3*
Low guilt	25	48	2.9*
Young mother	21	41	2.6*
Unemployed mother	25	45	2.5*
Low socioeconomic status	26	45	2.4*
Predicting Victims	**Percentage of Controls (1421)**	**Percentage of Offenders (29)**	**Odds Ratio**
Broken family	62	90	5.2*
Low achievement (CAT)	24	60	4.8*
Low guilt	25	59	4.3*
Old for grade	25	52	3.2*
Father behavior problems	17	38	3.0*
Hyperactivity	17	38	2.9*
Family on welfare	43	67	2.7*
Callous-unemotional	24	44	2.5*
Large family size	22	38	2.2*

Abbreviations: C, census measure; CAT, California Achievement Test measure.
$^*P < .05$

(likelihood ratio chi-squared [LRCS] = 13.22, $P = .0003$), old for the grade (LRCS = 9.21, $P = .002$), and a young mother (LRCS = 2.85, $P = .091$). In the final model, the partial ORs were 3.2 for a bad neighborhood, 3.1 for old for the grade, and 2.0 for a young mother.

An explanatory risk score was calculated for each boy based on the number of these three risk factors that he possessed. **Table 9.3** shows how much this risk score predicted homicide offenders; for example, 0.6% of boys with none of these three risk factors were homicide offenders, compared with 9.2% of boys with all three risk factors. Hence, there was significant predictability but also a high false-positive rate. Retrospectively, 19 of the 33 homicide offenders (58%) were among the one-fifth of the boys with two

TABLE 9.3 Explanatory Risk Scores

Predicting Offenders	**Number of Controls**	**Number of Offenders**	**Row (%)**	**Column (%)**
Score				
0	655	4	0.6	12.1
1	491	10	2.0	30.3
2	207	13	5.9	39.4
3	59	6	9.2	18.2
Predicting Victims	**Number of Controls**	**Number of Victims**	**Row (%)**	**Column (%)**
Score				
0	333	1	0.3	3.9
1	527	5	0.9	17.2
2	336	6	1.8	20.7
3–4	220	17	7.2	58.6

Predicting offenders: OR = 5.8 (01 versus 23); OR = 5.1 (012 versus 3); AUC = .771
Predicting victims: OR = 5.9 (01 vs. 234); OR = 7.7 (012 vs. 34); AUC = .796
Abbreviations: OR, odds ratio; AUC, area under the ROC (receiver operator characteristic) curve.

or three of these risk factors. Comparing boys with zero or one risk factor with boys with two or three risk factors, the OR was 5.8; comparing boys with zero to two risk factors with boys with three risk factors, the OR was 5.1. The receiver operator characteristic area under the ROC-AUC was .771.

A logistic regression analysis was also carried out to investigate which of the nine significant predictors of homicide victims were independent predictors. Four variables were significant or nearly significant in a stepwise analysis: low guilt (LRCS = 17.59, $P < .0001$), low achievement (LRCS = 9.56, $P = .002$), broken family (LRCS = 6.47, $P = .011$), and large family size (LRCS = 3.76, $P = .053$). In the final model, the partial ORs were 4.5 for broken family, 3.3 for low achievement, 3.8 for low guilt, and 2.3 for large family size.

An explanatory risk score was then calculated for each boy based on the number of these four risk factors that he possessed. Table 9-3 shows how much this risk score predicted homicide victims; for example, 0.3% of boys with none of these four risk factors became homicide victims, compared with 7.2% of boys with three or four of these risk factors. Retrospectively, 17 of the 29 homicide victims (59%) were among the one-sixth of the boys with three or four of these risk factors. Comparing boys with zero or one risk factor with those with two to four risk factors, the OR was 5.9; comparing boys with zero to two risk factors with those with three or four risk factors, the OR was 7.7. The AUC was .796. Hence, the prediction of homicide victims was at least as accurate as the prediction of homicide offenders.

WHY DOES RACE PREDICT HOMICIDE OFFENDERS AND VICTIMS?

It is plausible to suggest that race predicts homicide offenders and victims because African American and Caucasian boys differ on predictive risk factors. According to this hypothesis, race should not predict homicide offenders and victims after controlling for predictive risk factors. Indeed, after entering the eight significant explanatory risk factors in a logistic regression analysis, race did not significantly predict homicide offenders. After entering the nine significant explanatory risk factors in a logistic regression analysis, race was still a significant predictor of homicide victims (LRCS = 4.61, $P = .032$). However, the predictive power of race was considerably reduced after controlling for other risk factors. It might be concluded that race predicts homicide offenders and victims primarily because of racial differences in predictive risk factors. The most important risk factors that were significantly associated with race and that predicted homicide offenders and/or victims were a bad neighborhood, a broken family, the family on welfare, and a young mother.

Behavioral/Attitudinal Predictors

Ten of the 19 behavioral/attitudinal factors significantly predicted homicide offenders: a high risk score, suspended from school, disruptive behavior disorder, a favorable attitude toward delinquency, serious delinquency, a favorable attitude toward substance use, peer delinquency, truancy, covert behavior (concealing, manipulative, untrustworthy), and cruelty to people (**Table 9.4**). The strongest predictor was a high risk score in the screening assessment, based on antisocial behavior: 85% of homicide offenders had a high risk score, compared with 49% of controls (OR = 5.8).

Fourteen of the 19 behavioral/attitudinal factors significantly predicted homicide victims: serious delinquency, suspended from school, a high risk score, truancy, non-physical aggression, covert behavior, a bad relationship with peers, a bad relationship with parents, physical aggression, low school motivation, peer delinquency, cruelty to people, bad friends, and disruptive behavior disorder (Table 9.4). The strongest predictor was serious delinquency: 72% of homicide victims were serious delinquents, compared with 29% of controls (OR = 6.5).

Based on behavioral risk factors, homicide offenders were not more extreme than homicide victims. In fact, more significant results were found in these risk factors' ability to predict victims than to predict offenders.

TABLE 9.4 Behavioral Predictors of Homicide Offenders and Victims

Predicting Offenders	Percentage of Controls (1421)	Percentage of Offenders (33)	Odds Ratio
High risk score	49	85	5.8*
Suspended	43	79	5.0*
Disruptive behavior disorder	23	55	4.0*
Favorable attitude to delinquency	23	55	4.0*
Serious delinquency	29	61	3.8*
Favorable attitude to substance use	24	52	3.4*
Peer delinquency	24	52	3.4*
Truant	38	61	2.6*
Covert behavior	24	42	2.3*
Cruel to people	25	42	2.3*
Predicting Victims	**Percentage of Controls (1421)**	**Percentage of Victims (29)**	**Odds Ratio**
Serious delinquency	29	72	6.5*
Suspended	43	75	4.0*
High risk score	49	79	3.9*
Truant	38	69	3.7*
Nonphysical aggression	25	55	3.7*
Covert behavior	24	54	3.6*
Bad relation with peers	26	55	3.5*
Bad relation with parent	24	52	3.3*
Physical aggression	27	52	3.0*
Low school motivation	37	63	2.9*
Peer delinquency	24	45	2.6*
Cruel to people	25	45	2.5*
Bad friends	25	45	2.4*
Disruptive behavior disorder	23	39	2.1*

*$P < .05$

Behavioral Risk Scores

A logistic regression analysis was carried out to investigate which of the 10 significant predictors of homicide offenders were independent predictors. Five variables were significant or nearly significant in a stepwise analysis: a favorable attitude toward delinquency (LRCS = 14.77, $P < .0001$), disruptive behavior disorder (LRCS = 12.01, $P = .0005$), suspended from school (LRCS = 9.49, $P = .002$), a high risk score (LRCS = 4.89, $P = .027$), and a favorable attitude toward substance use (LRCS = 3.07, $P = .080$). In the final model, the partial ORs were 3.0 for high risk score, 2.8 for suspended from school, 2.2 for disruptive behavior disorder, 2.1 for a favorable attitude toward delinquency, and 2.1 for a favorable attitude toward substance use.

A behavioral risk score was then calculated for each boy based on the number of these five risk factors that he possessed. **Table 9.5** shows how much this risk score predicted homicide offenders; for example, 0.6% of boys with zero or one of these five risk factors were homicide offenders, compared with 11.1% of boys with four or five risk factors. Again, significant predictability was found, but also a high false-positive rate. Retrospectively, 22 of the 33 homicide offenders (67%) were among the quarter of the boys with three or more of these risk factors. Comparing boys with zero to two risk factors with those with three to five risk factors, the OR was 5.9; comparing boys with zero to three risk factors with those with four or five risk factors, the OR was 9.5. The AUC was .797.

A logistic regression analysis was also carried out to investigate which of the 14 significant predictors of homicide victims were independent predictors. Three variables were significant in a stepwise analysis: serious delinquency (LRCS = 23.01, $P < .0001$),

TABLE 9.5 Behavioral Risk Scores

Predicting Offenders	Number of Controls	Number of Offenders	Row (%)	Column (%)
Score				
0–1	703	4	0.6	12.1
2	355	7	1.9	21.2
3	230	6	2.5	18.2
4–5	128	16	11.1	48.5
Predicting Victims	**Number of Controls**	**Number of Victims**	**Row (%)**	**Column (%)**
Score				
0	589	33	0.5	10.3
1	458	55	1.1	17.2
2	290	1212	4.0	41.3
3	82	99	9.9	31.0

Predicting offenders: OR = 5.9 (012 versus 345); OR = 9.5 (0123 versus 45); AUC = .797
Predicting victims: OR = 7.4 (01 versus 23); OR = 7.3 (012 versus 3); AUC = .788
Abbreviations: OR, odds ratio; AUC, area under the ROC (receiver operator characteristic) curve.

bad relationship with parents (LRCS = 5.14, P = .023), and truancy (LRCS = 4.21, P = .040). In the final model, the partial ORs were 4.4 for serious delinquency, 2.4 for a bad relationship with parents, and 2.3 for truancy.

A behavioral risk score was then calculated for each boy based on the number of the three risk factors that he possessed. Table 9.5 shows how much this risk score predicted homicide victims; for example, 0.5% of boys with none of these three risk factors became homicide victims, compared with 9.9% of boys with all three of these risk factors. Retrospectively, 21 of the 29 homicide victims (72%) were among the quarter of the boys with two or more of these risk factors. Comparing boys with zero or one risk factor with those with two or three risk factors, the OR was 7.4; comparing boys with zero to two risk factors with those with all three risk factors, the OR was 7.3. The AUC was .788. The behavioral prediction of homicide victims was not significantly less accurate than the behavioral prediction of homicide offenders.

Comparing Predictors of Homicide Offenders and Victims

In comparing ORs, natural logarithms of ORs (LORs) were used, so as to convert this ratio variable into a linear scale. The LORs for explanatory predictors of homicide offenders and victims were significantly correlated (r = .45, P = .041). Eight significant explanatory predictors of homicide offenders and nine significant explanatory predictors of homicide victims were used in the analysis. Four factors significantly predicted both statuses: old for the grade, low guilt, the family on welfare, and a broken family.

The LORs for behavioral predictors of homicide offenders and victims were almost significantly correlated (r = .40, P = .089). There were 10 significant behavioral predictors of homicide offenders and 14 significant behavioral predictors of homicide victims. Eight factors significantly predicted both statuses: a high risk score, truancy, serious delinquency, covert behavior, suspended from school, cruel to people, peer delinquency, and disruptive behavior disorder. Over all 40 predictors (explanatory and behavioral), the LORs for homicide offenders and victims were significantly correlated (r = .44, P = .004). Thus the strongest predictors of homicide offenders tended also to be the strongest predictors of homicide victims.

It was expected that homicide offenders would be more antisocial and deviant than homicide victims, and hence that homicide offenders would be predicted more accurately than homicide victims. However, this was not the case. For explanatory predictors, the geometric mean OR was 2.1 for predicting victims and 1.9 for predicting offenders. For behavioral predictors, the geometric mean OR was 2.7 for predicting victims and 2.3 for predicting offenders. These ORs were not significantly different, but they indicate that homicide victims were predicted at least as well as homicide offenders. Not surprisingly, behavioral predictors were somewhat more accurate than explanatory predictors.

Of the 21 explanatory predictors, 9 predicted homicide offenders more strongly and 12 predicted homicide victims more strongly. Of the 19 behavioral risk factors, 6 predicted homicide offenders more strongly and 13 predicted homicide victims more strongly. Hence, young homicide offenders were not more extreme than young homicide victims in their possession of either explanatory or behavioral risk factors.

CONCLUSION

The research described here is the first prospective longitudinal study of characteristics of homicide offenders and victims. However, it has a number of limitations. In particular, the numbers of homicide offenders and victims are small (making it difficult to compare them directly), and results obtained in the city of Pittsburgh may not be generalizable to the rest of the United States. Also, the results are to some extent provisional, because the numbers of homicide offenders and victims (especially from the youngest sample) will continue to increase over time. The results apply to young homicide offenders and victims.

As described here, explanatory and behavioral risk factors measured in the first year of the study significantly predicted homicide offenders and victims between 4 and 15 years later. Importantly, young homicide victims were at least as deviant as young homicide offenders, and victims were generally predicted as accurately as offenders. For example, 61% of homicide offenders and 72% of homicide victims had already committed serious delinquency (i.e., burglary, vehicle theft, robbery, assault, or rape), compared with 29% of controls.

The strongest predictors of homicide offenders tended also to be the strongest predictors of homicide victims. However, among the significant explanatory predictors, homicide offenders tended to be predicted by sociodemographic factors and homicide victims tended to be predicted by individual factors. Thus victims tended to be individually deviant while offenders tended to be socially deprived. African American boys disproportionately tended to be homicide offenders and victims. These race differences were much reduced after controlling for risk factors associated with race, however—notably, a bad neighborhood, a broken family, the family on welfare, and a young mother.

Risk scores showed how much homicide offenders and homicide victims could be predicted in the first year of the study. Typically, fewer than 1% of the least at-risk boys, compared to approximately 10% of the most at-risk boys, became homicide offenders (and similar results were obtained in predicting homicide victims). These risk scores overestimate the degree of prospective predictive accuracy in a new sample, because they retrospectively choose the best predictors in this particular sample. However, the predictability of homicide offenders and homicide victims would be increased by including later features such as guns, gangs, and drugs.

One possible interpretation of these results is that multiple early risk factors (including socioeconomic deprivation and low school attainment) cause an antisocial lifestyle involving guns, gangs, and drugs, that increases the risk of being involved in a homicide either as an offender or as a victim. Further research is needed to test this theory and to investigate why sociodemographic factors are more important for homicide offenders, while individual factors are more important for homicide victims.

ACKNOWLEDGMENTS

Work on this chapter was supported by Grants 96-MU-FX-0012 and 2005-JK-FX-0001 from the Office of Juvenile Justice and Delinquency Prevention, Grant 50078 from the National Institute of Mental Health, and Grant 411018 from the National Institute on Drug Abuse. Points of view or opinions in this chapter are those of the authors and do not necessarily represent the official position or policies of the U.S. Department of Justice, the National Institute of Mental Health, or the National Institute on Drug Abuse.

REFERENCES

Achenbach, T. M., & Edelbrock, C. S. (1983). *Manual of the Child Behavior Checklist and Revised Child Behavior Profile*. Burlington, VT: University of Vermont Department of Psychiatry.

Cook, P. J., & Laub, J. H. (1998). The unprecedented epidemic in youth violence. In M. Tonry & M. H. Moore (Eds.), *Youth violence* (pp. 27–64). Chicago: University of Chicago Press.

Costello, A., Edelbrock, C. S., Kalas, R., et al. (1982). *The Diagnostic Interview Schedule for Children, Parent Version* (revised). Worcester, MA: University of Massachusetts Medical Center.

Edelbrock, C. S., & Achenbach, T. M. (1984). The teacher version of the Child Behavior Profile: Boys aged 6 though 11. *Journal of Consulting and Clinical Psychology, 52*, 207–217.

Farrington, D. P., & Loeber, R. (2000). Some benefits of dichotomization in psychiatric and criminological research. *Criminal Behaviour and Mental Health, 10*, 100–122.

Farrington, D. P., Loeber, R., & Stouthamer-Loeber, M. (2003). How can the relationship between race and violence be explained? In D. F. Hawkins (Ed.), *Violent crime: Assessing race and ethnic differences* (pp. 213–237). Cambridge, UK: Cambridge University Press.

Farrington, D. P., Loeber, R., Yin, Y., & Anderson, S. J. (2002). Are within-individual causes of delinquency the same as between-individual causes? *Criminal Behavior and Mental Health, 12*, 53–68.

Federal Bureau of Investigation. (2003). *Crime in the United States, 2001.* Washington, DC: U.S. Department of Justice.

Fleiss, J. L. (1981). *Statistical methods for rates and proportions* (2nd ed.). New York: Wiley.

Fox, J. A., & Zawitz, M. W. (1999). *Homicide trends in the United States.* Washington, DC: Bureau of Justice Statistics.

Frick, P. J., O'Brien, B. S., Wootton, J. M., & McBurnett, K. (1994). Psychopathy and conduct problems in children. *Journal of Abnormal Psychology, 103,* 700–707.

Heide, K. M. (2003). Youth homicide: A review of the literature and a blueprint for action. *International Journal of Offender Therapy and Comparative Criminology, 47,* 6–36.

Hoyert, D. L., Arias, E., Smith, B. L., et al. (2001). *Deaths: Final data for 1999.* Hyattsville, MD: National Center for Health Statistics.

Lauritsen, J. L., Laub, J. H., & Sampson, R. J. (1992). Conventional and delinquent activities: Implications for the prevention of violent victimization among adolescents. *Violence and Victims,* 7, 91–108.

Loeber, R., DeLamatre, M., Tita, G., et al. (1999). Gun injury and mortality: The delinquent backgrounds of juvenile victims. *Violence and Victims, 14,* 339–352.

Loeber, R., Farrington, D. P., Stouthamer-Loeber, M., & van Kammen, W. B. (1998). *Antisocial behavior and mental health problems: Explanatory factors in childhood and adolescence.* Mahwah, NJ: Lawrence Erlbaum.

Loeber, R., Farrington, D. P., Stouthamer-Loeber, M., & White, H. R. (2008. *Violence and serious theft: Development and prediction from childhood to adulthood.* New York: Routledge.

Loeber, R., Pardini, D., Homish, D. L., et al. (2005). The prediction of violence and homicide in young men. *Journal of Consulting and Clinical Psychology, 73,* 1074–1088.

Loeber, R., Stouthamer-Loeber, M., van Kammen, W. B., & Farrington, D. P. (1989). Development of a new measure of self-reported antisocial behavior for young children: Prevalence and reliability. In M. W. Klein (Ed.), *Cross-national research in self-reported crime and delinquency* (pp. 203–225). Dordrecht, Netherlands: Kluwer.

Rivara, F. P., Shepherd, J. P., Farrington, D. P., et al. (1995). Victim as offender in youth violence. *Annals of Emergency Medicine, 26,* 609–614.

Snyder, H. N., & Sickmund, M. (1999). *Juvenile offenders and victims: 1999 national report.* Washington, DC: Office of Juvenile Justice and Delinquency Prevention.

CHAPTER 10

Generality of Deviance and Predation: Crime-Switching and Specialization Patterns in Persistent Sexual Offenders

Patrick Lussier
Simon Fraser University

Benoit Leclerc
University of Montreal
Philippe-Pinel Institute of Montreal

Jay Healey
Simon Fraser University

Jean Proulx
University of Montreal
Philippe Pinel Institute of Montreal

Contrary to public perception, empirical studies have constantly shown that persistent sex offenders constitute a small subgroup of the sex offender population. Indeed, only a small subpopulation of sex offenders tends to persist in committing these crimes over time, as recidivism rates tend to be approximately 10% to 15% over a period of about five years after release (Lussier, 2005). This small subgroup has attracted a lot of attention from the criminal justice system, which in turn has led to the development of various risk assessment tools designed to help practitioners in screening for persistent offenders. Many characteristics have been identified, and theoretical models have been proposed (Beech & Ward, 2004; Hanson & Morton-Bourgon, 2005). In the meantime, however, the behavior of persistent sexual offenders has been overlooked to a great extent. One could reasonably question what those risk assessment tools are really predicting, as the criterion used to develop those instruments is "sexual recidivism," which includes much heterogeneity in its manifestations. Of interest is the fact that many predictors of sexual recidivism are related to offending characteristics, such as having offended against an extrafamilial, male, prepubescent victim.

The purpose of the study described in this chapter is to build on previous studies to further understand the sexual criminal activity of persistent sexual offenders (Guay, Proulx, Cusson, & Ouimet, 2001). The emphasis here is on the tendency for that subgroup of offenders to switch from one sex crime category to another. Building on the criminological literature, this investigation focuses on the sexual criminal versatility of persistent sex offenders and the associated risk factors.

SEXUAL POLYMORPHISM AND CRIME-SWITCHING PATTERNS IN SEX OFFENDERS

Sexual polymorphism refers to crime-switching patterns along several dimensions, such as victim's age, gender, relationship to the offender, and nature of acts committed by the offender. Few studies have examined the level of sexual polymorphism and crime-switching patterns in the sexual criminal activity of sex offenders. Based on the current state of knowledge, three broad conclusions can be drawn in regard to the offending patterns of persistent sexual offenders.

Specialization in Sexual Offending

Soothill and colleagues (2000) came to the conclusion that while sex offenders are generalists in their criminal offending, they tend to specialize in their sexual offending, confining themselves to one type of victim. Similarly, Radzinowicz (1957) also found specialization in victim choice, in that only 7% of his large sample of sex offenders had convictions for crimes against both male and female victims—a finding consistent with those of Gebhard and colleagues (1964). Cann, Friendship, and Gonza (2007) found that only 25% of their sample of incarcerated sex offenders was versatile when considering victim's age and gender as well as the offender–victim relationship.

Conversely, crime-switching patterns may vary as a function of the dimension of the sexual polymorphism considered. For instance, while they also found much stability as to the victim's gender, Guay et al. (2001) reported considerable versatility among those offenders targeting adolescents. Although offenders targeting children and those targeting adults remained in the same category, those offending against adolescents were likely to switch either to adults or to children. Guay and colleagues hypothesized that adolescents may be a sex surrogate choice when the preferred partner was not available.

Tendency Toward Sexual Polymorphism

Empirical studies conducted in clinical settings have yielded a divergent picture of sex offenders' crime-switching patterns. Weinrott and Saylor (1991) argued that official data hide an enormous amount of sex crimes. Using official data only, these authors found that only 15% of their sample of offenders was versatile when considering only three categories: adult females, extrafamilial children, and intrafamilial children. When results from a self-reported computerized questionnaire were considered, however, that percentage rose to 53% (Weinrott & Saylor, 1991). Similarly, Heil and his colleagues (2003) reported that incarcerated offenders in treatment were not versatile as to victim's age (7%) and gender (8%) when assessed with official data, but did meet this criterion when interviewed using a polygraph (70% and 36%, respectively). Less dramatic differences were reported for parolees, which might be explained by sampling differences (i.e., incarcerated offenders were more serious offenders) and the fact that admitting a crime was a prerequisite to enter treatment.

Abel and Rouleau's (1990) well-publicized study, which was conducted under strict conditions of confidentiality, revealed that 42% of their sample targeted victims in more than one age group, 20% targeted victims of both gender, and 26% committed both hands-on and hands-off crimes, such as exhibitionism. This study and its findings, however, have been criticized on methodological grounds (Marshall, 2007). Nevertheless, similar results have been reported elsewhere in a sample of sex offenders assessed in a forensic psychiatric institution. For example, 30% of rapists admitted to acts of heterosexual pedophilia (Bradford, Boulet, & Pawlak, 1992).

The overlapping nature of different forms of sexually deviant acts found in the clinical studies runs counterintuitive to current typological models of sex offenders based on the characteristics of the offense. The victim's gender, the victim's age, the offender–victim relationship, the level of sexual intrusiveness, and the level of force used during the commission of the crime are some examples of criteria that have been used over the years to classify sex offenders (Groth & Birnbaum, 1979; Knight & Prentky, 1990). The limited evidence does not allow for drawing firm conclusions as to whether those results invalidate current typological models. In fact, it remains unclear whether clinical studies found a general pattern of sexual polymorphism among most sex offenders or whether a subgroup of generally deviant sex offenders were found. In any case, this divergence raises the possibility of a common cause for different forms of deviant sexual acts.

One possible candidate that has been identified in recent years is the construct of sexualization or hypersexuality (Kafka, 1997; Knight & Sims-Knight, 2003). High sexualization refers to a disinhibited sexuality characterized by sexual preoccupation (e.g., excessive time spent thinking about sexual matters), sexual compulsivity (e.g., overwhelmed by sexual fantasies), and impersonal sex (e.g., preference for partner variety). Individuals characterized by a high sexualization might experience more difficulties in controlling their sexual urges. Thus it seems reasonable to propose that those sex offenders characterized by a disinhibited sexuality might have different means of sexual expression to fulfill their sexual needs. An individual with high sexualization, therefore, might be more likely to seek out sexual gratification in different contexts and different places. High sexualization has been shown to be related to frequency of sexual offending in a sample of incarcerated sex offenders (Lussier, Leclerc, Cale, & Proulx, 2007).

General Deviance

Smallbone and Wortley (2004) came to the conclusion that diversity in paraphilic activities may be a function of general deviance. Indeed, looking at different types of paraphilia (e.g., voyeurism, frotteurism, sexual sadism) in a sample of child molesters, these researchers found that a scale measuring the versatility of sexual deviance correlated significantly and positively with nonsexual offending. In other words, as the frequency of offending increases, so does the versatility in paraphilic interests and behaviors.

Similarly, Lussier, LeBlanc, and Proulx (2005) found in a sample of recidivists, all of whom had committed a sex crime, that versatility in sex offending was strongly related to versatility in nonsexual, nonviolent offending as well as versatility in nonsexual violent crime. Furthermore, using structural equation modeling, they found that such a pattern of general versatility was related to an early-onset and persistent antisocial behavior. In other words, sex offenders characterized by a life-course with persistent antisocial tendency were more likely to show much versatility in their sexual offending.

In that regard, Guay et al. (2001) hypothesized that crime switching in sexual offending might be partly explained by low self-control. Gottfredson and Hirschi (1990) have described individuals with low self-control as impulsive, characterized by a tendency to pursue easy, risky, and immediate gratifications requiring no special abilities in spite of more long-term negative consequences. A few emerging studies have shown that low self-control increases the risk of committing a sex crime. Thus the sexual behavior of sex offenders lacking control over their behavior might take different expressions given the opportunity to do so (Lussier, Proulx, & LeBlanc, 2005).

CURRENT FOCUS

The current studies conducted to date provide two different pictures of the persistent sexual offenders: one group specializing in one sex crime category and another being characterized by much versatility. The study described here addresses such contradictory findings by looking at the sexual offending patterns of persistent sexual offenders. While earlier studies have looked mainly at victim's characteristics to assess crime specialization/versatility, this investigation somewhat expands the scope by also looking at the nature of the sexual acts committed. In that regard, using kappa coefficients of agreement, Sjöstedt et al. (2004) found that the stability of the acts committed (e.g., violence, penetration, physical contact) was fair at best (i.e., kappa $< .40$) across offenses. The analytic strategy chosen here is in keeping with the long tradition of empirical studies in criminology examining specialization in general offending (Farrington, Snyder, & Finnegan, 1988).

To date, very few studies have looked at individual differences related to crime-switching patterns in sexual offending. In fact, to our knowledge, only one empirical study has looked at the individual differences related to sexual polymorphism. Cann et al. (2007) found that versatile sex offenders showed elevated scores on the Static-99, a risk assessment procedure designed to assess the risk of sexual recidivism specifically for sex offenders. Although interesting, these results are problematic considering that most of the items included in the Static-99 instrument are related to characteristics of the victims (e.g., a male victim) and offenses (e.g., sexual crime without contact). In other words, those results are not surprising, as more versatile sexual offending patterns are more likely to tap into different indicators on the risk assessment scale. Furthermore, those results do not provide avenues of explanation for sexual polymorphism.

Based on the review of the literature, two competing hypotheses were developed and empirically tested:

$H_{(1)}$: Sex offenders with high sexualization are more versatile in their sexual offending.

$H_{(2)}$: Sex offenders with low self-control are more versatile in their sexual offending.

METHODOLOGY

Subjects

The initial sample consisted of 553 adult males who had been convicted of a sexual offense. A total of 216 adult males convicted of a sexual offense against a minimum of two victims and having received a prison sentence of at least two years were included in this study. All of these individuals were incarcerated at the Regional Reception Centre of Ste-Anne-des-Plaines (in the province of Quebec, Canada) at the time of the study. They represented consecutive admissions between April 1994 and June 2000 at the Regional Reception Centre, a maximum-security federal institution run by the Correctional Service of Canada. This facility admits all individuals sentenced to a minimum of 2 years for the purpose of risk and treatment-needs assessment. The average stay in this institution is approximately 6 weeks, permitting completion of correctional assessment procedures prior to the individual's transfer to an institution suited to his risk level and treatment needs. The individuals' criminal history revealed that, on average, they were first convicted at age 35.2 years (standard deviation [SD] =15.2; range = 18.1–73.9).

Procedures

The data used to create scales measuring behavioral antecedents were collected in a semi-structured interview with each subject. Each subject was interviewed only once by a member of the research team who was unaware of the research questions and hypotheses. Participation in this study was strictly voluntary. Subjects included in this study signed a consent form indicating that the information gathered would be used for research purposes only.

Interviewers were all graduate students in criminology and psychology trained by a licensed forensic psychologist to conduct semi-structured interviews using a computerized questionnaire. The quality of data they collected was controlled by completing interrater agreement. Interrater agreement was measured on the basis of 16 interviews conducted jointly by two raters (the principal research assistant and the second author). Ratings were done independently following these interviews, which were conducted by one interviewer in the presence of the other. The mean kappa was .87, which represents very strong agreement. Note that interrater agreement analysis was not conducted for the developmental behavioral indicators (i.e., self-control and sexualization). Moreover, as participants granted access to their correctional files, official sources of information (e.g., police reports, victim statements, psychological assessments) were also used to validate information, when possible, obtained in interview. When disagreements were found between information gathered during the semi-structured interview and those collected from official files, official data were used.

Independent Variables

Age

The offender's age has been shown to be empirically related to sexual recidivism, in that younger offenders are more likely to sexually reoffend, but this effect varies greatly across studies (Hanson & Bussiere, 1998). More recently, some criminologists have argued that this relationship might be more complex than it appears. Thornton (2006) suggested that this trend might be due to the presence of differential recidivism rates between those offenders in their early twenties versus those in their sixties. Similarly, Doren (2006) raised the possibility that this effect might not hold true for all ages, as recidivism rates appear not to vary much between ages 40 and 60.

In the present study, age is defined as the age of the offender at the start of the current incarceration. On average, sex offenders included in this sample were 43.1 years old (SD = 12.6; range = 20–75).

Marital Status

A meta-analysis has shown that being single (i.e., never married) is significantly related to sexual recidivism. In the present study, 41% of the sample had never been married, 35% were in a relationship (e.g., common law, married), and 24% were either divorced or separated or widowed. The variable was coded as follows: (0) has been/being in a relationship or (1) being single/never married.

Ethnic Origin

Although ethnic origin has been extensively studied in the field of criminology, it has been somewhat overlooked in the field of sexual aggression. In total, 88% of the participants were Caucasian, 6% were African American, 4% were Native American, and nearly 2% were Hispanic. This variable was dichotomized (0 = non-Caucasian; 1 = Caucasian).

Educational Achievement

Educational achievement refers to the highest level of schooling completed. In the sample, 26% of participants had some elementary-level education, 61% had some high school education, and 12% had either some college- or university-level education. For the purpose of this study, this variable was dichotomized (0 = more than elementary level education; 1 = elementary level education or higher).

Number of Convictions

To examine an association between a diversity index of activity paraphilia and frequency of general offending, the total number of convictions was included in the analysis. Members of the sample had, on average, 3.9 convictions (SD = 4.0; range = 1–20). In total, only 28.4% of the sample had a prior record for a sex crime.

Low Self-Control

The construct of low self-control was operationalized using four behavioral indicators of general deviance:

- The scale of authority-conflict (α = .70) includes four items related to being defiant at home and at school: being disruptive in class, running away from home, being rebellious against an authority figure, and being short-tempered.
- The scale of reckless and imprudent behavior (α = .63) includes three items related to alcohol abuse, substance abuse, and dangerous behaviors.
- The scale of covert behavior (α = .61) is composed of five items that relate to being dishonest, deceitful, and committing concealing acts: repetitive and frequent lying, theft, selling drugs, fraud, and other property crime.
- The scale of overt behavior (α = .61) includes five items related to vandalism and acts of nonsexual aggression: major violence, serious violence, assault, cruelty against animals, and vandalism.

These four scales included items related to childhood (ages 0 to 12) and adolescence (ages 13 to 17) measured using a three-point scale: (0) did not commit the behavior; (1) committed the behavior either in childhood or adolescence; and (2) committed the behavior in both childhood and adolescence.

The scores of the four scales were standardized and summed. A higher score on the scale reflects lower self-control.

Sexualization

Following our previous studies, the construct of sexualization was operationalized using three behavioral indicators:

- The scale of impersonal sex (α = .64) includes four items: age at first sexual contact, age at first sexual intercourse, number of sexual partners (divided by age), and use of the services of prostitutes (0 = no; 1 = yes).
- The scale of sexual compulsivity (α = .68) is based on seven items: (1) age at first masturbation; (2) compulsive masturbation; (3) the average frequency of masturbation per week prior to incarceration, of which we isolated the 25th percentile with the highest frequency; (4) being overwhelmed by deviant sexual fantasies, lifetime; (5) being overwhelmed by nondeviant sexual fantasies, lifetime; (6) having deviant sexual fantasies one year prior to the sex crime for which they were incarcerated; and (7) the presence of a paraphilia (e.g., bestiality, fetishism) using the *DSM-IV* criteria developed by the American Psychiatric Association.
- The scale of sexual preoccupation (α = .63) includes three items relating to the use of pornographic magazines, use of pornographic movies, and practice of frequenting strip clubs.

The scores on the three scales were standardized and summed. A high score on this scale indicated a high sexualization.

Dependent Variables

Police records were consulted to determine the criminal activity in adulthood. The number of victims refers to the total number of victims for the index offenses. On average, sexual aggressors included in this sample had been charged with crimes committed against 3.4 victims (SD = 4.5; range = 2–65).

Crime Categories

There is no general consensus as to how sex crime should be categorized. Earlier work has focused mostly on distinguishing the age and the gender of the victim as well as the nature of the offender–victim relationship. Two main dimensions of the sexual criminal activity were considered in the present study—that is, the victim's characteristics and the characteristics of the sexual act. The victim's characteristics included three key dimensions: (1) the victim's gender (1 = male; 2 = female); (2) the victim's age (1 = between 0 and 12 years; 2 = between 13 and 17 years; 3 = 18 years or older); and (3) the relationship between the offender and the victim (1 = stranger; 2 = intrafamilial; 3 = known and familiar; 4 = known but unfamiliar). The three dimensions related to the characteristics of the sexual acts were as follows: (1) the presence or absence of sexual contact (1 = hands-off; 2 = hands-on); (2) the level of sexual intrusiveness (1 = fondling, rubbing, and masturbation; 2 = oral sex; 3 = anal, vaginal penetration); and (3) the level of physical force (1 = no force used; 2 = minimal force to gain compliance of the victim; 3 = excessive force used).

Considering their importance as distinct dimensions related to persistence in sexual reoffending, we investigated these variables' relevance in regard to specialization and crime switching. Transition matrices were performed for each of the six dimensions described here.

Transition Matrices

Crime-switching patterns across sex crimes were investigated to examine the tendency for sex offenders to specialize in a particular type of crime. To do so, we analyzed the sex crime transition for the first five victims. A focus on the first five victims was selected as a criterion because of the dropping out of offenders and the small sample size as more victims were considered in the analysis. Due to missing data, it was possible to chronologically order the victims for only 210 individuals. Because 210 offenders had two victims, but only 36 had at least five victims, we therefore considered only the first four crime transitions. Hence, 210 offenders had one crime transition (from crime #1 to crime #2), 100 had a second transition (from crime #2 to crime #3), 64 had a third transition (from crime #3 to crime #4), and 36 had four transitions (from crime #4 to crime #5).

Diversity Index

Following the work of Agresti and Agresti, diversity indexes were computed for different dimensions of the sexual offending activity (Agresti & Agresti, 1978). The application of the diversity index in criminology is well known (Piquero et al., 1999; Sullivan, McGloin, Pratt, & Piquero, 2006). This metric refers to the probability that any two offenses randomly selected from an individual's criminal history are in different categories. To compute the diversity index (D), the number of crimes categories (k) is determined, where p_i equals the proportion of crimes for each of the $i = 1, 2, \ldots, k$ categories identified. Then, D can be computed as follows:

$$D = 1 - \sum_{i=1}^{k} p_i^2$$

The D index can be characterized by the following set of properties. First, D does not consider the chronological order of the offenses. As a consequence, the level of specialization may be measured while taking all offenses into consideration simultaneously. Second, the minimum score of D is 0, indicating perfect or complete specialization (e.g., all offenses fall into one category). Third, D is "a function of both the number of categories and the dispersion of the population among the categories" (Agresti & Agresti, 1978, p. 206). Therefore, a D index indicating complete versatility will vary according to the number of offense categories considered—that is, $D_{max} = (k - 1)/k$. Hence, if four offense categories are considered, D will vary from .00 (complete specialization) to .75 (complete versatility).

RESULTS

The next sections contain the results of the statistical models relating to the diversity and specialization in offending.

Crime-Switching Patterns

Transition Matrices

In **Table 10.1**, crime transitions were analyzed using three victims' characteristics.

First, looking at the gender of the victim, we found much evidence for stability and specialization in both offenders against female victims and offenders against male victims. For offenders against female victims, the probability of repeating against the same victim remained consistently greater than .90 across the four transitions; this probability remained consistently greater than .80 for offenders against male victims.

Second, examination of the age of the victim provided more evidence of crime-switching patterns. Those offending against children tend to limit themselves to children, as the probability coefficient remained greater than .80 across offenses. If switching did occur, child molesters were more likely to revert to adolescents (range = .09–15) than to adults (range = .00–.05). Offenders against adolescents showed a relatively lower rate of specialization, as the probabilities of reoffending against the same age-category varied between .47 and .71 across offenses. When switching did occur, it was mainly for a child victim (range = .28–.53). Finally, among those individuals who had victimized adults, probabilities of specialization varied between .54 and .90 across offenses. When switching did occur, offenders against adult victims were more likely to revert to adolescents (range = .10–.36) rather than to child victims (range = .00–.11).

Third, the nature of the relationship between the offender and the victim was considered for each crime transition. Individuals having offended against strangers (range = .80–1.00) and those having offended known-unfamiliar victims (range = .73–.93)

TABLE 10.1 Crime-Transitions Based on the Victim's Characteristics

Victim's Gender[a]	Female (T_{+1})	Male (T_{+1})		
Female	T_1 =.92 (131) T_2 =.90 (55)	T_1 =.08 (12) T_2 =.10 (6)		
	T_3 =.94 (32) T_4 =.95 (18)	T_3 =.06 (2) T_4 =.05 (1)		
Male	T_1 =.20 (13) T_2 =.10 (4)	T_1 =.80 (53) T_2 =.90 (35)		
	T_3 =.10 (3) T_4 =.19 (3)	T_3 =.90 (26) T_4 =.81 (13)		
Victim's Age[b]	**Child (T_{+1})**	**Adolescent (T_{+1})**	**Adult (T_{+1})**	
Child	T_1 =.82 (116) T_2 =.81 (51)	T_1 =.15 (21) T_2 =.14 (9)	T_1 =.03 (4) T_2 =.05 (3)	
	T_3 =.86 (31) T_4 =.91 (20)	T_3 =.11 (4) T_4 =.09 (2)	T_3 =.03 (1) T_4 =.00 (0)	
Adolescent	T_1 =.28 (11) T_2 =.30 (7)	T_1 =.53 (21) T_2 =.65 (15)	T_1 =.20 (8) T_2 =.04 (1)	
	T_3 =.53 (9) T_4 =.29 (2)	T_3 =.47 (8) T_4 =.71 (5)	T_3 =.00 (0) T_4 =.00 (0)	
Adult	T_1 =.11 (3) T_2 =.07 (1)	T_1 =.36 (10) T_2 =.29 (4)	T_1 =.54 (15) T_2 =.64 (9)	
	T_3 =.00 (1) T_4 =.00 (0)	T_3 =.10 (1) T_4 =.33 (2)	T_3 =.90 (9) T_4 =.67 (4)	
Offender–Victim Relationship[c]	**Stranger (T_{+1})**	**Intrafamilial (T_{+1})**	**Known-familiar (T_{+1})**	**Known-unfamiliar (T_{+1})**
Stranger	T_1 =1.00 (21) T_2 =1.00 (11)	T_1 =.00 (0) T_2 =.00 (0)	T_1 =.00 (0) T_2 =.00 (0)	T_1 =.00 (0) T_2 =.00 (0)
	T_3 =.88 (7) T_4 =.80 (4)	T_3 =.00 (0) T_4 =.00 (0)	T_3 =.00 (0) T_4 =.00 (0)	T_3 =.13 (1) T_4 =.20 (1)
Intrafamilial	T_1 =.03 (2) T_2 =.00 (0)	T_1 =.73 (55) T_2 =.88 (21)	T_1 =.11 (8) T_2 =.04 (1)	T_1 =.13 (10) T_2 =.08 (2)
	T_3 =.06 (1) T_4 =.00 (0)	T_3 =.61 (11) T_4 =.75 (6)	T_3 =.11 (2) T_4 =.13 (1)	T_3 =.22 (4) T_4 =.13 (1)
Known-familiar	T1 =.00 (0) T2 =.04 (1)	T_1 =.13 (8) T_2 =.18 (5)	T_1 =.71 (45) T_2 =.64 (18)	T_1 =.16 (10) T_2 =.14 (4)
	T_3 =.00 (0) T_4 =.00 (0)	T_3 =.06 (1) T_4 =.00 (0)	T_3 =.75 (12) T_4 =1.00 (8)	T_3 =.19 (3) T_4 =.00 (0)
Known-unfamiliar	T_1 =.06 (3) T_2 =.03 (1)	T_1 =.02 (1) T_2 =.11 (4)	T_1 =.10 (5) T_2 =.14 (5)	T_1 =.82 (42) T_2 =.73 (27)
	T_3 =.09 (2) T_4 =.00 (0)	T_3 =.09 (2) T_4 =.07 (1)	T_3 =.00 (0) T_4 =.00 (0)	T_3 =.82 (18) T_4 =.93 (14)

Note: Probabilities are shown, with sample size in brackets.

[a] T_1 = Transition 1 ($n = 209$), T_2 = Transition 2 ($n = 100$), T_3 =Transition 3 ($n = 63$), T_4 = Transition 4 ($n = 35$).

[b] T_1 = Transition 1 ($n = 209$), T_2 = Transition 2 ($n = 100$), T_3 =Transition 3 ($n = 64$), T_4 = Transition 4 ($n = 35$).

[c] T_1 = Transition 1 ($n = 210$), T_2 = Transition 2 ($n = 100$), T_3 =Transition 3 ($n = 64$), T_4 = Transition 4 ($n = 36$).

showed a relatively higher level of crime specialization across offenses. In contrast, those who victimized a family member (range = .61–88) or a known-familiar individual (range = .64–1.00) showed relatively lower levels of specialization. For intrafamilial offenders, when switching did occur, it rarely involved stranger victims, but mostly known-familiar and unfamiliar victims. Similarly, when crime switching did occur for known-familiar victims, it almost never involved stranger victims, but rather intrafamilial and known-unfamiliar victims.

As shown in **Table 10.2**, crime transitions were also investigated using offense variables—that is, the nature of offense, the level of sexual behavior intrusiveness achieved in the offense, and the level of physical force adopted during the offense. The results indicate that offenders tended to remain stable in the nature of the offense from one transition to another, especially for hands-on offenses. For offenders who committed hands-on offenses, the probability of repeating the same pattern across crimes remained higher than .88. Although the number of offenders who committed hands-off offense was very small, 50% of them committed hands-on offense as their second crime, suggesting a low rate of specialization among those offenders. Thereafter, these offenders remained more stable from one crime to the next (range = .71–.75).

In contrast to the findings related to the offense nature, the level of intrusiveness achieved in the offense varied considerably over the sequence of crimes, suggesting more evidence of crime-switching patterns. Only 48% of offenders who fondled, rubbed, or masturbated their victim for their first crime did so in their second. Thereafter, the probability of subsequently demonstrating these sexual behaviors was 76%, 65%, and 83% in the second, third, and fourth transitions, respectively, showing a low rate of specialization. If switching did occur, those offenders were more likely to perform penetration (range = .08–.34) rather than oral–genital sex (range = .04–.18); thus they did not demonstrate a gradual escalation in the intrusiveness of sexual behaviors achieved. However, offenders who performed more intrusive sexual behaviors—that is, oral–genital sex (range = .50–.68) and penetration (range = .46–.67)—showed a higher rate of specialization and thus more stability across crimes than those who first adopted less intrusive sexual behaviors. It should be noted that for offenders who performed more intrusive sexual behaviors, percentages across transitions are quite low, suggesting that offenders are not necessarily achieving more intrusive sexual behaviors from one crime to the next.

Much like the case with the nature of offense, evidence for stability and specialization were found for the level of physical force adopted during the offense. Offenders who did not adopt physical force during their first crime repeated this pattern over the sequence of crimes (range = .86–.95). A similar pattern was observed for both offenders who adopted only the force necessary to commit the offense (range = .75–.92) and

TABLE 10.2 Crime-Transitions Based on Offense Characteristics

Presence of Physical Contact Between Offender and Victim[a]	**Hands-off (T_{+1})**	**Hands-on (T_{+1})**	
Hands-off	$T_1 = .50$ (6) $T_2 = .71$ (5)	$T_1 = .50$ (6) $T_2 = .29$ (2)	
	$T_3 = .71$ (5) $T_4 = .75$ (3)	$T_3 = .29$ (2) $T_4 = .25$ (1)	
Hands-on	$T_1 = .05$ (10) $T_2 = .02$ (2)	$T_1 = .95$ (188) $T_2 = .98$ (90)	
	$T_3 = .04$ (2) $T_4 = .12$ (4)	$T_3 = .96$ (53) $T_4 = .88$ (29)	
Level of Intrusiveness Achieved[b]	**Fondling, Rubbing, Masturbation (T_{+1})**	**Oral Sex (T_{+1})**	**Penetration (T_{+1})**
Fondling, rubbing, masturbation	$T_1 = .48$ (21) $T_2 = .76$ (19)	$T_1 = .18$ (8) $T_2 = .04$ (1)	$T_1 = .34$ (15) $T_2 = .20$ (5)
	$T_3 = .65$ (13) $T_4 = .83$ (10)	$T_3 = .10$ (2) $T_4 = .08$ (1)	$T_3 = .25$ (5) $T_4 = .08$ (1)
Oral sex	$T_1 = .21$ (7) $T_2 = .15$ (4)	$T_1 = .64$ (21) $T_2 = .50$ (13)	$T_1 = .15$ (5) $T_2 = .35$ (9)
	$T_3 = .33$ (4) $T_4 = .17$ (1)	$T_3 = .50$ (6) $T_4 = .68$ (4)	$T_3 = .17$ (2) $T_4 = .17$ (1)
Penetration	$T_1 = .23$ (25) $T_2 = .28$ (11)	$T_1 = .15$ (17) $T_2 = .05$ (2)	$T_1 = .62$ (69) $T_2 = .67$ (26)
	$T_3 = .10$ (2) $T_4 = .27$ (3)	$T_3 = .24$ (5) $T_4 = .27$ (3)	$T_3 = .67$ (14) $T_4 = .46$ (5)
Physical Force Adopted to Commit the Offense[c]	**No Force (T_{+1})**	**Only Force Necessary (T_{+1})**	**Excessive Force (T_{+1})**
No force	$T_1 = .90$ (101) $T_2 = .95$ (56)	$T_1 = .05$ (5) $T_2 = .05$ (3)	$T_1 = .05$ (6) $T_2 = .00$ (0)
	$T_3 = .95$ (38) $T_4 = .86$ (18)	$T_3 = .03$ (1) $T_4 = .14$ (3)	$T_3 = .03$ (1) $T_4 = .00$ (0)
Only force necessary	$T_1 = .06$ (4) $T_2 = .04$ (1)	$T_1 = .88$ (56) $T_2 = .88$ (22)	$T_1 = .06$ (4) $T_2 = .08$ (2)
	$T_3 = .00$ (0) $T_4 = .17$ (2)	$T_3 = .92$ (12) $T_4 = .75$ (9)	$T_3 = .08$ (1) $T_4 = .08$ (1)
Excessive force	$T_1 = .03$ (1) $T_2 = .07$ (1)	$T_1 = .24$ (8) $T_2 = .07$ (1)	$T_1 = .74$ (25) $T_2 = .86$ (12)
	$T_3 = .00$ (0) $T_4 = .00$ (0)	$T_3 = .10$ (1) $T_4 = .00$ (0)	$T_3 = .90$ (9) $T_4 = 1.00$ (3)

Note: Probabilities are shown, with sample size in brackets.

[a]T_1 = Transition 1 ($n = 210$), T_2 = Transition 2 ($n = 99$), T_3 = Transition 3 ($n = 62$), T_4 = Transition 4 ($n = 37$).

[b]T_1 = Transition 1 ($n = 188$), T_2 = Transition 2 ($n = 90$), T_3 = Transition 3 ($n = 53$), T_4 = Transition 4 ($n = 29$). Cases that included hands-off offenses were excluded from this analysis.

[c]T_1 = Transition 1 ($n = 210$), T_2 = Transition 2 ($n = 98$), T_3 = Transition 3 ($n = 63$), T_4 = Transition 4 ($n = 36$).

those who used excessive force (range = .74–1.00). In the latter case, however, when switching did occur, offenders who used excessive force were more likely to apply only the physical force necessary to commit the offense (range = .00–.24) rather than using no force (range = .00–.07). Overall, the results suggest that offenders are quite stable regarding the level of physical force adopted during the offense.

The Diversity Index

Bivariate Analyses

Table 10.3 presents the means and standard deviations (bottom row) of the five diversity indices computed as well as the bivariate relationships between those indices and the characteristics of the offenders. Overall, the diversity indices did not seem to vary according to the sociodemographic characteristics of the offenders. Younger offenders appeared to be more versatile in terms of the level of physical force used to commit the offense, suggesting that older offenders might be more specialized in that regard.

TABLE 10.3 Bivariate Associations Between Diversity Indexes and Individual Characteristics

	Diversity Indexes for Victim's Characteristics[c]			Diversity Indexes for Offense Characteristics[c]	
	Gender	Age	Relationship	Intrusiveness	Force
Age[a]	.01	–.12[‡]	.11	.04	–.16[†]
Ethnicity (Caucasian)[b]	–.92	–.82	–.65	–.74	–.36
Marital Status (Single)[b]	–.56	–.66	–1.12	–.47	–.17
Education (Elementary)[b]	–1.48	–1.86[‡]	–.94	–.78	–1.48
Number of Convictions[a]	.01	–.01	–.10	–.02	.10
Number of Victims[a]	.04	.10	.17[†]	.15[†]	–.03
Level of Sexualization[a]	–.04	.06	–.09	.19[*]	.05
Low Self-control[a]	.01	.14[†]	–.08	.05	.15[†]
N	214	213	214	216	216
Mean	.07	.17	.17	.26	.09
Standard deviation	.16	.23	.23	.25	.18
Range	.00–.50	.00–.67	.00–.67	.00–.75	.00–.63

[a]Continuous variables tested with Pearson's *r*.

[b]Categorical variables tested with Mann-Whitney *U*. *Z* scores are presented.

[c]Logarithmic transformations were used on each of the diversity indexes for bivariate analyses.

$^{*}P < .01$; $^{\dagger}P < .05$; $^{\ddagger}P < .10$

There was also a tendency for younger offenders to be more versatile in regard to the age of the victims selected, but the effect was only marginally significant. Moreover, less educated offenders tended to be more versatile in terms of the nature of relationship with their victims, but the effect was small and marginally significant.

Interestingly, the number of convictions was not significantly related to any of the diversity indices. In other words, the tendency to be more versatile in sexual offending was relatively independent of the tendency to reoffend for any type of crime. Conversely, versatility as to the nature of the relationship with the victim and sexual intrusiveness were both significantly related to the number of victims. In other words, as the frequency of sexual offending increases, offenders appeared to be more versatile in terms of the victims they selected (e.g., stranger, intrafamilial) and the nature of the sexual acts being committed (i.e., hands-off, fondling).

The level of sexualization was also significantly related to the diversity index of sexual intrusiveness, suggesting that offenders showing a higher level of sexualization are more likely to have a wider repertoire of sexual acts being committed across offenses. Sexualization was not significantly related to any other diversity index.

Finally, low self-control was significantly related to both the diversity index of age and the level of force used. Hence, offenders with a lower level of self-control were more likely to offend against more than one age category as well as to use a wider range of levels of physical force to commit their crimes.

Multivariate Analyses

Multivariate analyses were conducted to determine whether self-control and sexualization increased versatility in sexual offending independently of sociodemographic and offending characteristics. Considering that the skewed distribution of the five diversity indexes violated the assumption of normal distribution in ordinary least squares (OLS) regression, probit regression models were used for the analysis. **Table 10.4** presents results of the probit regression models for each of the five diversity indices. Probit coefficients and the associated standard errors are reported for each independent variable entered in the regression model. For each regression model, we controlled for the other diversity indices to avoid identification of spurious relationships.

Looking at the results from the probit regression models, different sets of predictors emerged for the five diversity indices. For the diversity index of gender, only one factor emerged as statistically significant—that is, the diversity index of sexual intrusiveness. For the age of the victim, six of the 12 factors entered in the model were statistically related to a higher score on the diversity index. Being Caucasian with a lower school achievement, a lower number of convictions, and a higher number of

TABLE 10.4 Prediction of Diversity Indexes Using Probit Regression Models

	Diversity Indexes for Victim's Characteristics[a]			Diversity Indexes for Offense Characteristics[a]	
	Gender	**Age**	**Relationship**	**Intrusiveness**	**Force**
Age	−.01 (.14)	−.01 (.01)	.00 (.04)	.00 (.01)	−.01 (.01)
Ethnicity (Caucasian)	.56 (.43)	.83 (.37)‡	.29 (.34)	−.17 (.34)	.19 (.41)
Marital Status (Single)	−.11 (.24)	.04 (.21)	−.05 (.21)	.12 (.21)	−.16 (.24)
Education (Elementary)	.32 (.26)	.47 (.23)‡	.28 (.22)	.40 (.23)**	−.32 (.28)
Number of Convictions	−.05 (.14)	−.32 (.12)†	.09 (.12)	−.10 (.12)	.07 (.14)
Number of Victims	.13 (.10)	.26 (.09)†	.22 (.09)‡	.21 (.09)‡	.01 (.10)
High Sexualization	−.06 (.06)	−.02 (.05)	−.04 (.05)	.19 (.05)*	−.05 (.06)
Low Self-control	.05 (.09)	.15 (.07)‡	−.05 (.08)	.02 (.07)	.04 (.08)
Gender Diversity Index (log)	–	−.57 (.74)	.09 (.68)	1.96 (.72)†	−.65 (.82)
Age Diversity Index (log)	−.46 (.62)	–	−.11 (.52)	−.14 (.53)	−.02 (.01)**
Relationship Diversity Index (log)	−.11 (.58)	−.18 (.51)	–	.66 (.50)	.63 (.57)
Intrusiveness Diversity Index (log)	1.50 (.58)‡	−.32 (.50)	.77 (.49)	–	1.24 (.57)‡
Force Diversity Index (log)	−.96 (.82)	1.26 (.64)‡	.66 (.63)	1.44 (.67)‡	–
N	209	209	209	209	209
Goodness-of-fit (df)	215.61 (196)	204.22 (196)	209.32 (196)	205.14 (196)	207.94 (196)

[a]Continuous variables tested with Pearson's *r*.
*$P < .001$; †$P < .01$; ‡$P < .05$; **$P < .10$
+P < .10, *P < .05, **P < .01, ***P < .001

victims, with higher scores on both the scale of low self-control and the diversity index of violence, were all characteristics statistically associated with a tendency to offend against more than one age category.

Only one factor was significantly related to the offender–victim relationship diversity index: Those offenders with more victims were more likely to offend against more than one category of victim.

Four of the 12 factors entered in the probit regression model were related to the diversity index of sexual intrusiveness. A higher number of victims, a higher level of sexualization, and higher scores on the diversity index of gender and force used to commit the sex crime were all related to more versatility regarding the nature of sexual acts committed against the victim across offenses.

Finally, only one of the 12 factors emerged as a significant predictor of the diversity index of force used to commit the sex crime: Higher scores on the diversity index of sexual intrusiveness were associated with higher scores on the diversity of violence.

DISCUSSION

Following in the footsteps of criminal career researchers, the analytical strategy used here helped frame six main conclusions about crime-switching patterns and specialization in sexual offending.

First, crime switching in sex offending is rather multidimensional. Clearly, the diversity indices were relatively independent from one another and associated with different individual characteristics. Contrary to Abel and Rouleau's (1990) observations, little evidence was found to support their conclusion that sex offenders offend against different types of victims in different contexts.

Second, the level of crime switching varies from one crime type to another. On one end of the continuum, the victim's gender and level of physical force were relatively stable across crime transitions, while on the opposite end of the continuum the victim's age and sexual intrusiveness predicted more crime switching. Interestingly, it was those two dimensions, for which more evidence of crime switching was found, that were more strongly related with individual characteristics of the offenders.

Third, the notion of preference is relevant and of importance in the understanding of persistence in sexual offending. We concur with Soothill and his colleagues to a certain extent in that the level of stability and specialization found for certain aspects of offending (e.g., targeting children) suggest that it is far from being purely random upon certain situational contingencies (Soothill et al., 2000). Although the notion of preference might first appear to contradict the impulsive nature of those crimes, the notion of self-control remains of importance, because one must recognize an opportunity as such before acting upon it.

Fourth, the concept of a sex surrogate might play a part in stimulating crime switching. This concept appears to be especially true for those having offended against adolescent victims, who might represent the second best option in the absence of the preferred victim type (i.e., children or adults). This situation appears to hold true for both child molesters and rapists. Of importance, and in keeping with the sex surrogate hypothesis, is the finding that very few child molesters also offended against adults, and vice versa.

Fifth, there is some commonality between crime-switching patterns, as some diversity indices tended to co-vary together. This finding raises the possibility that the same factors (not tested here) might be linked to different crime-switching patterns, such as force and intrusiveness.

Sixth, versatility appeared to increase as a function of repetition of the sexually deviant behavior, especially in terms of the victim's age, offender–victim relationship, and sexual intrusiveness. The more individuals offended against different victims, the more their sexual criminal repertoire tended to diversify. This relationship might partly explain discrepancies reported in earlier studies, as clinical samples including more serious and persistent offenders would be expected to report more evidence of crime switching.

To interpret these findings, we offer some hypotheses drawn from rational choice theory and its model of offending.

Crime Switching and Rational Choice

Rational choice approach theorists propose different offending models, each implying a different decision process influenced by a different set of factors.

For example, the continuing involvement model refers to the continuation stage of the offender's criminal career. Clarke and Cornish (1985) stipulate that as a result of generally positive reinforcement, the frequency of offending increases until it reaches some optimal level. Thus the rewards of crime (e.g., money, dominance over others, sexual gratification) are of special importance in continuance. Apart from situational variables such as opportunities, and inducements that trigger the decision to commit a crime, Clarke and Cornish (1985) also summarize three categories of variables of importance at this stage of the criminal career: (1) increase in professionalism or offending experience; (2) changes in lifestyles and values (e.g., enjoyment of this "delinquent" life, development of justifications for criminal behavior); and (3) changes in peer group (networks of delinquent peers).

Following this perspective, a key result emerging from the study described in this chapter is that each polymorphism indicator seems to involve a different continuation process as predicted by a different set of variables. Indicators that showed much variability across offenses are the age of the victim and the level of sexual behavior intrusiveness achieved in the offense. More specifically, the tendency to offend against more than one age category across sexual offenses could be viewed as a "general deviance" type of continuation, whereas the tendency to achieve a different level of sexual intrusiveness from one crime to the next may be considered as a "predation" type of continuation in crime.

The tendency to offend against more than one age category is driven principally by factors such as a high level of low self-control, and the tendency to be versatile across offenses by the level of force adopted to commit the crime. Offenders who sexually abuse victims regardless of their age (children, adolescents, women) may be versatile offenders overall—that is, offenders who commit a wide range of criminal acts with no strong inclination to specialize in a specific type of crime or pattern of criminal acts. Because they exhibit low self-control, they may not be able to resist crime opportunities so as to satisfy their immediate and various needs (e.g., money, sexual gratification, dominance over others), but rather exploit every target or victim regardless of their characteristics. The decision to sexually abuse regardless of the age of the victim might be spontaneous—the consequence of crime opportunities that offenders with low self-control encounter in the course of their "career."

This interpretation is somewhat simplistic, however. These offenders are versatile in the level of force they employ to commit their crime, suggesting that skills are at least required in some offenses. More importantly, offenders who abuse an adult or a child initially tend to switch to adolescent victims as a sex surrogate, suggesting that the phenomenon of crime switching regarding the age is not totally random. Offenders may decide to look first at whether they can satisfy their needs with a victim who most closely resembles their initial victims in terms of physical attractiveness. Thus offenders are versatile enough to switch to adolescent victims, but not totally versatile, as they still follow a certain path of stability. In other words, low self-control would not by itself account for the tendency to offend against more than one age category, but rather contributes to this type of continuance. Nonetheless, such sexual offending could be referred to as a "general deviance" type of continuation.

Offenders who achieve a different level of sexual intrusiveness across offenses are principally characterized by high sexualization, the tendency to abuse victims of both genders, and demonstration of different levels of physical force across offenses. It could be argued that these offenders are also versatile, but low self-control is not predictive of the intrusiveness diversity. Recall that high sexualization refers to a disinhibited sexuality characterized by sexual preoccupation, sexual compulsivity, and impersonal sex. Having a higher level of sexualization, these offenders instead represent a group more oriented toward committing sex offenses—a contention that is supported by the association found between high sexualization and the frequency of sexual offending within the same sample of offenders. Thus those individuals might be expected to experience more difficulties in controlling their sexual urges and to exploit different opportunities to satisfy these urges, which might also explain why they tend to abuse victims of both genders.

Furthermore, given the predictive nature of the level of physical force diversity (i.e., no force, minimum force, excessive force), those offenders might also be expected to acquire different and effective means to commit their crimes and satisfy their sexual needs. Recent evidence suggests that modus operandi strategies are purposeful—that is, the more strategic offenders are the most efficient in increasing victim participation and sexual intrusiveness during sexual episodes (Leclerc & Tremblay, 2007). The capacity to adopt efficient strategies for the purpose of achieving sexual gratification emerges as of special significance in that continuance process. In fact, across offenses, successive trials may lead offenders toward more intrusive sexual behaviors such as penetration. This progression may, in turn, reinforce sexual offending as a means to satisfy sexual needs especially for those characterized by a high sexualization. In this continuance process, individuals move within an interdependence cycle of offending between high sexualization (the propensity), force diversity (the means), and sexual intrusiveness diversity (the outcome). Disinhibited sexuality combined with force diversity would favor the commission of a wider variety of sexual behaviors in terms of intrusiveness, which in consequence would contribute to increase further disinhibition, and so on. Although we cannot adequately verify this hypothesis in the study described here, the continuance phenomenon resulting from positive reinforcement highlighted by Clarke and Cornish (1985) is clearly of importance here. This pursuit of sexual offending in the course of a criminal career might be referred to as a "predation" type of continuation in crime.

Finally, it should be noted that evidence of sexual intrusiveness versatility raises doubts concerning the possibility that sexual offenders may limit themselves to a specific type of sexual behavior across offenses. Recall that gender diversity predicts the intrusiveness diversity index. If one assumes that offenses are purposive and that positive reinforcement might occur, this possibility is rather unlikely.

Like any study, the present investigation is not without methodological limitations. As an exploratory study, its findings should be interpreted with caution. The study involved a small sample of Canadian federal inmates, who may not be representative of all sex offenders. Furthermore, it is possible that the retrospective and self-reported nature of the data used might have introduced some biases into the reporting of sex crimes. Likewise, the use of official data might have reduced the impact of those biases. Also, this study is based on offending data that are censored because the offenders included in the study are still active. Because some of these offenders will eventually reoffend, it is important to consider that this study presents only a snapshot of the sexual criminal activity of adult sex offenders.

REFERENCES

Abel, G. G., & Rouleau, J. L. (1990). The nature and extent of sexual assault. In W. L. Marshall, D. R. Laws, & H. E. Barbaree (Eds.), *Handbook of sexual assault: Issues, theories and treatment of the offender* (pp. 9–22). New York: Plenum Press.

Agresti, A., & Agresti, B. F. (1978). Statistical analysis of qualitative variation. *Sociological Methodology, 9*, 204–237.

Beech, A. R., & Ward, T. (2004). The integration of etiology and risk in sexual offenders: A theoretical framework. *Aggression and Violent Behavior, 10*, 31–63.

Bradford, J. M., Boulet, J., & Pawlak, A. (1992). The paraphilias: A multiplicity of deviant behaviors. *Canadian Journal of Psychiatry, 37*, 104–108.

Cann, J. Friendship, C., & Gonza, L. (2007). Assessing crossover in a sample of sexual offenders with multiple victims. *Legal and Criminological Psychology, 12*, 149–163.

Clarke, R. V., & Cornish, D. B. (1985). Modeling offenders' decisions: A framework for research and policy. In M. Tonry & N. Morris (Eds.), *Crime and justice: An annual review of research* (Vol. 6, pp. 147–185). Chicago: University of Chicago Press.

Doren, D. (2006). What do we know about the effect of aging on recidivism risk for sexual offenders? *Sexual Abuse: Journal of Research and Treatment, 18*, 137–157.

Farrington, D. P., Snyder, H. N., & Finnegan, T. A. (1988). Specialization in juvenile court careers. *Criminology, 26*, 461–488.

Gebhard, P. H., Gagnon, J. H., Pomeroy, W. B., & Christensen, C. V. (1964). *Sex offenders: An analysis of types*. New York: Harper & Row.

Gottfredson, M., & Hirschi, T. (1990). *A general theory of crime*. Stanford, CA: Stanford University Press.

Groth, A. N., & Birnbaum H. J. (1979). *Men who rape*. New York: Plenum Press.

Guay, J. P., Proulx, J., Cusson, M., & Ouimet, M. (2001). Victim-choice polymorphia among serious sex offenders. *Archives of Sexual Behavior, 30*, 521–533.

Hanson, R. K., & Bussiere, M. T. (1998). Predicting relapse: A meta-analysis of sexual offender recidivism studies. *Journal of Consulting and Clinical Psychology, 61*, 646–652.

Hanson, R. K., & Morton-Bourgon, K. E. (2005). The characteristics of persistent sexual offenders: A meta-analysis of recidivism studies. *Journal of Consulting and Clinical Psychology, 73*, 1154–1163.

Heil, P., Ahlmeyer, S., & Simons, D. (2003). Crossover sexual offenses. *Sexual Abuse: Journal of Research and Treatment, 15*, 221–236.

Kafka, M. P. (1997). Hypersexual desire in males: An operational definition and clinical implications for males with paraphilias and paraphilia-related disorders. *Archives of Sexual Behavior, 26*, 505–526.

Knight, R. A., & Prentky, R. A. (1990). Classifying sexual offenders: The development and corroboration of taxonomic models. In W. L. Marshall, D. R. Laws, & H. E. Barbaree (Eds.), *Handbook of sexual assault: Issues, theories and treatment of the offender* (pp. 23–54). New York: Plenum Press.

Knight, R. A., & Sims-Knight, J. E. (2003). Developmental antecedents of sexual coercion against women: Testing of alternative hypotheses with structural equation modeling. In R. A. Prentky, E. S. Janus, & M. Seto (Eds.), *Sexual coercive behavior: Understanding and management* (pp. 72–85). New York: New York Academy of Sciences.

Leclerc, B., & Tremblay, P. (2007). Strategic behavior in adolescent sexual offenses against children: Linking modus operandi to sexual behaviors. *Sexual Abuse: Journal of Research and Treatment, 19*, 23–41.

Lussier, P. (2005). The criminal activity of sexual offenders in adulthood: Revisiting the specialization debate. *Sexual Abuse: Journal of Research and Treatment, 17*, 269–292.

Lussier, P., LeBlanc, M., & Proulx, J. (2005). The generality of criminal behavior: A confirmatory factor analysis of the criminal activity of sex offenders in adulthood. *Journal of Criminal Justice, 33*, 177–189.

Lussier, P., Leclerc, B., Cale, J., & Proulx, J. (2007). Developmental pathways of deviance in adult sex offenders. *Criminal Justice and Behavior, 34*, 1441–1462.

Lussier, P., Proulx, J., & LeBlanc, M. (2005). Criminal propensity, deviant sexual interests and criminal activity of sexual aggressors against women: A comparison of explanatory models. *Criminology, 43*, 249–281.

Marshall, W. L. (2007). Diagnostic issues, multiple paraphilias, and comorbid disorders in sexual offenders: Their incidence and treatment. *Aggression and Violent Behavior, 12*, 16–35.

Piquero, A., Paternoster, R., Mazerolle, P., et al. (1999). Onset age and offense specialization. *Journal of Research in Crime and Delinquency, 36*, 275–299.

Radzinowicz, L. (1957). *Sexual offences: A report of the Cambridge Department of Criminal Justice.* London: Macmillan.

Sjöstedt, G., Långström, N., Sturidsson, K., & Grann, M. (2004). Stability of modus operandi in sexual offending. *Criminal Justice and Behavior, 31*, 609–623.

Smallbone, S. W., & Wortley, R. K. (2004). Criminal diversity and paraphilic interests among adult males convicted of sexual offenses against children. *International Journal of Offender Therapy and Comparative Criminology, 48*, 175–188.

Soothill, K., Francis, B., Sanderson, B., & Ackerley, E. (2000). Sex offenders: Specialists, generalists or both? *British Journal of Criminology, 40*, 56–67.

Sullivan, C. J., McGloin, J. M., Pratt, T., & Piquero, A. (2006). Rethinking the "norm" of offender generality: Investigating specialization in the short-term. *Criminology, 44*, 199–233.

Thornton, D. (2006). Age and sexual recidivism: A variable connection. *Sexual Abuse: Journal of Research and Treatment, 18*, 123–135.

Weinrott, M. R., & Saylor, M. (1991). Self-report of crimes committed by sex offenders. *Journal of Interpersonal Violence, 6*, 286–300.

CHAPTER 11

Comparing Women and Men Who Kill

Jennifer Schwartz
Washington State University

Homicide is a rare occurrence, especially among women. But how rare is it? Are some women more likely to kill than other women? Have women increased their involvement in homicide as they have increased their involvement in paid work and other previously male domains? When women kill, do they go about it differently than men? How does being a woman, or being a man, shape the ways in which people commit homicide? What accounts for the large difference in male and female homicide offending? Do women and men kill for the same reasons or do sources of homicide differ by gender? This chapter reviews the literature on homicide offending and presents data from the *Supplementary Homicide Reports* (SHR) to address these questions about similarities and differences in female and male homicide offending.

Fewer than 1% of all crimes committed last year were homicide offenses. Even so, this crime is the focus of much attention from the media and from criminologists who systematically study crime, in part because of the severity of the offense. Criminologists also study homicide because it is the most accurately measured offense and the offense for which the most statistical data are available at the national level. Moreover, the characteristics of homicide events are very similar to those for aggravated assault and other forms of violence. Therefore, studying homicide offending yields lessons about the causes and contexts of violence more generally. By exploring female and male homicide patterns, criminologists develop a more holistic picture of violent offending and get a sense of how similar or different the behaviors of the two sexes are. It should be noted, however, that gender differences are more apparent for offenses that are more serious in nature; gender similarities are greatest for minor sorts of violence, such as simple assault.

The Supplementary Homicide Reports are official, police-recorded statistics on almost all murders and non-negligent manslaughter incidents in the United States. Police record information on a voluntary basis on over 90% of the homicides of which they are aware. The FBI has accurately and consistently compiled these reports since the late 1970s. The SHR include demographic characteristics of the victims and offenders, their relationship to one another, and the situational features of the homicide incident, such as weapon used, motives, or circumstances. Because the SHR data include information on homicides still under investigation, information is incomplete for approximately 25% of the cases. Therefore, the involvement of young males is probably understated because they are more typically the perpetrators in the more difficult-to-solve stranger homicide cases. Using advanced statistical procedures to gain precision allows us to "guess" the characteristics of offenders based on their victims, minimizing the missing data problem (Fox, 2004). Therefore, SHR data present a fairly detailed nationwide portrait of homicide incidents, offenders, and victims. These data also allow researchers to describe changes in homicide offending over time.

In this chapter, the SHR data are used to detail the extent of female and male homicide offending. Then, in tandem with more detailed case analyses, we generate a portrait of the female and male homicide offenders.

EXTENT OF FEMALE AND MALE HOMICIDE OFFENDING

Of the 16,667 homicide victims identified in the United States in 2005, 3545 (21%) were women. Male and female victimization rates per 100,000 population were 8.7 and 2.4, respectively. As demonstrated by these statistics, homicide is a rare phenomenon. In comparison, deaths resulting from heart disease, accidents, and pneumonia are all far more common, but in all of these scenarios—including homicide—the death rate for men far exceeds the death rate for women. This gender disparity is even larger for homicide offending. In 2005, police were able to identify 19,127 homicide offenders, 17,301 of whom were men and 1826 of whom were women. Homicide offending rates for 2005 were 11.9 per 100,000 men and 1.2 per 100,000 women. Females represent only 9.5% of homicide offenders. Thus the large majority of perpetrators and victims of homicide are men.

Among both female and male homicide offenders, young adults (ages 18 to 24) have the highest rates of offending. Classified by gender and race, black males have higher homicide rates than white males, and black females have higher homicide rates than white females. Taken together, the offending rates of adolescent white girls (ages 14 to 17) are exceptionally low—0.7 per 100,000 is arrested for homicide. Black males aged 18 to 24 have the highest homicide offending rates (203.3 per 100,000). In between the two extremes are homicide rates of white males, followed by rates of black females. Within race groups, women make up a similar proportion of all homicide offenders—11% of white homicide offenders and 8% of black homicide offenders are women.

Homicide offending in the United States peaked in the early 1990s, but has since declined so that current rates are lower than those in the early 1980s (Fox & Zawitz, 2011). Rates of offending by black males dropped sharply in the mid-1990s and continue to decline, albeit at a slower pace. The trends of white males mirrored those of black males, but the declines were not as steep and seem to have leveled off by the early 2000s. The homicide rates of black and white females match one another and are characterized by steady declines since the 1980s. Driving the downward trend is a large drop in women's rates of intimate-partner homicide (trends in various sorts of homicide are discussed in more detail later in this chapter). The homicide rates of 14- to 17-year-olds were slightly elevated in the early 1990s; adolescent rates are less driven by intimate-partner homicide, in part because of these youths' lower exposure to intimate partners and lesser access to guns compared to other subgroups.

Generally, female trends in homicide mirror male trends in homicide, although women's involvement has been and remains very low in comparison to men. In fact, female representation among homicide offenders is somewhat lower today than it was almost 25 years ago (approximately 13% in 1980 compared to approximately 11% today). This consistency holds across age groups. Female representation dropped most sharply in the early 1990s when male rates were rapidly increasing, but the continued lower representation of females indicates female rates moved in tandem with male rates. The gender gap for 14- to 17-year-olds temporarily narrowed in the late 1990s and early 2000 because female declines were not as dramatic as male declines in that age group. The main conclusion, however, is that the gender gap in homicide has *widened* somewhat since the 1980s for all age and race groups.

The gender gap was declining even when effects of the women's movement might have been strongest in the 1980s and when women's arrest rates for assault were rising in the early 1990s. This stability in the gender gap is at odds with the widely held perception that women are becoming more violent, as suggested by the narrowing gender gap in arrests for assault offenses (Steffensmeier, Schwartz, Zhong, & Ackerman, 2005). Some commentators have interpreted increases in women's assault arrests as the ill effects of changing gender roles that have made female behaviors more masculine, but overall trends in homicide suggest otherwise.

In this chapter, we further explore the idea that changing gender roles might have altered the context of women's offending. On the one hand, if this is the case, we would expect women to increasingly kill strangers in felony-related homicides—in the past, a distinctly male scenario. Alternatively, perhaps female–female violence would be more likely as women start to interact with one another more like men do, including using violence to solve problems. On the other hand, if gender roles have not changed, or if they have not changed in a way that affects homicide and violent crime, stability in the context of women's and men's offending would be expected.

To examine these issues, we generate and examine statistics that indicate whether the extent and type of homicide committed by women and men have changed from 1980 to the present. Aside from addressing the theoretical debate regarding the effects of social change on women's homicide offending patterns, there are other good reasons to study female homicide, despite its statistical rarity.

WHY STUDY FEMALE HOMICIDE?

Some might question why the study of female homicide is warranted. A number of reasons can be cited for why such study is necessary.

First, despite its low incidence, women's homicide offending may have wider-reaching consequences for future crime trends than male offending because women are typically the primary caregivers for children in U.S. society. Incarcerated female homicide offenders are more likely to leave children behind, both compared to other types of female offenders and compared to male offenders of all types. Although obviously it is not desirable for children to have homicidal mothers, female offenders tend to direct their aggression against abusive partners rather than their children, and most do not have any prior arrests.

Second, no other violent crime is measured as accurately and precisely as homicide offending, making homicide a good barometer for violent offending in general, despite homicide's comparatively low frequency. Indeed, many criminologists consider homicide to be an overly successful assault. Therefore, similar distributional patterns in victim–offender relationship, motive, and so forth likely hold for assault offending. Note, however, that the gender gap systematically narrows as less serious violence is considered, so that women are far more involved in simple assault (accounting for approximately 25% of arrests), which includes minor harm such as scratching or shoving, versus aggravated assault (20%) or homicide (10%).

Third, studying female homicide offending, and comparative research more generally, can help clarify our current understanding of causes of violent offending. If women's homicide patterns do not fit with dominant theories of crime, these discrepancies should challenge criminologists to refine their explanations of criminal offending. Moreover, studying female homicide offending is useful in and of itself for better understanding and demonstrating the pervasive influence of gender on behavior, even in extreme actions such as taking the life of another. Men and women kill in ways that reflect their gender roles. We now explore patterns of female and male homicide to demonstrate this point.

WOMEN'S AND MEN'S OFFENDING PATTERNS: A QUANTITATIVE COMPARISON

Criminologists may characterize female and male homicide offending as either very similar or very different depending on the lens they use to view homicide offending. Many approaches to studying gender differences in homicide offending are possible, ranging from interviews with convicted offenders, to in-depth analyses of legal documents generated in the criminal justice system, to secondary data analysis of police records. Each methodology produces a slightly different picture of gender and homicide, making it important to look at multiple sources of evidence. No matter which methodology is used, the findings must be interpreted, and the researcher's theoretical orientation to gender differences—whether women and men are viewed as fundamentally alike or essentially different—is likely to color this interpretation. The truth probably lies somewhere in the middle: There are both gender similarities and differences in homicide offending.

Consider the following prototypical examples of a male homicide versus a female homicide:

The offender, Stephen, and the victim, Mark, have each spent the evening drinking in bars with friends before happening to go to the same nightclub. By this time, about 1 A.M., both men were drunk. Stephen and Mark had seen each other around before, but didn't really know each other. For reasons that are not obvious, Mark, with his friends watching nearby, makes repeated derogatory comments to Stephen. Mark had a reputation for getting into fights. Stephen, with his friends looking on, tells Mark to "[expletive] off" and, on a subsequent occasion, warns him he'll hit him if Mark provokes him again. Stephen has been arrested before, but has not served time for a violent offense. Mark makes a verbal threat, Stephen lands one solid blow to Mark's face, and Mark falls to the floor unconscious. Mark dies of a brain hemorrhage (Brookman, 2005, p. 128).

During five years of marriage, both the offender and her husband drank heavily. Things were OK until the last year of marriage, when he would drink and become violent. He would beat her, sometimes to the point of where she had to go to the emergency room of a local hospital. . . . One night, after he had been drinking to excess, he heard her talking on the telephone. He thought it was another man and became enraged. He began to strangle her with the telephone cord. She freed herself and ran into the kitchen. He started to strangle her with his hands. She reached into the sink where she found a knife that she used to stab him to death (Brownstein et al., 1994, p. 105).

Based on these two cases, how alike or different are these two homicide cases? The two events are similar in that neither homicide was planned but rather resulted from an argument, spurred in part by the use of alcohol, which escalated into physical violence. Both incidents took place late at night, probably on a weekend. Both victims were male and, to some extent, provoked their attack.

The circumstances surrounding these two deaths, however, might also be characterized as very different. The male's victim was a casual acquaintance or stranger, whereas the female offender was married to her victim. Although the female offender could claim self-defense and had no prior record, the male offender was not protecting himself, and both he and the victim had previous involvements with the law. Finally, the location of the two events differed—the female offender killed at home, and the male offender in public at a bar. We more systematically explore these differences in delineating a portrait of male and female homicide offenders, by drawing on the *Supplementary Homicide Reports*.

Victim-Offender Relationship

Perhaps the most crucial gender difference in homicide offending relates to who women and men kill. Overwhelmingly, females kill family members. In fact, almost 60% of female homicide offenders kill an intimate partner, child, or other family member. Men, in comparison, kill a family member approximately 20% of the time. In almost one-third of female homicides, her victim is a boyfriend, husband, or former partner. Children are the next most common targets of women's homicide (19%). In comparison, approximately 13% of men's homicides are directed against intimates and 3% against children. Consequently, the large majority of women's victims, roughly

75%, are men. Likewise, 75% of males' victims are men, but male offenders' targets are acquaintances approximately half the time. Men kill strangers (29%) more often than they kill family members (22%). In contrast, women rarely kill strangers (10%) and only sometimes kill acquaintances (33%).

Importantly, there is *no marked shift among women in the relational aspects* of their homicide offending profile. If gender roles were shifting in ways that affected homicide offending patterns, one might expect to see a shift toward stranger homicide or, perhaps, acquaintance homicide. In fact, no such shift is evident. The distribution of women's victims has remained essentially the same over the past 25 or so years—approximately 60% family, 30% acquaintances, and 10% strangers. One notable change, however, is the decrease in the percentage of victims who are intimate partners (from 49% of women's victims to 29%).

Women's rates of intimate-partner homicide have dropped precipitously since 1980; by comparison, women's intimate-partner homicide rates in 1980 were four times higher than current rates. Men's rates of intimate-partner violence have dropped as well, but not as much as women's rates (men's rates in 1980 were roughly twice as high as present rates). In fact, women's rates of homicide are down since 1980 for every victim–offender relationship category and family subcategory, though by far the greatest decline is in intimate-partner homicide. Declines in male acquaintance and stranger homicides are also notable: Rates of each have dropped by more than half since 1990; 1990 rates were marginally higher than rates in 1980. Since 1990, the declines in male rates of stranger and acquaintance homicide outpaced female declines.

The gender gap in homicide offending is greatest for stranger homicides. Females have consistently accounted for only 5% to 7% of those individuals identified as perpetrators of a stranger homicide. Women's representation in homicides against acquaintances is also low and unchanged: Between 1980 and the present, 7% to 10% of these offenders were women. The gender gap is narrowest for homicides against family members (including intimates). More than 25% of homicides against family are now committed by female offenders, which represents quite a drop from the 40% rate noted in 1980. Notably, women are still underrepresented as offenders, given that they make up half the population.

Women are not underrepresented as homicide offenders in terms of those individuals who victimized young children (ages 0 to 12). Women account for almost half of all offenders arrested for child homicide. Female involvement in child homicide declines with victim age, however, so that the gender gap is 50% for infanticides, 33% for toddlers, and 21% for older children. These gender gap percentages are nearly identical to those identified in 1980 and 1990. In terms of victim selection, women more often offend against infants (37% of child victims), whereas males offend against older children (27% of child victims). Toddlers are the most vulnerable age group, accounting for roughly half of both women's and men's non-adult victims.

Motives and Circumstances

The immediate motivation for the majority of both women's and men's homicides are arguments and fights. Nearly half of all homicides, committed by men or women, occur because of some sort of argument or fight, such as a conflict over money or property, anger over one partner cheating on the other, severe punishment of a child or abuse of a partner, retaliation for an earlier dispute, or a drunken fight over an insult or other affront. In addition, qualitative analyses (discussed in more depth later in this chapter) show that many female homicides that result from an argument are often directed against the violence or abuse of a partner and may be viewed as "extra-legal" self-help, or even self-defense. Incidents of severe child abuse are also categorized as arguments however. Less common are arguments with friends, neighbors, or acquaintances that end in homicide (Jurik & Winn, 1990).

The second most common male homicide circumstance is felony related (Wilbanks, 1983). Approximately 25% of men's homicides, compared to 15% of women's homicides, occur within the context of committing another felony. Almost half the time these sorts of homicides are related to a robbery. Other felony-related homicides are related to drug dealing (one in five felony homicides for men; one in eight for women) or, less commonly, burglary, arson, sex offenses, or theft. Of note, female felony homicide offenders often co-offend with male partners (Schwartz & Steffensmeier, 2007).

In the SHR data, some 70% of female felony homicide offenders had a partner, compared to 50% of male felony homicide offenders. In contrast, roughly 25% of offenders of either sex had a co-offender when the homicide resulted from a fight or argument. Felony-related homicides are more likely to be committed with a partner, particularly when women are involved. Qualitative accounts suggest that the crime partners of women are often boyfriends and that the women play a more secondary role in these events. It is more likely that the male planned the offense—to the extent that planning occurred—and that he brought the weapon, especially if it was a gun. That women's homicide is often related to men's violence—either as a co-offender or in response to abuse—is consistent with the findings from past studies of women's assault and use of other forms of violence (Miller, 1998). Women's homicides, if not related to a fight, are next most often categorized as "other." Since the 1990s, women's homicides increasingly have been categorized as "other," to the extent that one-third of women's homicides are now classified as having an unknown origin. This trend is also noted among men (increasing from 15% to 19% of homicides), but "other" homicides remain less common among men compared to felony-related homicides.

The smallest proportions of homicides for both men and women are those that are gang related. Only 1% to 2% of women's homicides are gang related—a low level of involvement that has not changed over time. Fewer than 50 women in the entire United States were identified as being participants in gang-related homicides between 2001 and 2003. For men, the percentage of gang-related homicides increased from 3% to 9% over the past 20 years, with the bulk of this increase occurring in the late 1980s. The proportion of male homicides motivated by gang involvement has increased, in part, because all other types of homicides declined more precipitously than did gang-related homicides. In sum, both sexes most often kill in the context of an argument or fight, but men's homicide more often is felony related whereas women are next most often involved in "other" sorts of homicides.

The gender gap in homicide offending is narrowest for uncategorized "other" sorts of homicide offenses and homicides stemming from an argument. The gender gap is largest for gang- and felony-related offenses. This pattern holds across each decade examined in this chapter, indicating that any changes in women's homicide motivation generally have mirrored changes in men's motivations. Since at least 1980, female involvement in gang-related homicides has remained at 2%, felony-related homicides at 7%, and "other" homicides at 17% to 18%. Rates of involvement in these three types of homicide declined evenly for women and men. Felony-related and "other" homicide rates are currently roughly half what they were in the 1980s. Rates of gang-related violence increased in the 1990s for both women and men; though declining, these rates in 2003 remained somewhat higher than 1980s rates. All other offense types are at lower levels now than in 1980.

The greatest declines have occurred in homicides motivated by arguments, and these declines have been more substantial among women than men. Women's relative involvement in homicides stemming from an argument dropped sharply, by nearly half since 1980. Women's rates are only one-fifth what they were in 1980; men's rates have been cut by more than half. The sustained decline in female-partner violence rates mainly is responsible for the widening of the gender gap in argument-related homicides.

To summarize these trends, women's and men's homicides are most likely to be sparked by arguments or fights, but gender differences are more notable in the extent of involvement in felony- and gang-related homicides. Although the SHR data lack detail in terms of offender motivations, case studies and qualitative accounts show important gender differences in the content of arguments, with women more often engaging in self-defensive violence. There has been little, if any, shift in the immediate circumstances surrounding women's homicide: Women (and men) most often kill in the context of a heated argument. Changes in the circumstances surrounding female homicide generally follow a similar pattern as those of men. Notably, however, women's involvement in the most common type of homicide—argument-motivated offenses—has declined more substantially than males' involvement owing to declining rates of female-partner violence.

Offense Characteristics

Some offense characteristics, such as the time and day of the incident, are similar for both females and males. For instance, homicides occur more often on weekends and late at night. Female homicide offending usually takes place in her and/or the victim's home; male homicides are more likely to occur in public places, such as bars or other public locations. This gender difference occurs primarily because the settings in which women kill intimate partners and children differ from the settings in which men kill acquaintances and strangers.

In terms of weapon used, men are far more likely than women to use guns. Females were involved in only 6% of the homicides perpetrated with a gun. Female representation in this category has declined since 1980, when 13% of gun offenders were women. In fact, a woman today is as likely to use a knife as a gun to commit her homicide, whereas in the past 60% of female homicides were committed with a gun. Note that this shift is mainly due to the steeper declines in women's rates of gun violence as compared to the declines in homicides committed with knives. Gun homicide rates of males have also declined, but only since 1990. Men's homicides still tend to be committed primarily with guns.

Interestingly, women and men have equal representation in homicides in which the victim is poisoned or drugged. Otto Pollak (1950) asserted that a large portion of the gender gap in offending might be due to women's more surreptitious style of committing crimes, citing poisoning as a means of women secretly engaging in homicide. This style of offending, he explained, was required by women's smaller physique but facilitated by their domestic social role as caregivers (e.g., via opportunities presented when preparing food). Criminologists today would likely concur with Pollak's contention that female crime is more hidden than male crime, but would attribute this difference to the typically less serious nature of women's offending compared to men's. Criminologists would also agree that homicide is the most accurately measured offense and women's overall involvement in this crime, as compared to men's engagement, is very low. Whether Pollak's assertion was true at the time, today fewer than 100 women each year are *detected* killing a victim with poison or drugs. Only a fraction of female-committed homicides involved poisoning or drugs (e.g., ranging from less than 1% to 3%). Of course, with approximately 20,000 "unintentional" poisonings occurring each year in the United States, there is always the potential that untold numbers of women are getting away with murder.

Offender Background

Like many offenders, female and male homicide offenders tend to come from economically and educationally disadvantaged backgrounds and communities. Most female offenders have low educational attainment—on average, reaching only the eleventh-grade level. Both sexes of homicide offenders, if employed, tend to work in a menial occupation, although for women it is in the service sector and for men it is more likely in a blue-collar occupation. Women's rates of homicide tend to be high in settings where male rates are also high, such as urban areas (Scott & Davies, 2002).

In comparison to male homicide offenders, fewer women had a prior record or felony offense. More than 75% of male homicide offenders are believed to have been previously arrested, many repeatedly and for felony offenses. Previous research suggests that most (70%) female homicide offenders do not have a prior felony record, although many (approximately 60%) have prior arrests for misdemeanors. Many of these arrests are likely to be for assault, possibly related to domestic abuse; status violations as a juvenile (such as running away from abuse in the home); or minor property crimes, such as shoplifting, credit card fraud, or check forgery. Very few women who commit homicide have ever been previously incarcerated, compared to one-fifth of the men arrested for homicide (Jurik & Winn, 1990). Clearly, female homicide offenders are not immersed in criminal subcultures to the same extent as male homicide offenders (Suval & Brisson, 1974).

Compared to other types of female offenders, female homicide offenders were more likely to be from a solid family background, to have children, and to a have stable living arrangement prior to the offense. Female murderers appear to have somewhat higher levels of problem drinking, but lower levels of drug use compared to other female offenders. These patterns again indicate that female homicide offenders are not often part of the criminal subculture. Such differences also reflect the influence of domestic violence on women's offending patterns.

In sum, the background characteristics of female and male offenders are similar: They tend to be characterized by poverty, low educational attainment, and no or low-quality employment. Yet female homicide offenders are less likely to have a prior record, especially for more serious offenses, and are less likely to be involved in a drug or criminal subculture. Gender differences in previous criminal involvements and opportunities for offending clearly shape the context, or type, of offending for women and men.

WOMEN'S AND MEN'S OFFENDING PATTERNS: A QUALITATIVE COMPARISON

Previous case studies shed additional light on the family and gendered dynamics of women's and men's homicide. This section draws on previous case studies and other in-depth modes of analysis to elucidate key gender differences (and similarities) in the ways in which men and women engage in homicide offending. We begin by considering the more unusual forms of female and male homicide offending—neonaticide for women and familicide for men. Next, we describe the more common (albeit still rare) forms of female and male homicide—child and partner homicide for women and acquaintance and stranger homicide for men.

Child Homicide

A predominantly female form of homicide is neonaticide, when a mother kills her newborn child. These offenders are often young, unmarried women who may conceal pregnancy, give birth alone, and commit homicide using more "delicate" methods such as exposure, suffocation, or strangulation. They may kill their newborns in fear of stigmatization should their pregnancy be detected, due to feeling unable to care for the child, or because of extreme stress or mental illness. Until an infant is one week old, the most immediate threat to the child's well-being is the neonate's mother (Gartner & McCarthy, 2006).

Almost uniquely male are murder-suicides and family massacres (familicides), where a man kills his children, his partner, and possibly himself for reasons such as possessive jealousy, anger or vindictiveness toward a partner, loss of children through separation, or sometimes the inability to financially support the family. The victims are usually biological children, are older than most child victims, and are killed with a gun. Like the women who commit neonaticide, these men may express feelings of powerlessness, of matters having gone beyond their control; they may also express feelings of pain or anger (Alder & Polk, 1996; Daly & Wilson, 1988).

Although both sorts of homicide sometimes demonstrate planning or forethought, some offenses seem to occur with little forethought, with the offenders showing signs of irrationality or imbalance, at least at the time of the incident if not previously. The more common situation of child homicide, for both women and men, arises not out of the intent to kill the child, but rather as the end result of harsh punishment (or discipline, from the parent's perspective). Consider the following two examples—the first a female description, the second a male description:

> *I was packing up my stuff [to move] and my son was acting up and I didn't know what to do, 'cause I don't understand nothing about disciplining a child, 'cause how I was raised by my own family, how they abused me and I didn't know what to do, so I took it out on my son and sent him to his room and I made him go to bed and he went to bed. I went near and he wasn't breathing, he stopped breathing, wouldn't breathe. I know he was sleeping and he didn't wake up. I hit him, I only hit him twice in the head with my hand. I don't know, with my shoe, my flat shoe in the head twice and that was it, and I sent him to his room 'cause I didn't want to hit him no more . . . It was very hard for me 'cause I didn't know what to do. The only thing I knew was to take him to the doctor when he needed to go to the doctor and feed him and keep him clean, that was it. I didn't know how to love him, 'cause I didn't have, didn't love myself, I didn't know how to love him.—Female offender (Crimmins, Langley, Brownstein, & Spunt, 1997, p. 58).*
>
> *Austin was sitting on the floor eating a packet of chips and he started crying. I picked him up and whacked him on the bum three or four times with an open hand. I put him down and he was still crying. I picked him up and shook him [to] shut him up . . . I didn't lose my cool, I was just annoyed . . . I was just annoyed because I couldn't hear the video. He was getting on my nerves.—Male offender (Alder & Polk, 1996, pp. 404-405).*

The homicides of preschool children, like the ones described in the preceding examples, tend to be more brutal than those of infants killed by their mothers. Such child victims likely have suffered from abuse by their mother or, especially, their father for much of their young lives. Female offenders are usually the primary caretakers in such cases. Many of the women report having felt socially isolated, trapped by their responsibilities at a young age. The male child homicide offenders share more in caretaking responsibilities than most men; they are often stepfathers or other men living with the family.

The mother, though often married or partnered, tends to commit the crime alone, in the bathroom or bedroom, using manual force (hitting, kicking, choking, or drowning). She is unlikely to be under the influence of drugs or alcohol at the time of the crime's commission. While women filicide offenders often claim innocence or state that the crime was an accident, more than 6 in 10 children (ages 2 to 5) have multiple wounds. Half the women arrested for child homicide have recorded child abuse histories, particularly those who kill a preschooler (ages 2 to 5). A large minority of the women were abused themselves as children. Conviction charges typically are lower than the initial murder charge in 80% of cases, suggesting lesser culpability or some mitigating circumstances are identified by prosecutors upon weighing the evidence (Mann, 1996).

Men may initially attempt to deny or cover up their role in the homicide, sometimes persuading their partners to support their story. Case files show male offenders saying, "I was just playing;" that the child was "accident-prone;" or that the child "fell downstairs." The child's death is not likely to be premeditated, but the physical evidence often shows signs of prior abuse and extreme aggressive acts may have precipitated the child's death.

Although these acts are inexcusable, in general, the families involved are typically enmeshed in stressful circumstances—money troubles and unemployment, frequent fights with their partner, high-poverty communities. Approximately 70% of murdered children reside in severe urban poverty. Higher rates of child homicide (and child maltreatment) occur in socially disorganized communities—communities characterized by poverty and social isolation, single-parent family structures, and conflicting behavioral norms regarding the use of violence as a means of problem solving. This suggests that child homicide is, in part, rooted in the social organization of society. Likewise, intimate-partner homicide rates tend to be higher in communities characterized by high poverty, social isolation, and similarly stressful factors.

Partner Homicide

Exploring homicide directed against intimate partners paints a very different portrait of female and male homicide offenders. Recall that women's homicidal behavior and violence is most often directed against a partner rather than against a child. In fact, a woman's decision to kill her partner may be motivated be a desire to protect her children (or herself). Angela Browne's (1987) landmark study of women who killed their abusive husbands compared to women who escaped abuse showed few differences among the two groups of women. Their victims, however, differed: Men who more frequently and severely assaulted or raped their partners, made more death threats, frequently used alcohol or drugs, and abused the children were more likely to be killed. Patterns of violence escalated more among the women who killed their husbands, prompting Browne to conclude "women's behavior seemed to be primarily in reaction to the level of threat and violence coming in." What appears to trigger the homicide, despite the long history of abuse, is the woman's feeling that her death or, importantly, the safety of her children is at stake based on an event out of proportion with past "normal" violent events (e.g., physical abuse of a child or discovery of sexual abuse of a daughter) (Browne, 1987). As many as 60% of women are being abused at the time they kill their partners, and women who kill their partners are unlikely to be the first to use force in the event precipitating the homicide (Johnson & Hotton, 2003).

Qualitative analyses suggest that women's homicide, and their violence more generally, often revolves around relational concerns. Female motives for violence often involve self-defensive acts against abusive male partners, risk taking to protect emotional commitments and valued relationships, or co-offending with male partners (the last condition is described later in this chapter). Moreover, a myriad of studies indicate that female violence mainly occurs only under conditions of extreme stress or repeated provocation, such as in the case of aggression against small children or in response to male-instigated violence including domestic abuse and assault (Bailey & Peterson, 1995; Dobash, Dobash, Wilson, & Daly, 1992). Indeed, some have characterized women's violence as a form of "extra-legal" self-help, in that women typically kill only in situations they perceive as life-threatening. Males, in contrast, are more likely to kill as a result of jealousy or rage, during trivial arguments, or in the course of committing another felony.

Male homicide offenses against partners are dominated by motives of possessiveness, jealousy, abuse and control, and arguments. Men's violence in these cases is aimed at preventing the woman from leaving, retaliating for her departure, or forcing her to return. Some studies indicate that women who are separated from their partners are at an elevated risk of violent victimization, including homicide. When men kill their partners, the act often represents the culmination of a prolonged history of abuse. Another motive related to possessiveness is sexual jealousy, such as over a suspected or known infidelity (e.g., love triangles).

Thus motives relating to perceived infidelity or termination of the relationship center on themes of male domination and control, whereas females are the majority of offenders when the motive is self-defense. A portrait of female homicide offenders as women acting in self-defense out of desperation would be one-dimensional, however, as there have always been some women who are motivated by material wealth and financial gain, revenge, involvement in a criminal subculture or gang, or the wish to continue an illicit affair (Weisheit, 1993).

Acquaintance and Stranger Homicide

Female-on-female violence most frequently is directed against neighbors, their intimates' other sex partner, or friends and acquaintances. It should be noted, however, that as many as one-third of female-on-female homicide victims are daughters of the offender. Offenders who kill another (unrelated) woman tend to be younger in age than the average female offender. Women may kill other women in fights over men, to prevent a romance from occurring, to preserve their sexual or social reputation, and as a result of an ongoing feud.

For example, two neighbors had an ongoing feud over neighborhood matters. On the occasion of the homicide, the two women were struggling in the kitchen and, as a result, the offender's three-year-old child was inadvertently injured. The offender responded by stabbing the other woman in the chest one time, killing her. Although certainly extreme, this incident also reflects female offenders' greater willingness to aggress to protect a loved one.

Other female–female violence may result from fights related to gossip or jealousy, disrespect (e.g., negatively evaluating another woman's appearance), and interactions with another woman's boyfriend. All of these causes may be regarded generally as reputational challenges (Miller & Mullins, 2006).

Importantly, the extent of female–female homicides has not risen over the past 25 years, as some might expect if women truly have become more violent. Female-on-female homicides continue to account for fewer than 2% of total homicides.

Although women rarely kill strangers, when this type of offense occurs, the victim is likely to be a man. Approximately half the time, such a crime occurs in the context of committing another felony offense—namely, robbery. In these situations, women often co-offend with men, typically filling an ancillary role in the crime. Many of the women charged with a robbery-related homicide are ultimately considered accessories to the homicide rather than major actors, and men usually have brought the weapon (80% of the time) and initiated the violence. Women who co-offend are probably romantically involved with their crime partners (approximately 75%); men, by comparison, mainly co-offend with other men and, therefore, are far less likely to be romantically involved (14%). Thus, even when women are not offending against their intimate partners, women's violence is intertwined with men because they are offending with their intimates.

After robbery-related homicides, female felony-related stranger homicides are most often perpetrated in the context of prostitution. To some extent, these homicides may be self-defensive because prostitutes may be severely victimized by their clients (e.g., beaten, robbed, raped). Prostitutes are also known to sometimes rob their clients, perhaps with the help of "their man," so "prostitution-related" homicides may also reflect robberies "gone bad."

Approximately one-third of stranger homicides by women result from a fight or argument. Most of these assaults involve a man, but 30% of the time they result from an argument with a woman.

By far the most common homicide situation is male-on-male violence. Indeed, men are the most likely victims of homicide (and violent crime generally) as well as the more common perpetrators. As with female offenders, most male homicide is often the result of an argument. The nature of male disputes with other males, however, differs markedly in character from the types of domestic arguments that spur women to violence. Male–male homicide events often result from what appears to be a minor or trivial provocation, such as a shove, insult, or the "wrong look" among friends and acquaintances, or, less often, strangers. These events typically occur in a place of leisure, such as bars, parties, barbecues, and parks, or on the streets, where groups of young men may congregate. Insults or threats appear to escalate and develop "spontaneously" into violence, though this amplification process is often facilitated by the use of alcohol and the presence of young male peers and a social audience. At some point, both parties must interpret the exchange as requiring retaliation and mutually agree to aggression (Polk, 1998).

Why would such seemingly trivial slights with so little consequence cause a person to risk death? Events that seem absurdly trivial to some provide some males the opportunity to demonstrate masculinity through "honor contests." Lower-class males may develop a stronger allegiance to street norms and aggressive means of demonstrating masculinity than middle- and upper-class males, who more strongly prescribe to mainstream society's norms relating to problem solving. Whereas middle- and upper-class males may have myriad opportunities to "do masculinity," lower-class males living in concentrated urban poverty may, over time and as an adaptation to persistent economic strain, place stronger collective emphasis on more achievable goals than financial success, such as respect and prestige conferred to those with a street reputation. Violations of the "code of the street"—sustained eye contact or a disrespectful demeanor, for example—can be lethal (Anderson, 1999; Messerschmidt, 2004).

Male homicide offenders are more likely to be immersed in street culture. Compared to female homicide offenders, who are unlikely to have a prior felony record, male offenders (and sometimes victims) tend to have lengthy arrest and, for some, imprisonment histories. Moreover, many male offenders often have long histories of unemployment. Whereas male–male status contests are the predominant form of male homicide, two other distinctly male patterns are felony-related homicide and gang-related killings. These types of homicide also stem from male involvement in criminal subcultures. More than 95% of gang homicides are perpetrated by males against males, with both parties being likely to be members of (rival) gangs. Motives for gang homicide often are similar to those for non-gang homicide—retaliation to save face and to establish social position, although gang interactions tend to be more conflict oriented than those involving other groups of youth. Felony-related homicide is also viewed by some as a means of displaying masculinity via demonstrating willingness to engage in risky behavior with the potential for violence (Short & Strodtbeck, 1965).

CONCLUSION

While the motivational and situational contexts of homicide offending may differ markedly by gender, males and females appear to be subject to the same social and cultural influences on homicide offending. That is, when it comes to forces outside the individual, men and women are more alike than different in the factors that increase the likelihood of homicide and other violent offending. For example, places with high levels of male homicide tend to have correspondingly high levels of female homicide, although male homicide rates greatly exceed female rates in all places. A number of studies have also shown that women's homicide trends tend to mirror men's trends in homicide offending. Both of these facts provide indirect evidence that women and men experience similar pushes toward violence.

The considerably smaller set of studies that have directly compared causes of women's and men's homicide have concluded that similar social forces are at work in producing men's and women's violence—namely, concentrated poverty, inequality, and single-parent family structures that weaken allegiance to mainstream norms regulating the use of violence and inhibit the capability of residents to exert informal social control. Of note, these conditions are associated with all sorts of homicide offending, including partner homicide, child homicide, gang killings, and so on. These "milieu effects" are felt by females and males alike, but males more readily respond to these conditions with violence than do females, explaining some gender differences in offending levels (Kubrin, 2006; Schwartz, 2006a, 2006b; Steffensmeier & Haynie, 2000a, 2000b).

The stronger influence on men of their immediate surroundings only partially explains why females are so much less likely than males to kill, why women need a higher level of provocation before turning to serious violence, and why women so infrequently kill a rival female, use a gun, or engage in felony offenses or gang violence. These differences in the extent and context of female and male homicide offending might be attributed to differences in the organization of gender. Gender norms and stereotypes surrounding femininity and masculinity are less or more compatible with violence (and involvement in the criminal underworld). Violence by females is more stigmatizing, whereas violence by males is more status enhancing. Women's violence tends to be restrained by an "ethic of care," which carries the expectation that women will maintain and establish social relationships and be more responsive to the needs of others. By comparison, men are conditioned more toward status seeking, competition, and distributive justice. Gender socialization works to limit female motivations for violence, but it heightens male motivations. Further, males have greater opportunities to commit violence, because they tend to be less supervised than females, allowing greater and more frequent contact with male peers and engagement in violence-likely social situations (e.g., parties, "hanging out," involvement in criminal enterprises). Females, by comparison, are more closely watched, limiting their exposure to violence-likely situations (Bottcher, 1995). There also is a good deal of evidence that sexism in male peer groups and in the underworld limits the extent and nature of female involvement in crime groups, regardless of a woman's ability or desire to engage in criminal violence (Steffensmeier & Terry, 1986).

Homicide is rare, particularly for women, who account for only some 10% of homicide offenders. Relative to men, female involvement in this type of crime has not changed much over time. Whether change in women's propensity toward violence is measured using overall rates of homicide or "masculine" homicide classifications such as stranger homicide or intra-sex homicide, little evidence suggests that women have become more violent over the past 25 years. This conclusion meshes with other studies on girls' and women's assault patterns, which show that females today are not more likely to commit assault (although they do appear more prone to arrest for assault). Importantly, no marked shift has occurred among women in terms of their homicide offending profile over time.

Women's homicide patterns are characterized by victimization of those persons who are closest to them—partners and children. Only rarely do women kill strangers. When a woman kills, it is likely she was under extreme pressure or was provoked by fear for her life or for the sake of someone close to her (e.g., children), although sometimes her motives are less altruistic (e.g., committing another felony, often with her romantic partner, or fighting with another woman). Men's relational patterns of homicide are more heavily weighted by victimization of friends, acquaintances, and strangers; such events typically take place in a public place, perhaps where alcohol and an audience are present. Men often kill over matters that appear to be trivial—minor insults or minimal physical contact—yet these challenges are viewed by participants as requiring a response to defend one's masculinity, particularly if the offenders are involved in street culture. When a man kills his partner, the cause is rarely mortal fear; instead, the crime usually occurs in response to jealousy or some other control motive.

Men and women kill in ways that uniquely reflect their gender roles and related social positions. As such, the context of homicide differs substantially by gender. Contexts of offending, such as the target of aggression, specific motivation, and commission of the crime, are profoundly shaped by gender roles and opportunities, which differ markedly for men and women. Thus women's aggression tends to be directed at intimate partners, who may precipitate their own death, and at children, with whom these women spend much of their time. Men's homicide occurs more often in the context of the criminal underworld and in public spaces.

The backgrounds and social structural conditions underpinning homicide, however, are quite similar for women and men. Communities marked by social disorganization render both women and men more susceptible to dealing with problems via the use of violence (even if the targets of aggression differ). Thus temporal and spatial patterns of female and male homicide tend to be the same.

Because female homicide is so intertwined with male homicide, solutions that address the structural conditions associated with violence should apply equally well to females and males. At the same time, there is also a need to take into account the gendered ways in which people offend as part of interventions to prevent future homicides from occurring.

REFERENCES

Alder, C., & Polk, K. (1996). Masculinity and child homicide. *British Journal of Criminology, 36*, 396–411.

Anderson, E. (1999). *Code of the street.* New York: W. W. Norton.

Bailey, W., & Peterson, R. D. (1995). Gender inequality and violence against women: The case of murder. In J. Hagan & R. D. Peterson (Eds.), *Crime and inequality* (pp. 174–205). Stanford, CA: Stanford University Press.

Bottcher, J. (1995). Gender as social control. *Justice Quarterly, 12*, 33–57.

Brookman, F. (2005). *Understanding homicide.* London: Sage.

Browne, A. (1987). *When battered women kill.* New York: Free Press.

Brownstein, H. H., Spunt, B. J., Crimmins, S., et al. (1994). Changing patterns of lethal violence by women. *Women and Criminal Justice, 5*, 99–118.

Crimmins, S., Langley, S., Brownstein, H. H., & Spunt, B. J. (1997). Convicted women who have killed children. *Journal of Interpersonal Violence, 12*, 49–69.

Daly, M., & Wilson, M. (1988). *Homicide.* Hawthorne, NY: Aldine de Gruyter.

Dobash, R., Dobash, R. E., Wilson, M., & Daly, M. (1992). The myth of sexual symmetry in marital violence. *Social Problems, 39*, 71–91.

Fox, J. A. (2004). Missing data problems in the *Supplementary Homicide Reports*: Imputing offender and relationship characteristics. *Homicide Studies, 8*, 214–254.

Fox, J. A., & Zawitz, M. W. (2011). *Homicide trends in the United States.* Washington, DC: U.S. Department of Justice, Bureau of Justice Statistics. Retrieved January 13, 2011, from http://bjs.ojp.usdoj.gov/content/homicide/homtrnd.cfm

Gartner, R., & McCarthy, B. (2006). Killing one's children: Maternal infanticide and the dark figure of homicide. In K. Heimer & C. Kruttschnitt (Eds.), *Gender and crime: Patterns in victimization and offending* (pp. 91–114). New York: New York University Press.

Johnson, H., & Hotton, T. (2003). Losing control: Homicide risk in estranged and intact intimate relationships. *Homicide Studies*, 7, 58–84.

Jurik, N., & Winn, R. (1990). Gender and homicide: A comparison of men and women who kill. *Violence and Victims, 5*, 227–242.

Kubrin, C. (2003). Structural covariates of homicide rates: Does type of homicide matter? *Journal of Research in Crime and Delinquency, 40*, 139–170.

Mann, C. R. (1996). *When women kill.* Albany, NY: State University of New York.

Messerschmidt, J. (2004). *Flesh and blood: Adolescent gender diversity and violence.* Totowa, NJ: Rowman, Littlefield.

Miller, J. (1998). Up it up: Gender and the accomplishment of street robbery. *Criminology, 36*, 37–66.

Miller, J., & Mullins, C. W. (2006). Stuck up, telling lies, and talking too much: The gendered context of young women's violence. In K. Heimer & C. Kruttschnitt (Eds.), *Gender and crime: Patterns of victimization and offending* (pp. 41-66). New York: New York University Press.

Polk, K. (1998). Males and honour contest violence. *Journal of Homicide Studies, 3*, 6–29.

Pollak, O. (1950). *The criminality of women.* Philadelphia: University of Pennsylvania Press.

Schwartz, J. (2006a). Effects of diverse forms of family structure on women's and men's homicide. *Journal of Marriage and the Family, 68*, 1292–1313.

Schwartz, J. (2006b). Family structure as a source of female and male homicide in the United States. *Homicide Studies, 10*, 253–278.

Schwartz, J., & Steffensmeier, D. (2007). The nature of female offending: Patterns and explanation. In R. Zaplin (Ed.), *Female offenders: Critical perspective and effective interventions* (pp. 43-76). Sudbury, MA: Jones & Bartlett.

Scott, L., & Davies, K. (2002). Beyond the statistics: An examination of killing by women in three Georgia counties. *Homicide Studies, 6*, 297–324.

Short, J. F., & Strodtbeck, F. L. (1965). *Group process and gang delinquency.* Chicago: University of Chicago Press.

Steffensmeier, D. J., & Haynie, D. (2000a). Gender, structural disadvantage, and urban crime: Do macrosocial variables also explain female offending rates? *Criminology, 38*, 403–438.

Steffensmeier, D. J., & Haynie, D. (2000b). The structural sources of urban female violence in the United States. *Homicide Studies, 4*, 107–134.

Steffensmeier, D., Schwartz, J., Zhong, H., & Ackerman, J. (2005). An assessment of recent trends in girls' violence using diverse longitudinal sources: Is the gender gap closing? *Criminology, 43*, 355–406.

Steffensmeier, D., & Terry, R. (1986). Institutional sexism in the underworld: A view from the inside. *Sociological Inquiry, 56*, 304–323.

Suval, E. M., & Brisson, R. C. (1974). Neither beauty nor beast: Female criminal homicide offenders. *International Journal of Criminology and Penology, 2*, 23–34.

Weisheit, R. (1993). Structural correlates of female homicide patterns. In A. W. Wilson (Ed.), *Homicide: The victim–offender connection* (pp. 191–206). Cincinnati: Anderson.

Wilbanks, W. (1983). The female homicide offenders in the U.S. *International Journal of Women's Studies, 6*, 302–310.

CHAPTER 12

Female Sexual Offending

Danielle A. Harris
San Jose State University

It is well known that men commit more crimes than women, but they commit an even greater proportion of sexual crimes. Compared to their male counterparts, women who do commit sexual offenses are usually subject to extreme media coverage and a fierce double standard. Female sexual offenders receive differential treatment at every level of the criminal justice system from law enforcement through legal processing and corrections. When compared to men convicted of the same offenses, women often appear to be held to a different standard not only in the eyes of the law, but also in the eyes of the media and by the larger community. Understanding crime by women requires criminology to think outside of its typically male perspective. As one can imagine, describing female sexual offending requires an even bigger paradigm shift.

This chapter provides an overview of the nature and extent of sexual offending by women. It reviews the many typological approaches that have been developed to explain female sexual offenders, including descriptions of what might be considered "typical" examples of each classification. In addition, the chapter focuses more specifically on the differential treatment that male and female sexual offenders receive in the news media and in the criminal justice system owing to their crimes.

It is necessary to begin any crime-specific discussion with an overview of the nature and extent of the problem. Very roughly speaking, research indicates that female offenders account for fewer than 5% of all reported incidents of child sexual abuse and fewer than 10% of all known sexual offenders (Finkelhor & Williams, 1988; Strickland, 2008). Incidence rates differ when the question is asked from the victim's perspective: As many as 75% of various samples of anonymous respondents report having been abused by women. Specifically, Strickland (2008) found that 15% of college students and 39% of victims receiving treatment identified female perpetrators as their offenders. Finally, more than half (59%) of male rapists reported having been molested by a woman during childhood.

Regardless of the specific study or sample, the actual impact of female sexual offending is substantial. It is clear that the "dark figure of crime" (the amount of crime that remains unreported) is especially high in cases of sexual abuse. Also, the amount of unreported crime is amplified when the perpetrator is female, when the victim is male, when the offender is in a position of authority, and when the abuse seemingly occurs within the context of a relationship (Briggs, 1995; Turner, Miller, & Henderson, 2008). As described later in this chapter, these conditions make it all the more likely that the true extent of female-perpetrated sexual offending is unknown. Further, the news media tend to skew these data in the direction of moral panic by sensationalizing sexual crime. All of these variables paint an incomplete picture of the true extent of sexual crime by women.

There is a strongly held—but erroneous—assumption that the absence of a penis renders female–child abuse less traumatic, less violent, and less physically damaging than sexual abuse by a male (Hislop, 2001). However, victims of female perpetrators consistently describe similarly or *more* severe psychological and behavioral symptoms as victims of male perpetrators. For survivors of abuse by both men and women, it is often against the woman that they feel more anger, because the expectations of maternity and caretaking create a far greater sense of betrayal and powerlessness when the abuser is female (Saradjian, 1996).

FEMALE SEXUAL OFFENDER TYPOLOGIES

The population of known sexual offenders constitutes an extremely heterogeneous group (Parkinson, Shrimpton, Oates, Swanston, & O'Toole, 2004). Typologies have developed as a way of making sense of this heterogeneity by zooming in on individual differences, and generating categories of like individuals based on demographic or other criteria that are related to offending (Blackburn, 1993; Gannon & Rose, 2008; Harris, 2010; Matravers, 2008). Due to very small samples, most of the typologies that have been constructed for women have so far been essentially descriptive and not grounded in theory. Nevertheless, they provide some illustration of the vast spectrum of female sexual offenders and have assisted clinicians who deal directly with these women.

Typologies are constructed by arranging individuals into subtypes based on various differentiating characteristics. These features might include offender characteristics (e.g., age, history of childhood abuse, substance abuse, presence of a co-offender), offense characteristics (e.g., location of offense, motivation, recidivism, general criminal history), and victim type (e.g., age, gender, relationship). Numerous female sexual offender typologies have emerged over the past two decades. The eight examples that were reviewed for this chapter appear in **Table 12.1** (Faller, 1987; Finkelhor & Williams, 1988; Mathews, Matthews, & Speltz, 1989; Matravers, 2008; McCarty, 1986; Sandler & Freeman, 2007; Sarrel & Masters, 1982; Vandiver & Kercher, 2004).

Each column in Table 12.1 represents a separate typology that was developed using a specific sample of female sexual offenders. For example, Sarrel and Masters' typology resulted from dividing their sample of female sexual offenders into four distinct groups (or "types"). All of the typologies in the table are arranged in rows, to allow for comparison between approaches based on four specific themes regarding victim characteristics (including age and relationship). Thus Sarrel and Masters (1982) examined a sample that included women who had abused adults, or adolescents, or their own children. Further, as can be seen from the table, Sarrel and Masters' typology and Vandiver and Kercher's typology are the only approaches that feature women who have adult victims. By comparison, McCarty's entire sample included women who had abused their own children, and the women in Finkelhor and Williams' study abused unrelated adolescents, exclusively. Nearly all of the typologies emphasize the presence of a co-offender. In Table 12.1, women who offended in the company of men are indicated by shaded cells. That is, five of the eight typologies included women who were accompanied or coerced by men.

Although these approaches are clearly diverse (proposing anywhere from two to six separate categories), some themes are common to each description. This section describes the four most obvious areas of similarity across all typologies: women who abuse adolescent boys exclusively, women who abuse their own or other young children, women who co-offend with men, and women who sexually assault or coerce adults. As indicated by Table 12.1, various studies have revealed even more specific offending circumstances exist, but these four categories are the broadest and most inclusive and provide a useful overview of the nature of sexual offending by women (Harris, 2010).

Women Who Abuse Adolescent Boys

Mary Kate LeTourneau was a 34-year-old married mother of four when she was convicted in 1997 of the statutory rape of a 12-year-old student at the elementary school in Washington where she was a teacher. She was originally given a suspended sentence, which was revoked when she became pregnant by the same boy. LeTourneau

TABLE 12.1 Typologies of Female Sexual Offenders (shaded cells indicate women who offend in the company of men)

	Sarrel & Masters (1982)	McCarty (1986)	Mathews, Matthews, & Spelt (1993)	Matravers (2008)	Faller (1987)	Finkelhor & Williams (1988)	Vandiver & Kercher (2004)	Sandler & Freeman (2007)
Unrelated adult victims only	Dominant women abuse						Aggressive homosexual Homosexual criminal	
Unrelated adolescent victims only	Babysitter Forced assault		Experimenter/exploiter Teacher/lover	**Group offender**	Adolescent abusers	Lone abusers **Multiple perpetrators**	Female sexual predators Heterosexual nurturers Noncriminal homosexual	Non-habitual offender High risk chronic offender Criminally limited hebephile Criminally prone hebephile
Incest *and* unrelated child victims			**Male-coerced**	Lone **Partner offender**	Psychotic abusers **Poly-incestuous**		Young adult child exploiters	Young adult child molester Homosexual child molester
Incest only	Incestuous assaults	Independent abuser **Co-offending mother** **Accomplice-colluded abuser**	Intergenerationally predisposed Psychologically disturbed		Single-parent abusers Non-custodial abusers			

gave birth to two children while serving seven years in custody, and now lives with the boy (whom she married upon release) and their children.

Six of the eight typologies reviewed for this chapter identify a specific group of female sexual offenders who abuse unrelated male adolescents exclusively. Referred to most often as "teacher/lovers" (Mathews et al., 1989) other typologies have labeled such women as "heterosexual nurturers" (Vandiver & Kercher), "adolescent abusers" (Faller, 1987), or "criminally limited hebephiles" (Sandler & Freeman, 2007).

Teacher/lover female sexual offenders typically promote an adolescent boy to adult status and tend not to see the abusive behavior as criminal (Atkinson, 1996; Cortoni, 2009). Instead, they believe the victim is a willing participant in a consensual relationship, or that it is their responsibility to "teach" the child about sexuality, thereby saving the child from apparently inevitable abuse later on (Hunter & Mathews, 1997). This type of offender typically acts from a position of authority that is achieved either through her age or her role in the boy's life (e.g., as his teacher or babysitter). Although the initial sexual encounter is often not premeditated, it might later become the result of calculation and careful planning. Although a large majority of female offenders are survivors of abuse, clinical experience indicates that teacher/lovers are less likely than other subtypes to have been victimized sexually as children. However, research does indicate that they experience verbal or emotional abuse in their childhood and may have had a distant (or absent) father.

Women Who Abuse Young Children

A comparably small but still identifiable group of female sexual offenders target younger, prepubescent children. Some typologies have further separated these women by victim gender (i.e., showing a preference for boys or girls) or by their relationship to the victim (i.e., abusing their own or other children). A key feature of this group is the label that they are "predisposed" to offend in some way. A typical predisposed sexual offender would act alone and victimize her own children, although some may select victims outside the home in addition to their own offspring (Atkinson, 1996; Faller, 1987; Vandiver & Kercher, 2004).

In their roles as wives and mothers, women typically have substantially more domestic opportunities than men to commit undetectable crimes. Performing basic tasks associated with care, control, and socialization, women who abuse young children characteristically do so by disguising their behavior in routine caretaking activities such as bathing, clothing, or feeding a child (Hislop, 2001).

Predisposed offenders are especially likely to have experienced severe childhood trauma or to have long histories of sexual abuse. Their adult relationships are frequently unstable, unhealthy, or abusive, and they tend to experience additional difficulties such as low self-esteem, extreme anger, cognitive distortions, or substance abuse (Cortoni, 2009; Mathews et al., 1989). Of the four types of female sexual offenders described in this chapter, women who abuse young children are the most likely to report having deviant fantasies about their sexual offending, display other sexually deviant behaviors, and use violence during their offense.

Studies of adult survivors indicate that victims of predisposed offenders, who are very young during the abuse, tend to report the greatest difficulty in coping, sense of loss, betrayal, and pain (when compared with other victims of female offenders). This outcome is usually explained by the fact that the offender is so often the victim's mother or other caregiver, or at least someone with whom the victim had an expectation of trust (Hislop, 2001).

Women Who Have Co-offenders

A large proportion of identified female criminals commit their offenses in the company of a male co-offender. Such relationships are especially likely to occur with female sexual offenders, whether coerced or accompanied (Gannon, Rose, & Ward, 2008). Sexual co-offending typically involves women who are in intimate relationships with men who influence or force them to engage in an offense. Although teacher/lovers are clearly overrepresented in the popular media coverage on female sexual offenders, women who co-offend with men tend to account for the largest proportion of female sexual offenders in most typologies (Atkinson, 1996).

The extent to which the woman willingly accompanies the man or is coerced or forced by him exists on a continuum. Recognizing this continuum, some classification schemes describe male-coerced offenders separately from male-accompanied offenders (Syed & Williams, 1996). The main distinction between the two categories is that the male-coerced offender is believed to typically abuse others *only* in the presence of her male coercer. In these situations, the woman's behavior is believed to be motivated by extreme emotional dependency or by a fear of physical punishment or sexual assault at the hands of her partner. By contrast, the accompanied offender tends to participate more actively in the offense and might, over time, come to initiate the sexual abuse herself (Atkinson, 1996).

As might be expected, co-offending women are extremely non-assertive and emotionally dependent upon their male coercers. They often subscribe to extremely traditional gender roles that promote a man's relational and societal dominance over a woman. These women might participate directly in the abuse, or simply facilitate the procurement of victims for sexual activity. The woman's daughter is the most typical victim in these cases, but instances involving their sons or unrelated children are not unheard of (Mathews et al., 1989; Matravers, 2008; McCarty, 1986).

The differences between solo female offenders and female co-offenders (either coerced or accompanied) are unknown, but it has been proposed that meaningful distinctions might exist between them. For example, women who co-offend sexually with men have been found to have a greater number of arrests for nonsexual crimes than those who exclusively engage in solo offending. This finding suggests that women who co-offend with men might be prone toward more general criminality (Vandiver, 2006). So far, it is unknown whether the nonsexual crimes committed by women who co-offend with men are also committed in the company of a male co-perpetrator. This is an interesting gap in the existing body of knowledge regarding female offending and versatility.

Women Who Abuse Adults

Women who target adult victims constitute a small minority of female sexual offenders. For example, only 8.3% of Vandiver and Kercher's (2004) total sample of 471 women were found to offend against adults. "Homosexual criminals" (n = 22) and "aggressive homosexual offenders" (n = 17) were the oldest groups in their typology, with almost all of them (88%) having female victims exclusively. "Homosexual criminals" were described as a more general group of offenders for whom sexual offenses (such as forcing females into prostitution) constituted only a small part of their overall offending behavior. In contrast, "aggressive homosexual offenders" were the most likely to be arrested for a contact sexual assault against an adult. Schwartz and Cellini (1995) have suggested that these women are much less likely than other subtypes to come to the attention of authorities, which might account for their small representation in extant typologies.

An interesting gap that is left by existing typologies relates to women who sexually coerce or assault adult men. A considerable reluctance to acknowledge this phenomenon exists, likely perpetuated by the myths that males cannot be sexually victimized and that men's sexual desire is unremitting. In fact, "Aggressive female seduction [of men] is increasingly portrayed as positive for both the sexually assertive and sophisticated female aggressor and as a reflection on the male partner's desirability and sexual prowess" (Kernsmith & Kernsmith, 2009, p. 604).

Women are, indeed, capable of rape, but substantial obstacles stand in the way of a man disclosing rape or sexual assault by a woman. Whether they are met with disbelief by trusted friends, encounter suspicion by police officers, or simply experience a personal unwillingness or inability to recognize the act as abuse, men are considerably less likely than women to report being victimized. Thus very little is known about the characteristics of female sexual offenders who have adult male victims. This is certainly an area in need of further research.

To summarize, typologies are intended to classify individuals into theoretically useful categories that will facilitate a deeper understanding of their behavior. In terms of clinical utility, descriptions of offender subtypes are perceived as helping to guide therapeutic interventions, management techniques, and treatment alternatives. Although existing typologies provide descriptions of female sexual offenders, they are limited by their use of small, convenient clinical samples. Clear areas of overlap within and between subtypes have been identified, but it bears repeating that each individual will present with unique characteristics that might not be fully represented by a clear-cut classification scheme and further elements are needed to understand female sexual offenders.

SOCIETAL RESPONSE TO FEMALE SEXUAL OFFENDING

The criminal justice system has a long history of dealing inconsistently with criminal women, where they are viewed dichotomously as either not *real* criminals or not *real* women (Worrall, 1990). At one end of the spectrum, many acts committed by women are downplayed or dismissed as not *real* crimes. Exhibitionism by a woman, for example, might more readily be labeled promiscuity, whereas exhibitionism by a man would be labeled "indecent exposure" and would attract a criminal charge (Fehrenbach & Monastersky, 1988). At the other end of the spectrum, particularly heinous or unusual offenses by women draw considerable attention, reinforcing the assumption that the women who *do* offend have transgressed societal expectations in such an outrageous fashion that they can no longer be viewed as *real* women, and are instead portrayed as monsters.

The first perspective can be understood by the "chivalry hypothesis" or paternalism. It is generally attributed to the belief that women are dependent and childlike and need to be protected. Here, women are generally afforded more leniency and shorter sentences than their male counterparts by a criminal justice system that consists largely of men (Cullen & Agnew, 2003).

Although paternalism can result in less severe sanctions, it has also led to harsher penalties, especially for sex-related offenses (Naffine, 1996). Harsh treatment for women in these circumstances is usually either justified as maintaining their socially submissive role or explained to be "for their own good" (Vold, Bernard, & Snipes, 2002) A type of double standard exists within the paternalism of the criminal justice system where there is a simultaneous reluctance to incarcerate women but an eagerness to commit them for treatment (Schwartz & Cellini, 1995). Whereas indeterminate psychiatric hospitalization is a last resort for male offenders, imprisonment is the end of the road for women. Viewing women as in need of "care" rather than "punishment" clearly reinforces the paternalism of the criminal justice system.

The second perspective—that criminal women are not *real* women—is reminiscent of some of the earliest and most primitive explanations of offending. This line of thought suggests that a woman who engages in criminal behavior is doubly deviant; in other words, there is something inherently "wrong" (or at the very least, "unfemale") with her. Where men are regarded as the standard case, women become the unfortunate aberration (Naffine, 1996). In this explanation, female offenders are seen to have deviated from socially construed sexual norms and are trying to be men (Campbell, 1993). Such women have been described as possessing a strongly masculine psychological makeup, being characterized by the presence of physiological inactivity and amorality, or being inherently deceitful, manipulative, calculating, and secretive.

Neither perspective (criminal women as not *real* criminals or criminal women as not *real* women) truly appreciates the capacity of a woman to offend, or the seriousness of her offense. Under such circumstances, neither she nor her victims receive the attention, acceptance, or assistance they deserve. This omission is likely no more evident than in the case of female sexual offending.

Female Sexual Offending and the Media

Much of what the public understands about crime and the criminal justice system comes from popular media. Female sexual offending is no exception. In recent years, this phenomenon has received a growing and increasingly skewed amount of attention in the press. What is actually reported, however, reveals a widespread lack of understanding that both reflects and perpetuates the inconsistent manner in which female sexual offenders are treated by the criminal justice system. Society's general reluctance to accept that women can commit sex crimes against children is documented in articles that describe the prevailing sentiment that "It's not a big deal."

In a systematic review of newspaper coverage of cases of child abuse, Cheit (2003) concluded that the media tend to over-report cases of sexual abuse (when compared to physical or psychological abuse), especially those that involve stranger offenders or abuse outside the home. Further, he found that newspapers overemphasize cases where the offense takes place in a public space (such as a church or school) and when details of the event are considered bizarre, astonishing, or unexpected.

Given these findings, it should come as no surprise that much of the media coverage of female sexual offenders in the United States typically involves teacher/lover cases. Thanks to the skewed attention they receive in the news, when confronted with the concept of a "female sexual offender" many members of the community will instantly call upon the image of a demure and possibly emotionally fragile high school teacher who has slept with a male student.

It is worth noting the many double standards here, where even the language we use implies innocence in the "offender" and consent from the "victim." A female teacher is accused of having *slept with a student* or reported to have *made a mistake with an underage partner.* Of course, when the genders are reversed, male teachers are instead portrayed as having committed *abuse, molestation, or rape of a minor* (Frei, 2008).

This double standard is well demonstrated by the publicity that surrounded Mary Kate LeTourneau (mentioned earlier in this chapter). Her story is referred to in many subsequent cases involving female teachers and male students. In the news media at the time, LeTourneau was described as "a randy Miss" and her 12-year-old *victim* was described as "her grade school lover." The *couple* were reported to be "tangled in an intimate relationship" that was characterized by one journalist as a "steamy 18-month affair" (Flynn, 2002; Gold & Dirmann, 2002; Radue & Kriva, 2002). A consideration of this language again makes one wonder just how different the media's (and the criminal justice system's) response would have been had the genders been reversed.

This kind of media coverage creates an extremely unrealistic impression of the true nature and extent of sexual offending by women. Female sexual offenders are neither harmless nor monsters. Teacher/lovers are overrepresented in the media, and stories of their crimes perpetuate the message that young, attractive, confused, and nonthreatening female teachers are "engaging in affairs" with willing "underage lovers" who are not adversely affected by the experience (Frei, 2008).

The distorted exposure of female sexual offenders in the media has important implications for everyone affected by them. It fails to take seriously the extent of their offenses, it underestimates their capacity for harm, and it perpetuates a double standard regarding gender bias in similar circumstances. Such biased reporting perpetuates community-level denial, it increases the likelihood of disbelief or downplaying of the crime by law enforcement, and it shapes the offenders' capacity for denial during treatment. As Frei (2008) asks, "How can society properly address the issue of sex offending as a comprehensive social problem if the media and general public continue to avoid serious acknowledgment of female perpetration of sex offense as dangerous or abusive?" (p. 495).

Perhaps worst of all is the great disservice that this double standard does to victims. Portraying male sexual offenders as violent predators, while simultaneously failing to recognize the ability of a woman to offend in the same way, sends an unfortunate message to survivors of abuse. Victims of female-perpetrated sexual offending deserve validation, respect, and support that are commensurate with the gravity of their experience. The news media, along with law enforcement officers, treatment providers, educators, parents, and the community at large, would do well to find ways to more responsibly manage this complex population.

We now know that women engage in similar criminal acts as men. But, without overstating the obvious, men and women behave differently. More specifically, a woman's pathway to offending is uniquely female and, consequently, warrants a separate explanation from those that have been generated to explain the behaviors of men. Further, a continued failure to acknowledge the actual extent of male victimization by women serves to further the idea that males always want sex and that coercion of males is acceptable. These myths discourage men from seeking help when they have been victimized because they may fear that they will not be believed or that services are not available for them.

CONCLUSION

This chapter examined the current state of knowledge regarding female sexual offenders. It discussed the nature and extent of this important problem and reviewed eight typological approaches that have assisted in our understanding of these women. Finally, it provided an overview of the double standard applied to these women by the media, the criminal justice system, and the community. The differential treatment that female sexual offenders receive, in comparison to their male counterparts, fails to meet the needs of the community, these women, and, most of all, their victims. As criminologists, clinicians, and community members, we must develop a much deeper understanding of female sexual offenders to enhance our efforts to identify, treat, manage, and understand this small but complicated population.

REFERENCES

Atkinson, J. L. (1996). Female sex offenders: A literature review. *Forum on Corrections Research, 8*, 39–42.

Blackburn, R. (1993). *The psychology of criminal conduct: Theory, research and practice*. Chichester, UK: John Wiley and Sons.

Briggs, F. (1995). *From victim to offender: How child sexual abuse victims become offenders*. Sydney: Allen and Unwin.

Campbell, A. (1993). *Men, women, and aggression*. New York: Basic Books.

Cheit, R. (2003). What hysteria? A systematic study of newspaper coverage of accused child molesters. *Child Abuse and Neglect, 27*, 607–623.

Cortoni, F. (2009). Violence and women offenders. In J. Barker (Ed.), *Women and the criminal justice system: A Canadian perspective* (pp. 175–199). Toronto: Edmond Montgomery.

Cullen, F., & Agnew, R. (2003). *Criminological theory: Past to present, essential readings* (2nd ed). Los Angeles: Roxbury.

Faller, K. (1987). Women who sexually abuse children. *Violence and Victims, 2*, 263–276.

Fehrenbach, P., & Monastersky, C. (1988). Characteristics of female adolescent sexual offenders. *American Journal of Orthopsychiatry, 58*, 148–151.

Finkelhor, D., & Williams, L. M. (1988). Perpetrators. In D. Finkelhor, L. M. Williams, & N. Burns (Eds.), *Nursery crimes: Sexual abuse in day care* (pp. 27– 69). Newbury Park, CA: Sage.

Flynn, B. (2002, April 5). "See me after history class . . . and I'll strip!" *The Sun http://wn.com/See_me_after_class.*

Frei, A. (2008). Media consideration of sex offenders: how community response shapes a gendered perspective. *International Journal of Offender Therapy and Comparative Criminology, 52*, 495–498.

Gannon, T. A., & Rose, M. R. (2008). Female child sexual offenders: Towards integrating theory and practice. *Aggression and Violent Behavior, 13*, 442–461.

Gannon, T. A., Rose, M. R., & Ward, T. (2008). A descriptive model of the offense process for female sexual offenders. *Sexual Abuse, 20*, 352–374.

Gold, S., & Dirmann, T. (2002, May 3). Teacher held in kidnapping: Police discover 33-year-old woman in a Las Vegas hotel room with a 15-year-old boy from her science class. *Los Angeles Timeshttp://articles.latimes.com/2002/may/03/local/me-wanted3.*

Harris, D. A. (2010). Theories of female sexual offending. In T. R. Gannon & F. Cortoni, *Female sexual offenders* (pp. 31-52). London: Wiley and Sons.

Hislop, J. (2001). *Female sex offenders: What therapists, law enforcement and child protective services need to know.* Ravensdale, WA: Idyll Arbor.

Hunter, J. A., & Mathews, R. (1997). Sexual deviance in females. In R. D. Laws & W. O'Donohue (Eds.), *Sexual deviance: Theory, assessment and treatment* (pp. 465– 480). New York: Guilford Press.

Kernsmith, P. D., & Kernsmith, R. M. (2009). Female pornography use and sexual coercion perpetration. *Deviant Behavior, 30*, 589–610.

Mathews, R., Matthews, J. A., & Speltz, K. (1989). *Female sexual offenders: An exploratory study.* Orwell, VT: Safer Society Press.

Matravers, A. (2008). Understanding women who commit sex offenses. In G. Letherby, K. Williams, P. Birch, & M. Cain (Eds.), *Sex as crime?* (pp. 299–320). Devon, UK: Willan.

McCarty, L. M. (1986). Mother–child incest: Characteristics of the offender. *Child Welfare, 65*, 447–458.

Naffine, N. (1996). *Feminism and criminology.* Philadelphia: Temple University Press.

Parkinson, P., Shrimpton, S., Oates, R., Swanston, H., & O'Toole, B. (2004). Nonsex offenses committed by child molesters: Findings from a longitudinal study. *International Journal of Offender Therapy and Comparative Criminology, 48*, 28–39.

Radue, A., & Kriva, C. (2002, May 6). Students address teacher–student relationships in light of area cases. *Milwaukee Journal Sentinel B6.*

Sandler, J. C., & Freeman, N. J. (2007). Typology of female sex offenders: A test of Vandiver and Kercher. *Sexual Abuse, 19*, 73–89.

Saradjian, J. (1996). *Women who sexually abuse children: From research to clinical practice.* Chichester, UK: Wiley.

Sarrel, P. M., & Masters, W. H. (1982). Sexual molestation of men by women. *Archives of Sexual Behavior, 11*, 117–131.

Schwartz, B. K., & Cellini, H. R. (1995). *The sex offender: Corrections, treatment and legal practice.* Kingston, New Jersey: Civic Research Institute.

Strickland, S. (2008). Female sex offenders: Exploring issues of personality, trauma, and cognitive distortions. *Journal of Interpersonal Violence, 23*, 474–489.

Syed, F., & Williams, S. (1996). *Case studies of female sex offenders in the Correctional Service of Canada.* Ottawa: Correctional Service Canada.

Turner, K., Miller, H., & Henderson, C. (2008). Latent profile analysis of offense and personality characteristics in a sample of incarcerated female sexual offenders. *Criminal Justice and Behavior, 35*, 879–894.

Vandiver, D. M. (2006). Female sex offenders: A comparison of solo offenders and co- offenders. *Violence and Victims, 21*, 339–354.

Vandiver, D. M., & Kercher, G. (2004). Offender and victim characteristics of registered female sexual offenders in Texas: A proposed typology of female sexual offenders. *Sexual Abuse, 16*, 121–137.

Vold, G. B., Bernard, T. J., & Snipes, J. B. (2002). Theoretical criminology (5th ed.). New York: Oxford University Press.

Worrall, A. (1990). *Offending women.* London: Routledge.

CHAPTER 13

Gang Involvement and Predatory Crime

Jean Marie McGloin
University of Maryland

Recent decades have witnessed major growth in street gang presence and activity. The National Youth Gang Center estimates that there are approximately 760,000 street gang members and 26,000 street gangs across nearly 3000 jurisdictions in the United States. Their proliferation has not been limited to urban centers—rural and suburban areas across the nation have also seen a dramatic increase in street gangs. At the same time, gangs have shown an expansion of members' age and ethnicity, as well as an increase in female membership (Crane, Boccara, & Higdon, 2000; Curry & Decker, 2003; Howell, 1998; Howell, Moore, & Egley, 2002; Miller, 2001). Today, membership in street gangs is no longer reserved solely for young, minority males.

These apparent quantitative and qualitative shifts in the character of street gangs and gang membership are interesting, but they do not compel the attention of criminologists in isolation. If gangs are simply self-formed groups of individuals who adopt a common name, have some sense of longevity, and have a basic organization, then they would arguably not be in the purview of criminological focus. After all, we rarely find ourselves in the business of studying sports teams or fraternities.

Instead, it is the *criminal* element that invites our attention. Research has long suggested that predatory crime is somehow part, if not a product, of gang life. Indeed, as the prevalence of street gangs increased across the United States, Chicago saw a 500% increase in its number of gang-motivated homicides from 1987 to 1994, while the increase in homicides from 1999 to 2001 in California was fully attributable to an upswing in gang homicides in Los Angeles County (Block, Christakos, Jacob, & Przybylski, 1996; Tita & Abrahamse, 2004).

For such reasons, scholars and law enforcement practitioners alike invest resources in understanding the relationship between gang involvement and offending behavior. This chapter reviews these linkages, paying particular attention to the theoretical mechanism that underlies it, and it considers what may be uniquely criminogenic about gang membership. It also discusses the implications of this linkage, focusing on policy and intervention considerations.

GANG INVOLVEMENT AND THE INCREASED RISK OF OFFENDING

A number of studies have found that gang members tend to be more serious offenders than their non-gang counterparts and that they are responsible for a disproportionate amount of crime. For example, data from the Rochester Youth Development Study reveal that gang members are significantly more likely to report involvement in violence than non-gang members (Thornberry, Krohn, Lizotte, & Chard-Wierschem, 1993). At the same time, 30% of youth in this sample self-reported gang membership, but they were responsible for 65% of the reported delinquent acts over a four-year period (Thornberry, 1998; Thornberry & Burch, 1997). This disproportionate involvement emerged across levels of crime seriousness as well as for violent and property crime.

Similarly, approximately 14% of the subjects in the Denver Youth Study over a four-year period reported being gang members, but they were responsible for 79% of the reported serious violence and 71% of the reported serious property crime (Huizinga, 1997). Gang members in this sample self-reported approximately two to three times more delinquency than non-gang members.

Sara Battin and her colleagues (1998) investigated this pattern within the Seattle Social Development Project. They found that gang members made up 15% of the sample, but were responsible for approximately 58% of the general delinquent acts reported over a four-year period. As with the other studies, results from the Seattle data revealed that this disproportionate involvement spanned an array of crime types, including assault, robbery, and theft (Battin, Hill, Abbott, Catalano, & Hawkins, 1998; Hill, Lui, & Hawkins, 2001).

It is important to highlight three points about these findings. First, despite the varying geographic locations and the range of gang membership prevalence, gang members consistently were more serious delinquents and were responsible for the lion's share of crime. Second, the findings were based on self-reports, so it is not the case that this relationship simply reflects a systematic bias of official records being more likely to capture the behavior of gang members than that of non-gang members. Finally, this trend cuts across types of predatory crime. Thus it is not exclusive to a specialized type of crime, such as robbery, which suggests a general connection between gang membership and offending behavior.

Interestingly, this "risk" of gang involvement for offending appears to exist on a continuum. David Curry, Scott Decker, and Arlen Egley (2002) acknowledge that most gangs are loosely organized and that the process of affiliation or disaffiliation is rarely rapid or spontaneous. Full-fledged gang membership is often a gradual process; accordingly, distinctions can made be made regarding the extent to which individuals are affiliated or associated with street gangs. Thus some individuals may not be distinctly gang members or non-gang members, but their associations and linkages rather place them somewhere in the middle.

Relying on a survey of middle school students in St. Louis, Curry and his colleagues (2002) found that 15% reported being a past or current gang member and nearly 49% reported at least some association with gangs (e.g., had gang members as friends, wore gang colors, flashed gang signs, or hung out with gang members) but did not identify themselves as gang members. Both self-reported and official recorded delinquency showed a rank-order relationship with the level of gang involvement. Gang members reported the most delinquency and were most likely to have court referral records, followed by youth associated with gangs and non-gang youth, respectively. In short, the increased risk of delinquency is not specific to gang membership, but also extends to individuals who associate with street gangs.

Just as the division between gang and non-gang members is not always clear-cut but rather exists on a spectrum, so not all gang members are the same. Research has long confirmed that gang members vary in the extent to which they are committed to or embedded in street gangs. Most often, scholars distinguish between core members, who are involved in many gang activities and have more stable memberships, and peripheral members, who are relatively less involved and have more transient stays in the gang. Some research has moved past this relatively simple dichotomy, however. For example, Jean McGloin (2005a) found that gang members in Newark, New Jersey, occupied a range of social positions and roles in street gang networks. For the purposes of this chapter, it is important to note that the more involved and embedded members tend to be more criminally active than the peripheral, less embedded members. Indeed, research has revealed that being a central member of a delinquent group serves to amplify the individual's criminal behavior (Haynie, 2001). This relationship, along with the previous findings, underscores the robustness of the linkage between gang involvement and predatory crime. Even so, these results do not necessarily reveal what this linkage means.

Some people may assume that gang involvement causes delinquent behavior, but this is not the only possible interpretation of the data. Individuals who gravitate toward gang involvement may be different somehow than individuals who are not attracted to these groups. If so, this linkage reflects enduring individual-level differences in criminal propensity, rather than some detrimental impact of gang involvement. Thus it would seem prudent for theorists, researchers, practitioners and policy makers to study this relationship in more depth.

WHY IS THERE A RELATIONSHIP BETWEEN GANG INVOLVEMENT AND CRIME?

Scholars have long debated the theoretical "meaning" of gang membership in much the same way as they have debated the relationship between having deviant peers and an increased likelihood of offending. Traditionally, this linkage has served as a place of intersection between learning and control theories of crime, which adopt decidedly different viewpoints. Other perspectives, such as opportunity theories, certainly have the ability to comment on the importance of deviant peers with regard to offending, but learning and control theories nonetheless have largely defined and structured this debate.

The *socialization or facilitation model* views gang involvement or membership as causally meaningful. Based primarily on propositions from differential association and social learning theories, it assumes that the normative processes of the gang/delinquent peer group create and sustain delinquents (Akers, 1998; Sutherland, 1947). As a consequence of gang involvement, individuals learn to commit crime because their primary social environment provides access to definitions favoring the commission of crime, to sources of reinforcement for delinquent/criminal behavior, and to a number of delinquent models for observational learning. Simply put, without this social–psychological context, the individual might not engage in delinquency. Thus the gang is criminogenically important and deserves our attention (see **Figure 13.1**).

In contrast, the selection model asserts that gang membership does not hold any causal significance with regard to delinquency. Instead, this model argues that the relationship is spurious; a common factor explains both gang membership and delinquency, as Figure 13.1 demonstrates. Drawing on findings by control theorists, it proposes that an individual will offend when social controls or self-control are weak (Glueck & Glueck, 1950; Gottfredson & Hirschi, 1990; Hirschi, 1969). Under the

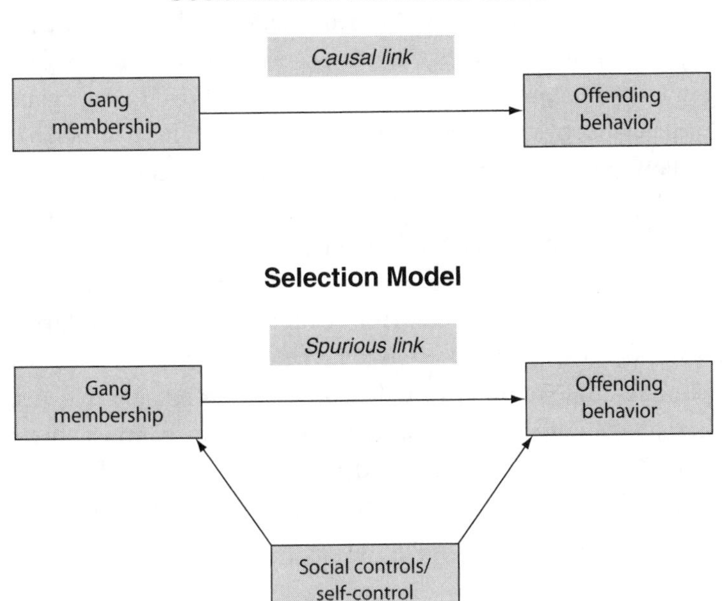

FIGURE 13.1 Models of gang delinquency.

premise of "birds flock together," more serious delinquents self-select into gangs. According to this model, the gang simply reflects the already established tendencies of its members; it does not cause delinquency. As part of their general theory of crime, Michael Gottfredson and Travis Hirschi (1990) state that "adventuresome and reckless children who have difficulty making and keeping friends tend to end up in the company of one another, creating groups made up of individuals who tend to lack self-control. The individuals in such groups will therefore tend to be delinquent, as will the group itself" (p. 158). In turn, Gottfredson and Hirschi assert that low self-control underlies both involvement with street gangs and delinquency.

Some researchers have recognized that couching this discussion in a single, unidirectional pathway is not the most productive enterprise. Rather, it might be more accurate to consider the possibility that controls, learning environments, and delinquency influence one another in a reciprocal manner. Accordingly, Terence Thornberry suggested an interactional perspective that accounts for bidirectional relationships among social controls, delinquent peers, and delinquency. The *enhancement model*, which is influenced by this perspective, proposes that people who are part of gangs may be different than their non-gang counterparts even before membership, but the group processes nonetheless encourage and amplify delinquent behavior

(Elliott & Menard, 1996; Thornberry, 1987). In short, it represents a middle ground, or compromise between the control and learning sides of this theoretical debate.

These perspectives all have conceptual merit and can potentially explain the linkage between gang involvement and a heightened risk of offending. To truly gain insight into which theoretical perspective has the most merit, researchers have turned to longitudinal data. Doing so allows a finer-grained investigation of the connection between the timing of gang membership and criminal behavior, about which the selection and socialization models offer divergent predictions. According to the selection model, when comparing gang members and non-gang members over time, the former should always be more delinquent and there should be no systematic change in this behavior during the time of gang membership. Because gang membership is simply a reflection of a person already being more delinquent due to other causal factors, such as self-control, being an active member of the gang should have no facilitating effect. In contrast, the socialization model argues that these two groups would look very similar before and after gang membership, but that during the time of membership, gang members would be more active in delinquency due to the criminogenic social context.

The Timing of Gang Membership and Delinquency

Finn Esbensen and David Huizinga (1993) investigated the temporal relationship between gang membership and delinquent behavior in an attempt to shed light on this theoretical debate. First, they confirmed the cross-sectional relationship between gang membership and delinquency, finding that gang members were more criminally active than their non-gang delinquent counterparts. Then, turning to four years of available data, Esbensen and Huizinga focused on the timing of the subjects' gang membership, if applicable, and delinquency. Their findings revealed that offending prevalence among gang members was higher than that among non-gang subjects even before membership began, which provides initial support for the selection model. The results also offered some support for the socialization model, however, because the offending prevalence among gang members declined subsequent to their departure from the gang.

When looking at individual-level patterns of offending, Esbensen and Huizinga (1993) discovered that offending rates among gang members were particularly pronounced during the time of membership—though these individuals engaged in more crime than their non-gang counterparts before joining the gang. This divergence became more pronounced during the time of active membership. These researchers' investigation provided the most support for the enhancement model. Although youth who are already more delinquent may gravitate toward gangs, it suggested, there is also something criminogenic about the gang that amplifies their behavior (Esbensen & Huizinga, 1993).

Thornberry and his colleagues (1993) also investigated this premise with male youth from the Rochester Youth Development Study. Their investigation largely provides support for the socialization model, given that gang members had higher rates of crimes against persons (compared to non-gang members) only when these individuals were actively part of the gang.

Gordon and colleagues (2004) relied on 10 years of data from the Pittsburgh Youth Study for their inquiry. They discovered that boys who joined gangs were more delinquent than their non-gang counterparts before they were active members, which provided some support for the selection model. They also found that criminal behavior increased significantly among gang members when they were actively part of the gang and diminished after they left the gang, which strongly supported the enhancement model. Interestingly, this pattern emerged across an array of dependent variables, including self-reports, parent reports, and teacher reports of delinquent behavior.

On the whole, studies vary in the extent to which they provide support for each model, although the enhancement explanation tends to have the strongest standing. It also seems clear that gangs do provide some facilitating effect with regard to delinquency, even if gang members were more criminally active prior to membership. The natural question that follows, then, is why gang membership acts as a facilitator. Given that researchers and theorists often group research results on gang membership with those on deviant peers, it is not unreasonable to ask whether the same mechanism is at work. Does this effect arise simply because gang members are exposed to a greater number of more serious delinquent peers? If so, then there is nothing uniquely criminogenic about the gang or gang membership per se, but rather gangs are simply another form of a delinquent peer group.

In an empirical investigation of this question, Gordon and colleagues (2004) found that exposure to delinquent peers was one socializing mechanism at work for gang members, although it did not completely explain the relationship between gang membership and heightened delinquency. Relying on a different data set, Battin and her colleagues (1998) also investigated whether delinquent peer associations were able to explain the effect gang membership has on criminal involvement, measured both via self-reports and official records. They found that being a gang member exerted an effect on delinquency above and beyond having delinquent friends. Collectively, these two investigations suggest that gang membership has some uniquely criminogenic influence.

WHAT IS UNIQUE ABOUT GANG MEMBERSHIP?

In considering which aspects of gang membership might cause and support increased criminal behavior, a number of possibilities emerge. This section elaborates on one primary possibility—namely, the interactive social processes that are part of gang life. Although a street gang might be argued to be simply one type of delinquent peer group, it is a particularly powerful social network that often embraces predatory and violent behavior. Street gangs have the power to constrain the behavior of their members so that individuals behave in a manner consistent with the values, norms, and symbolic nature of the group (Thornberry et al., 2003).

Individuals who gravitate toward gang involvement are often marginalized from acquiring status in general society via legitimate means (Cloward & Ohlin, 1960; Thrasher, 1927). To be sure, threats to individuals' status often shed insight into the origin of gangs and gang membership; even so, it is important to recognize that vying for status within the gang is also an important process. James Short and his colleagues have argued that studying the social process of negotiating social status within the gang often provides more insight into the behavior of gang members than does considering one's status in reference to society at large (Short & Strotdbeck, 1965; Hughes & Short, 2005). This perspective is somewhat similar to Elijah Anderson's discussion of the disadvantaged underclass, who adapt to their socioeconomic context, which includes the perceived continual threat of violence, by adopting a "code of the street." Part of this code entails the notion that failing to address transgressions or actions of disrespect will merely serve to invite them more often in the future. As in this subculture, the nature of gang life can structure certain expectations and support social processes that facilitate the offending behavior of gang-involved youth (Anderson, 1999).

Hughes and Short (2005) studied gang members in Chicago and noted that "several gang members lost status within the gang after becoming 'too committed' to school or family . . . the gang context was conducive to violence, and for gang youth, willingness to fight was often critical to status management" (p. 56). When Decker reviewed interview data from 99 gang members in St. Louis, he concluded that gang violence was an expressive result of interactive social processes. In particular, he suggested that violent behavior among gang members emerged out of collective behavior, whereby real or perceived "threats" set an escalating, dynamic interaction in motion, which culminated in violence through retaliation. Empirical data have demonstrated support for this argument (Decker, 1996; Decker & Van Winkle, 1996). For example, when Pizarro and McGloin (2006) reviewed more than five years' worth of homicide data in Newark, New Jersey, they discovered that escalating disputes embedded within group contexts predicted gang-related homicides.

In short, offending behavior often becomes an intimate part of acquiring and maintaining social status within the gang, which might help elucidate the mechanism whereby gangs facilitate predatory behavior. Gang membership comes with particular expectations about behavior and interactions—expectations that shape and influence individuals who are involved in street gangs. The initiation for some gang members is an inherently violent process, consisting of being "beat in," which sends a resonating message about the meaning and expected interactional processes of gang life (Vigil, 1996). It is not surprising that gang membership has been established as a correlate and a predictor of gun ownership, gun possession, and gun use (Bjerregaard & Lizotte, 1995; Lizotte, Howard, Krohn, & Thornberry, 1997; Lizotte et al., 2000; Luster & Oh, 2001). Indeed, Terence Thornberry et al. (2003) found that gang members in the Rochester Youth Development Study were approximately 10 times more likely to carry a firearm than non-gang members. Consistent with the previous discussion, the often-cited reasons for such activities include protection/self-defense from other gangs, intimidation, and status.

Unfortunately, these violent expectations often have dire consequences. In a rather grim commentary, Decker and Janet Lauritsen (2006) followed up 99 gang members five years after interviewing them. Of the 51 subjects found, 19 were dead, 2 were in prison, and 4 were in wheelchairs.

The label of "gang member" or "gang involved" may also spark assumptions of an individual's behavior from those outside the gang. For instance, Esbensen and his colleagues (1993) found that gang members and non-gang members look quite similar across an array of psychosocial variables, yet individuals in the former group are more likely to perceive having been labeled as bad or disturbed by their teachers. Accordingly, individuals involved in gangs may not only perceive expectations from within the gang about social interactions and behavior that are tied to status and respect, but might also perceive expectations from those in the larger society to act or behave in a violent fashion (Esbensen, Huizinga, & Weiher, 1993). Under the premise of labeling theory, this perception may set a self-fulfilling prophecy in motion. Taken together, external labels and internal expectations can interact and cumulatively push an individual toward predatory crime as he or she becomes involved in street gangs (Lemert, 1951). This discussion is only one example of what might make gang involvement uniquely risky, but it provides some points for consideration.

POLICY IMPLICATIONS

That gang members are responsible for a disproportionate amount of crime clearly warrants intervention strategies and resources directed toward these groups. The notion that something unique about gang involvement amplifies individual-level offending behavior further underscores this focus. A vast array of gang prevention and intervention programs exist, even though our sense of "what works" is fairly limited. It would seem relatively straightforward to determine which strategies are best suited to combat gangs and prevent gang membership, yet programs rarely define their problem of interest with clarity (e.g., intimidation of the public by gang members, an increased homicide rate attributed to gangs, an increasingly violent drug market related to an influx of gangs, an increase in commercial robberies attributed to a small group of gang members), let alone complete a process or outcome evaluation (McGloin, 2005b). The purpose of this chapter is not to review the vast array of intervention strategies, however, but rather to elaborate on the implications for anti-gang programs derived from the linkage between gang involvement and predatory crime.

The first implication is that prevention strategies are clearly warranted. Prevention programs are directed at the general population, which stands in stark contrast to intervention strategies that are directed at at-risk populations or groups. As this chapter has illustrated, gang membership has a crime-facilitating effect—one that extends to individuals who associate with gangs as well. Therefore, directing policies or programs only at those individuals who are already gang members or gang associates would miss the opportunity to prevent these linkages and reduce the risk of increased offending.

One relatively well-known prevention program is Gang Resistance and Education Training (G.R.E.A.T.), a school-based curriculum in which law enforcement officers act as instructors. The middle school portion of the curriculum comprises a 13-week course in which lessons center on preventing gang involvement and delinquency, while promoting life skills and prosocial ties within families and the community. Results from an early cross-sectional evaluation indicated that this program had some moderate success, but a longitudinal evaluation was less promising (Esbensen et al., 2001). Although the efficacy of G.R.E.A.T. is not clear, it is important to note that programs that attempt to prevent the development of linkages to gangs do and should exist.

Not all individuals who are linked to a gang are equal in their connections. Some people are merely associates. Even within the category of "gang member," the extent to which people are involved and embedded in the gang varies. With this point in mind, as well as the positive relationship evidenced between the extent of gang involvement and offending behavior, it would be unwise to subject all gang-involved youth to the same intervention strategy. For example, peripheral members might be persuaded to terminate their gang involvement more easily, perhaps through a social services–based strategy, when compared to core/central gang members, who may require more law enforcement resources to break the gang's hold on them. With regard to the former group, social services providers could intervene and potentially prevent individuals from becoming further enmeshed in the gang, thereby reducing the heightened risk of becoming more serious offenders. At the same time, levels of involvement in the gang may reflect gradations in the "dangerousness" posed to public safety, further suggesting a benefit to differentiating among the types and extent of linkages to street gangs.

Mark Fleisher (2002) has argued that "gangs are social networks composed of individual gang members, and that gang member behavior is determined in part by a gang member's location in the structure of the social network. That location in the social network structure determines opportunities and constraints that expand or limit a gang member's choices" (p. 200). Not surprisingly, then, recent work has argued that intervention and suppression techniques should be cognizant of social positions within the gang and not treat all individuals in the same manner. Programs would be well served by seeing gangs as social networks in which positions and the extent of "gang involvement" vary.

One manner in which to do so is to appeal to a particular method during the problem analysis phase—namely, network analysis. The primary purpose of network analysis is to reveal the presence of any regular patterns in social relationships: Rather than focusing on attributes of individuals, it focuses on the associations or linkages among people (Knoke & Kuklinski, 1982; Wasserman & Faust, 1994).

McGloin (2005a) recently described a network analysis of gang members in Newark, New Jersey. The data included information on various social connections among gang members, including who was related to whom, who was incarcerated in the same correctional facility, and who hung out and committed crimes together. As part of this analysis, it became clear that individuals exist on a spectrum of involvement with the gang, across a range of linkages. This kind of problem analysis could directly inform policy and prevention decisions about how to allocate resources. For example, a core member who is central to the street gang and heavily involved in serious crime warrants law enforcement attention. This need can be further underscored if this core person is the primary linkage to the gang for a number of peripheral members. Under such hypothetical circumstances, focusing on this person would address an individual who is responsible for a disproportionate amount of crime and, coincidentally, would cleave other persons' linkage to gang involvement, potentially ending the facilitating context for delinquency (McGloin, 2005a).

It is also important to be cautious when designing such programs so as to ensure that interventions do not have the unintended consequence of strengthening an individual's linkage to the gang. Law enforcement attention on gangs, through arrest, patrol, surveillance, and other suppressive tactics, can serve as a powerful external source of cohesion. It provides the group with a common point of conflict and solidified identity, setting a self-fulfilling prophecy in motion (Klein & Crawford, 1968). Such a risk is not limited to suppressive tactics. For example, the Los Angeles Group Guidance Project offered tutoring, counseling, and recreational activities to local gangs in an attempt to integrate street gang members into the prosocial community. Unfortunately, this program provided the gang members with more opportunities to socialize and become more linked to one another, producing a more cohesive gang that was involved in more crime. In short, it is terribly important for policy makers and those part of an intervention to consider the level of gang involvement when planning a strategy and to continually assess whether that involvement is changing as a consequence of the tactics (either for better or for worse).

CONCLUSION

This chapter discussed the positive correlation between gang involvement and predatory crime. Although scholars have long known that such a link exists, in terms of both an individual's increased likelihood of offending and gang members' responsibility for a disproportionate amount of crime, they have contested the meaning of this relationship. Whereas the socialization/facilitation model argues that gang membership plays a causal role in offending behavior, the selection model maintains that this relationship merely reflects more serious offenders self-selecting into gang membership.

The evidence presented in this chapter suggests that gangs do have a facilitating effect on delinquency, above and beyond simple exposure to delinquent peers. The unique criminogenic nature of gang involvement compels us to carefully consider intervention strategies that are cognizant of the gradations that exist regarding gang involvement. When tailored appropriately, such interventions may disproportionately affect the crime rates, as well as evidence greater success through the intelligent allocation of resources.

REFERENCES

Akers, R. L. (1998). *Social learning and social structure: A general theory of crime and deviance.* Boston: Northeastern University Press.

Anderson, E. (1999). *Code of the streets.* New York: W. W. Norton.

Battin, S. R., Hill, K. G., Abbott, R. D., Catalano, R. C., & Hawkins, J. D. (1998). The contribution of gang membership to delinquency beyond delinquent friends. *Criminology, 36,* 93–115.

Bjerregaard, B., & Lizotte, A. J. (1995). Gun ownership and gang membership. *Journal of Criminal Law and Criminology, 86,* 37–58.

Block, C. R., Christakos, A., Jacob, A., & Przybylski, R. (1996). *Street gangs and crime: Patterns and trends in Chicago.* Chicago: Illinois Criminal Justice Information Authority.

Cloward, R. A., & Ohlin, L. E. (1960). *Delinquency and opportunity: A theory of delinquent gangs.* New York: Free Press.

Crane, J., Boccara, N., & Higdon, K. (2000). The dynamics of street gang growth and policy response. *Journal of Policy Modeling, 22,* 1–25.

Curry, G. D., & Decker, S. H. (2003). *Confronting gangs: Crime and the community* (2nd ed.). Los Angeles, CA: Roxbury.

Curry, G. D., Decker, S. H., & Egley, A. (2002). Gang involvement and delinquency in a middle school population. *Justice Quarterly, 19,* 275–292.

Decker, S. H. (1996). Collective and normative features of gang violence. *Justice Quarterly, 13,* 243–264.

Decker, S. H., & Lauritsen, J. (2006). Leaving the gang (pp. 21–40). In A. Egley, C. L. Maxson, J. Miller, & M. W. Klein (Eds.), *The gang reader* (3rd ed., pp. 21–40). Los Angeles, CA: Roxbury.

Decker, S. H., & Van Winkle, B. (1996). *Life in the gang: Family, friends, and violence.* New York: Cambridge University Press.

Elliott, D. S., & Menard, S. (1996). Delinquent friends and delinquent behavior: Temporal and developmental patterns. In J. D. Hawkins (Ed.), *Delinquency and crime: Current theories* (pp. 28–67). New York: Cambridge University Press.

Esbensen, F., & Huizinga, D. (1993). Gangs, drugs, and delinquency in a survey of urban youth. *Criminology, 31,* 565–587.

Esbensen, F. A., Huizinga, D., & Weiher, A. W. (1993). Gang and non-gang youth: Differences in explanatory factors. *Journal of Contemporary Criminal Justice, 9,* 94–116.

Esbensen, F. A., Osgood, D. W., Taylor, T. J., et al. (2001). How great is G.R.E.A.T.? Results from a longitudinal, quasi-experimental design. *Criminology and Public Policy, 1,* 87–118.

Fleisher, M. S. (2002). Doing field research on diverse gangs: Interpreting youth gangs as social networks. In C. R. Huff (Ed.), *Gangs in America, III* (pp. 199–218). Thousand Oaks, CA: Sage.

Glueck, S., & Glueck, E. (1950). *Unraveling juvenile delinquency.* Cambridge, MA: Harvard University Press.

Gordon, R. A., Lahey, B. B., Kawai, E., et al. (2004). Antisocial behavior and gang membership: Selection and socialization. *Criminology, 42,* 55–87.

Gottfredson, M. R., & Hirschi, T. (1990). *A general theory of crime.* Stanford, CA: Stanford University Press.

Haynie, D. L. (2001). Delinquent peers revisited: Does network structure matter? *American Journal of Sociology, 106,* 1013–1057.

Hill, K. G., Lui, C., & Hawkins, J. D. (2001). *Early precursors of gang membership: A study of Seattle youth.* Washington, DC: Office of Juvenile Justice and Delinquency Prevention.

Hirschi, T. (1969). *Causes of delinquency.* Berkeley, CA: University of California Press.

Howell, J. C. (1998). *Youth gangs: An overview.* Washington, DC: Office of Juvenile Justice and Delinquency Prevention.

Howell, J. C., Moore, J. P., & Egley, A. (2002). The changing boundaries of youth gangs. In C. R. Huff (Ed.), *Gangs in America, III* (pp. 3–18). Thousand Oaks, CA: Sage.

Hughes, L. A., & Short, J. F. (2005). Disputes involving youth street gang members: Micro-social contexts. *Criminology, 43,* 43–76.

Huizinga, D. H. (1997). *Gangs and the volume of crime.* Paper presented at the annual meeting of the Western Society of Criminology, February 14–16, 1997, Honolulu, HI.

Klein, M. W., & Crawford, L. Y. (1968). Groups, gangs, and cohesiveness. In F. Short (Ed.), *Gang delinquency and delinquent subcultures* (pp. 63–75). New York: Harper & Row.

Knoke, D., & Kuklinski, J. H. (1982). *Network analysis.* Thousand Oaks, CA: Sage.

Lemert, E. M. (1951). *Social pathology: Systematic approaches to the study of sociopathic behavior.* New York: McGraw-Hill.

Lizotte, A., Howard, G. J., Krohn, M. D., & Thornberry, T. P. (1997). Patterns of illegal gun carrying among young urban males. *Valparaiso Law Review, 31*, 375–394.

Lizotte, A. J., Krohn, M. D., Howell, J. C., et al. (2000). Factors influencing gun carrying among young urban males over the adolescent–young adult life course. *Criminology, 38*, 811–834.

Luster, T., & Oh, S. M. (2001). Correlates of male adolescents carrying handguns among their peers. *Journal of Marriage and the Family, 63*, 714–726.

McGloin, J. M. (2005a). Policy and intervention considerations of a network analysis of street gangs. *Criminology and Public Policy, 4*, 607–636.

McGloin, J. M. (2005b). *Street gangs and interventions: Innovative problem solving with network analysis.* Washington, DC: Office of Community Oriented Policing Services.

Miller, W. B. (2001). *The growth of youth gang problems in the United States: 1970–1998.* Washington, DC: Office of Juvenile Justice and Delinquency Prevention.

Pizarro, J., & McGloin, J. M. (2006). Gang homicides in Newark: Collective behavior or social disorganization? *Journal of Criminal Justice, 34*, 195–207.

Short, J. F., & Strotdbeck, F. L. (1965). *Group process and gang delinquency.* Chicago: University of Chicago Press.

Sutherland, E. H. (1947). *The principles of criminology* (4th ed.). Philadelphia: Lippincott.

Thornberry, T. P. (1987). Toward an interactional theory of delinquency. *Criminology, 25*, 863–891.

Thornberry, T. P. (1998). Membership in youth gangs and involvement in serious and violent offending. In R. Loeber & D. P. Farrington (Eds.), *Serious and violent juvenile offenders: Risk factors and successful interventions* (pp. 147–166). Thousand Oaks, CA: Sage.

Thornberry, T. P., & Burch, J. H. (1997). *Gang members and delinquent behavior.* Washington, DC: Office of Juvenile Justice and Delinquency Prevention.

Thornberry, T. P., Krohn, M. D., Lizotte, A. J., & Chard-Wierschem, D. (1993). The role of juvenile gangs in facilitating delinquent behavior. *Journal of Research in Crime and Delinquency, 30*, 55–87.

Thornberry, T. P., Krohn, M. D., Lizotte, A. J., et al. (2003). *Gangs and delinquency in developmental perspective.* New York: Cambridge University Press.

Thrasher, F. (1927). *The gang: A study of 1,313 gangs in Chicago.* Chicago: University of Chicago Press.

Tita, G., & Abrahamse, A. (2004). *Gang homicide in L.A., 1981–2001.* Perspectives on Violence Prevention, Number 3. Sacramento, CA: California Attorney General's Office.

Vigil, J. D. (1996). Street baptism: Chicano gang initiation. *Human Organization, 5*, 149–153.

Wasserman, S., & Faust, K. (1994). *Social network analysis: Methods and applications.* New York: Cambridge University Press.

CHAPTER 14

The Heterogeneity of Predatory Behaviors in Sexual Homicide

Eric Beauregard
Simon Fraser University

The first obstacle to characterizing sexual homicide is the absence of any legal definition of this crime. When investigators attempt to solve a murder and decide whether it is sexual in nature, they rely on two types of information: (1) testimony, from either the murderer or someone else, and (2) physical evidence, of which the crime scene is the key source. According to Robert Ressler and his colleagues (1988), a murder can be considered sexual if at least one of the following conditions is present:

- The victim is found totally or partially naked.
- The genitals are exposed.
- The body is found in a sexually explicit position.
- An object has been inserted into a body cavity (anus, vagina, mouth).
- There is evidence of sexual contact.
- There is evidence of substitutive sexual activity (e.g., masturbation and ejaculation at the crime scene) or of sadistic sexual fantasies (e.g., genital mutilation).

The main obstacle to the exploitation of this type of evidence is that some police officers do not recognize its sexual character (Ressler, Burgess, & Douglas, 1988).

Another problem related to the definition of sexual homicide has been the failure of some researchers to take into account differences in the sex and age of the victims. In several cases, in fact, murders of women, men, and children have been considered to belong to a single category, despite the lack of evidence that these types of sexual homicide are part of a homogenous phenomenon. On the contrary, the absence of such evidence makes it essential that definitions of sexual homicide consider the age and sex of the victims.

This chapter addresses different issues related to the heterogeneity of sexual murderers and sexual homicides. The first section deals with the different types of sexual murderers of women and the existing classifications based on their criminal behaviors. The second section addresses the sexual murder and murderers of men. After looking at some of their differences and similarities with sexual murderers of women, this section ends by suggesting a classification specific to sexual murderers of men. Finally, the third section investigates sexual murderers of children and their crimes, especially the differences and similarities they share with sexual murderers of women.

SEXUAL MURDERERS OF WOMEN

To date, 10 typological studies of sexual murderers have been conducted (Beauregard & Proulx, 2002; Beech, Fisher, & Ward, 2005; Beech, Robertson, & Clarke, 2001; Clarke & Carter, 1999; Folino, 2000; Keppel & Walter, 1999; Kocsis, 1999; Meloy, 2000; Ressler et al., 1988; Revitch & Schlesinger, 1981). These typological studies reveal four types of sexual murderers, with each study describing between two and four types. Two types of sexual murderers are consistently reported in the various studies: angry and sadistic. In addition, some sexual murderers kill to eliminate witnesses. Finally, some studies include a residual category that includes types such as the "power-assertive," "power-reassurance," "mixed," and the "neuropsychological dysfunction."

Despite terminological variations from study to study, sadistic and angry murderers present a number of consistent characteristics. To illustrate this point, **Table 14.1** and **Table 14.2** list the primary characteristics of sadistic and angry sexual murderers, respectively.

Sexual murderers who kill the victim to eliminate any witnesses have as their primary intent the sexual assault of their victim; the murder is merely instrumental to this motive. This type of murderer is often described as less likely to have had long-term emotional relationships, and his phallometric (sexual interests measured by penile plethysmography) profile is deviant. His victims are unknown to him and often younger than age 30, and his sexual assault is characterized by coitus and some sadistic elements. The murder may or may not be premeditated, and may be panicky or cold-blooded, depending on the offender's prior criminal experience. Usually, the victim's wounds are restricted to a single site on the body, and the victim is found lying on her

TABLE 14.1 Summary of the Characteristics of the Sadistic Sexual Murderer

Characteristics of the Murderer

- mobility
- high intelligence
- post-secondary education
- preference for work placing him in contact with authority or death
- cruelty towards animals
- enuresis during childhood
- fire-setting
- tendency to be isolated
- antisocial, narcissistic, schizoid, and obsessive-compulsive personality disorders
- severe psychopathy
- lack of empathy
- history of breaking and entering
- history of sexual crimes
- fascination for objects related to police work
- possession of violent pornography and detective magazines
- paraphilias: cannibalism, vampirism, necrophilia, fetishism, masochism, transvestitism, voyeurism, exhibitionism, obscene telephone calls
- sexual sadism
- serial sexual crimes

Pre-Crime Phase

- impressive amounts of unexpressed aggression
- very elaborate sadistic fantasies
- fantasy world more important to him than reality
- situational stress
- often murders after a blow to his self-esteem
- "hunts" his prey
- selection of a specific victim following surveillance
- premeditation of crime
- consumption of alcohol prior to the crime

Modus operandi

- ruse and manipulation to approach the victim
- victim unknown to him
- isolated crime scene, chosen in advance, far from his residence
- vehicle used in commission of crime
- presence of instruments of torture or of a rape kit
- modus operandi reflects sadistic fantasies
- victim held captive for several hours, with recording of aggression
- consumption of alcohol during the crime

(continued)

TABLE 14.1 (Continued)

- submission of victim demanded
- victim tied up and gagged
- fellatio by victim
- anal or vaginal penetration of victim
- possible sexual dysfunction
- insertion of objects in various body cavities
- prolonged and ritualized torture
- mutilation of genital organs
- non-random pattern of wounds on the victim's body
- pre-mortem mutilation
- death by strangulation
- sexual arousal elicited by violent acts committed on the victim, culminating in the murder
- dismembering
- retention of souvenirs belonging to the victim or taken from her

Post-Crime Phase

- moving of the corpse
- hiding of the corpse
- modus operandi reflects control
- absence of weapons or evidence at the crime scene
- interest in media coverage of the crime
- possible change of job or city after the crime
- may volunteer to help during the investigation
- relatively normal behavior between crimes
- absence of remorse for his acts
- pleasure in describing the horror of his acts
- low profile during incarceration

back. Often, the crime is committed, and the victim's corpse found, at the site at which first contact between the murderer and the victim occurred.

The typological studies of sexual homicide conducted to date suffer from several limitations. First, most of the studies encompass a wide spectrum of sexual murderers (serial and non-serial murderers; sexual murderers of women, men, and children), completely ignoring the specific characteristics of each group. Moreover, some studies have limited themselves to a single source of data, such as clinical observations, self-reported information, or data from official sources, which precludes the possibility of using multiple sources of data to create a complete and detailed profile of the sexual murderer and his crimes. Additionally, while some studies have proposed models in which the murderer's motivation and personality have been inferred, none has identified a standardized method that permits adequate evaluation of these elements.

TABLE 14.2 Summary of the Characteristics of the Angry Sexual Murderer

Characteristics of the Murderer

- mid-20s
- residence and workplace near crime scene
- average intelligence
- married or in a stable relationship
- little work experience
- socially incompetent
- not socially isolated
- difficulties with authority
- impulsivity
- anger
- selfish
- various non-specific personality disorders
- low to moderate psychopathy
- presence of severe mental disorders such as mood disorders
- history of violence towards women
- history of substance abuse
- no consumption of pornography
- sexual dysfunctions
- absence of sadistic sexual fantasies

Pre-Crime Phase

- depressive mood
- suicidal ideation
- feelings of anger
- reduced alcohol consumption prior to the crime
- victim selected from murderer's daily activities
- victim selected from a familiar setting
- victim stance
- rage displaced towards victim
- desire to kill
- absence of premeditation

Modus operandi

- access to crime scene on foot
- victim is known to murderer
- victim older than murderer
- crime scene outdoors
- crime scene known to murderer
- use of a weapon found at the crime scene
- minimal use of physical restraints

(continued)

TABLE 14.2 (Continued)

- reduced consumption of alcohol during the crime
- anxiety during crime
- explosive and violent attack
- vengeance is displaced to a specific victim
- blows specifically directed towards victim's face
- wounds on several areas of the bodies
- humiliation and extreme violence
- possible sexual assault of the victim
- insertion of objects into various body cavities of the victim
- absence of sperm at the crime scene
- murder provoked by the victim's words or actions
- murder by strangulation
- post-mortem sexual acts and mutilation
- overkill

Post-Crime Phase

- feelings of relief
- corpse of victim left at the crime scene
- corpse left in view
- body found on its back
- crime scene reflects lack of organization
- no interest for media coverage of the crime

Furthermore, most of the studies—regardless of whether their approaches are rooted in clinical psychology or law enforcement—pay little attention to pre-crime-phase variables, including situational factors that may have precipitated the sexual homicide. Finally, several authors have constructed their typology on the basis of clinical judgment, which precludes any verification of the validity of their analyses.

In light of the limitations just described, Beauregard and Proulx (2002) developed a typology of sexual murderers of adult women based on objective crime-scene criteria and multivariate analyses. This analysis yielded two subgroups of sexual murderers—namely, sadistic and angry.

Sadistic Sexual Murderers

Sadistic murderers premeditate their crimes, and they select victims previously unknown to them. During the crime, these murderers humiliate their victims, in many cases physically restrain them, and mutilate them. The crime lasts more than 30 minutes, which increases the risk of being caught by police. This type of murderer tends to move or hide the victim's corpse in an attempt to avoid its detection. Moreover, sadistic sexual murderers sometimes dismember and hide their victim's corpse, select victims with specific characteristics, oblige their victims to commit sexual acts, and torture their victims. Nevertheless, sadistic murderers are less likely to leave their victims on their backs than angry murderers.

Sadistic murderers present a positive affect prior to the crime, such as sexual arousal, calmness, or feelings of well-being, and they report having deviant sexual fantasies prior to their crime. Their positive affect may be related to deviant sexual fantasies and may be the expression of strong feelings associated with the thrill associated with the hunt and capture of the victim. Further, it may be hypothesized that premeditation and deviant sexual fantasies favor the execution of certain acts (e.g., torture, mutilation, dismemberment) that instantiate the imagined crime scenario.

Angry Sexual Murderers

Angry murderers do not premeditate their crimes, and their victims are circumstantial. They do not usually humiliate, physically restrain, or mutilate their victims. The crime lasts less than 30 minutes, which decreases the risk of being caught by the police. The victim's corpse is often left at the crime scene. Furthermore, angry murderers present an anger affect prior to crime and report having experienced problems of loneliness and idleness prior to crime.

The differences in the relational problems of the two types of sexual murderers reflect differences in the personality disorders they experience. Our results demonstrate that angry murderers typically have a borderline personality disorder, indicative of a desire for contact with other people concomitant with an inability to obtain and maintain intimacy (Beck & Freeman, 1990, p. 396). In contrast, sadistic murderers exhibit schizoid and avoidant personality disorders, suggesting that they have renounced contact with other people in favor of fantasy worlds. This being so, it is understandable that sadistic murderers do not report problems of loneliness prior to committing their crimes.

As this discussion suggests, different types of sexual murderers of women can be identified as illustrated by their different offending patterns (Beauregard, Proulx, Brien, & St-Yves, 2005). Most studies on the classification of sexual homicide have identified types congruent with the sadistic and the angry pathways. However, as mentioned previously, these studies have mixed together sexual murderers of women, men, and children. But are they really all the same? When looking at studies on sexual murderers, the sexual homicide of women was the most frequently investigated;

consequently, little research has been conducted on other types of sexual murderers, especially on sexual murderers of men.

SEXUAL MURDERERS OF MEN

Sexual murderers of men commit crimes that fit the same definition of sexual homicide given by Ressler et al. (1988). However, Drake (1999) identified other elements typically present in cases of homosexual homicide: death does not occur on the offender's territory, no signs of forced entry, body found in the bedroom, signs of overkill and arson, concomitant robbery, and victim lived alone.

Homosexual Homicide Typology

To date, only one typology of sexual homicide of men has been developed. Based on his investigative experience, Geberth (1996) suggested a six-type classification of homosexual homicide. The first type of crime identified in this model comprises "interpersonal violence-oriented disputes and assaults," which are the most common type of sexual homicide of men. These murders are mostly the result of disputes between partners, ex-partners, or love triangles. This type of homosexual homicide may include instances where "ground rules" are not respected by one of the parties involved in the sexual activity (e.g., an older male attempting to carry the sexual activities beyond what has been negotiated). Often, these homicides are committed in a context of prostitution, where the prostitute or hustler denies being a homosexual and responds with extreme violence to this threat to his masculinity.

The second type of homosexual homicide identified by Geberth comprises "murders involving forced anal rape and/or sodomy." These murderers are usually sexually motivated, but they do not experience any sexual gratification associated with the killing. Death occurs mainly from the amount of force used to overcome the victim's resistance or to prevent identification.

The third type identified is the "lust murder." Such homicides often present evidence of sadism and mutilation to the victim's genitals. The crime is meticulously premeditated according to the deviant sexual fantasies of the offender. According to Geberth, the offender will exhibit several characteristics in keeping with Hare's description of a psychopath, such as deception, superficial charm, and callousness.

"Homosexual serial murders" represent the fourth type identified by Geberth. These criminals hunt for vulnerable victims who are easy to control, such as children and prostitutes. Their homicides involve lust murders, thrill killings, and child killings as well as robbery homicides that are homosexually oriented. The crimes may be characterized by acts of mutilation and dismemberment of the victim's body, performed to shock those who will find the corpse, to facilitate its transportation, or simply to prevent the victim's identification.

In the "robbery/homicide of homosexuals" (the fifth type in Geberth's model), offenders hunt for potential victims engaging in high-risk behaviors (e.g., cruising) in locations known to be frequented by some homosexuals (e.g., gay bars, saunas). Others, either alone or in a group, will use homosexual prostitution as a vehicle to assault or rob a gay customer who is willing to pay to have sex. Finally, "homophobic assaults and gay bashing" incidents (the sixth type) are performed by individuals showing an intense hatred for homosexuals.

The typology by Geberth is interesting, particularly from an investigative viewpoint, because the main characteristics concern the crime scene. However, this perspective neglects the pre-crime phase and the characteristics of the offenders. To better understand sexual murderers of men, Beauregard and Proulx (2007) have described a group of men who have killed another man in a sexual context and proposed a new classification for this particular type of sexual murderer.

Characteristics of Sexual Murderers of Men

Victimology

In cases of sexual homicide of men, the victims are generally older than their offenders. The murderers in the sample of Beauregard and Proulx (2007) were, on average, 27.8 years old at the time of the crime, compared to 33 years old for the victims. These findings are in agreement with other studies on homosexual violence, which report that victims of homosexual homicides are usually older than the offender (Sagarin & Macnamara, 1975; Tremblay, Boucher, Ouimet, & Biron, 1998).

Two-thirds of the sexual murderers in Beauregard and Proulx's sample killed a victim who was of homosexual orientation. Even if this particular result is not reported in other studies on sexual homicide, it suggests that the routine activities of some victims may put them more at risk for such crimes. A single homosexual man may look for sexual contact with unknown partners in gay bars, thereby increasing his risk of being victimized by a predator. Moreover, in half of the cases in the sample, victims were living alone. According to Marcus Felson (2002), the absence of a "capable guardian" is a factor that increases the probability of a crime being committed by a motivated offender. Knowing that the victim lives alone assures the offender that there will be no witnesses and no one to interfere during the crime.

Finally, in almost half of the cases, the victims had used drugs or alcohol prior to the crime. This consumption of drugs or alcohol, combined with a desire to have sexual contact with another man, may have reduced the victims' ability to assess the potential dangers of bringing home a stranger. Furthermore, it is possible that victims under the influence of drugs or alcohol may be more vulnerable because they are less able to defend themselves during an attack.

Pre-crime Factors

The majority of sexual murderers of male victims used alcohol and/or drugs before committing their crimes. It can be hypothesized that the consumption of drugs or alcohol serves as a disinhibitor, bringing to the fore anger and violence in some sexual murderers of men. Interestingly, this consumption of drugs or alcohol is similar to the behavior demonstrated by some sexual murderers of adult female victims from Beauregard and Proulx's (2007) study. Other sources have reported that substance abuse in individuals who commit homicide and sexual assault is more serious than in the general population and could contribute to the crime by unleashing rage as well as stimulating sexual desire (Langevin et al., 1988; Yates, Barbaree, & Marshall, 1984).

Interestingly, Beauregard and Proulx's (2007) study showed that in almost half of the cases, financial problems prior to the crime were reported by sexual murderers of men. This result, which is particular to this study, may be related to a motivation apart from sexual or vindictive motives for these sexual murderers. For some sexual murderers in the sample, sex was used to attract a victim for an intended theft crime, but because of unplanned situational elements (e.g., inability to control the victim), the planned theft degenerated into murder. Burglary has been linked to sexual homicide by other authors (Schlesinger & Revitch, 1999).

Finally, only a minority of sexual murderers in the sample admitted to deviant sexual fantasies prior to the crime. This result corresponds with observations of sexual murderers of female victims, in that only a minority of these crimes are sexually motivated. For instance, Langevin and his colleagues (1988) found that 31% of sexual murderers reported sexual gratification as a motivation, whereas 69% identified a fusion of anger and sexuality. In Beauregard and Proulx's (2002) study, 33% of sexual murderers of female victims reported having deviant sexual fantasies before committing the crime.

Crime Characteristics

Concerning crime characteristics, it is noteworthy that almost all crimes committed by sexual murderers of male victims were premeditated. For instance, Ressler and his colleagues (1988) found that 86% of organized sexual murderers premeditated their offense compared to 44% for disorganized offenders. In Grubin's (1994) study, one-third of his sample planned the offense; in contrast, in Beauregard and Proulx's (2002) study, only 5% of sexual murderers who used the anger pathway premeditated the crime compared to 81% of sexual murderers who used the sadistic pathway.

The use of a weapon is a crime characteristic that distinguishes sexual murderers of male victims from sexual murderers of female victims. The majority of the former offenders used a weapon to commit their crime in Beauregard and Proulx's research. Similarly, in the 37 cases of homosexual serial homicides reported by Geberth (1996), 35 victims were killed either by a firearm or a knife. According to Miller and Humphreys (1980), a knife was used in 54% of cases of homosexual homicides, and in 19% of cases a firearm was used, often after being stabbed and assaulted. In sexual murderers of women, the preferred method used to kill the victim is strangulation.

Two possible explanations could account for such a difference. First, male victims possess greater physical strength than female victims, and offenders may want to prevent opposition or resistance. Second, the use of a weapon may be directly related to the premeditation of the crime, with the weapon being used primarily to threaten or control the victim during the burglary or sexual assault.

Mutilations rarely occurred in the male sexual murders documented in Beauregard and Proulx's study—a result that contrasts drastically with the literature on homicide of men, where mutilations, and especially overkill, are typical characteristics. Moreover, a study by Bell and Vila (1996) revealed that overkill was greater among homosexual victims of homicide compared to heterosexual victims. One possible explanation for this discrepancy is that mutilation and overkill have not been well documented in the available data.

Cases of sexual homicide usually present a certain degree of organization. Along with premeditation of the crime and use of a weapon, 60% of sexual murderers reported having moved or hidden the body after the crime. Hiding or moving the victim's body after the crime usually represents an attempt to delay body discovery and conceal evidence. Moreover, it may provide more time for the offender to flee the crime scene, come up with a good alibi, or even move to another city or another state. In cases of sexual homicides of men, it is possible that high organization is related to the crime location, such as the offender's or the victim's residence. In those cases, it is crucial to move the victim's body to avoid being associated with the crime.

A PROPOSED TYPOLOGY OF SEXUAL MURDERERS AGAINST MEN

Descriptions of sexual murderers of adult male victims provide some understanding of this particular type of sexual homicide and its main features. Separate descriptive analysis of each variable, however, does not offer a clear picture of the context in which the crimes are committed. To remedy this shortcoming, the classification of sexual murderers who killed men proposed here takes into account the entire criminal event of the sexual homicide, including the study of the offender, the victim, and the context of the crime (Meier, Kennedy, & Sacco, 2001).

The Avenger

Individuals corresponding to the avenger type of sexual murderer are usually involved in prostitution activities and can be of homosexual, heterosexual, or bisexual orientation. The consumption of drugs and alcohol is an important feature of these murders. Most have been convicted of property and violent crimes before the murder. Most also have experienced psychological, physical, and sexual abuse during childhood. Moreover, the type of sexual activity requested by the client in a prostitution context, or a triggering event occurring during or after the sexual exchange, may elicit memories from the abuse, unleash the rage of the offender, and lead to the homicide. In this sense, the offender is avenging himself directly on his current partner for all the grievances (present or past) of which he feels he has been a victim. The homicide is preceded by anger; it is usually committed by strangulation or by using a weapon of opportunity (kitchen knife, a pillow, phone cord). Because of the rage of this offender, evidence of expressive violence can be found on the victim, who is usually older than the offender.

The Sexual Predator

The second type of sexual murderer of male victims identified is the sexual predator. This offender is mainly motivated by deviant sexual fantasies. Homosexual in sexual orientation, he presents criminal antecedents of sexual crimes, especially against male children or adolescents. For these sexual murderers, the sexual assault and the homicide are premeditated. Most often, the targeted victim is an adolescent or a young man, unknown to the offender, and not necessarily of homosexual orientation. The offending process starts with the abduction and confinement of the victim, and sadistic acts (mutilations, sodomy, and humiliation) are performed during the crime. Evidence of expressive violence is found on the victim's body. The crime usually lasts more than 30 minutes, but it can endure for as long as 24 hours in some cases.

The Nonsexual Predator

The nonsexual predator is motivated by neither anger nor deviant sexual fantasies. The homicide is not planned, but rather is more accidental or instrumental in nature. The principal motivation for the crime is to rob the victim, and sex is used solely as a means of entrapment. In most cases, this type of sexual homicide can be described as a robbery that degenerated into the death of the victim because of his resistance.

Predation is related to the choice of victim: Often, this offender uses the visibility and the homosexual orientation of the victim to seduce him and to bring him to an isolated area (usually the victim's residence) where he will be able to commit his crime without interference. His hunting field is established according to the availability of targets and potential victims. Thus gay bars, "cruising" bars, and places reputed to be frequented by homosexuals are attractive to this offender. The victim is chosen for his vulnerability (e.g., lives alone, reluctant to report a robbery to the police, feels guilty after being manipulated), his easy access (lives in or goes often to the gay district), and his visibility (gay bars and gay district are usually frequented by homosexuals, some of whom are very open about their sexual orientation).

The nonsexual predator murderer may be heterosexual or homosexual; he may act alone or with the help of an accomplice. The crime may be committed with a weapon that is found at the crime scene or one that is brought by the offender. Violence is instrumental in the sense that it serves to commit the burglary or to overcome the victim's resistance. Usually, the victim is not sexually assaulted, but sexual contact may occur between the offender and the victim so as to manipulate the victim; that is, the sexual contact serves to trap the victim. The crime phase is generally of short duration, with the offender leaving the crime scene right after the homicide. Sexual murderers of this type may use alcohol or drugs prior to the crime. Often, they have a diversified criminal career with an emphasis on crimes against property.

Beauregard and Proulx (2007) have suggested that homosexual men are victimized mainly because of their situational vulnerability. Life habits or routine activities of these men may have increased their risk of becoming a victim of sexual homicide. Routine activities theory demonstrates that the probability that a crime will occur depends on three factors: an attractive target, the absence of a capable guardian, and the presence of a motivated offender. As can be seen from Beauregard and Proulx's results as well as in other studies, male victims of sexual homicide generally live alone, they have used alcohol or drugs before the crime, they are reputed to not report certain crimes to the police, and some of them go to gay bars to get acquainted with a stranger or to "cruise." In turn, certain types of criminals choose these hunting fields to select an attractive and vulnerable target.

Sexual murderers of male victims exhibit a variety of motivating factors. Some are motivated by revenge, others by profit or sadistic fantasies. Interestingly, revenge and profit are motivations that have not been found in the sexual homicide of women.

Moreover, these motivations are directly related to specific contexts (e.g., financial difficulties, prostitution)—a relationship that once again highlights the importance of looking at the entire criminal event in sexual homicide.

Sexual murderers of men share some similarities with those who kill women. Their offending patterns and motivations vary, however, and are specific to this type of murderer, once again pointing to the heterogeneity of sexual murderers and sexual homicides. Individuals who kill children in a sexual context seem also to exhibit a specific offending pattern mainly influenced by different routine activities and a sadistic motivation.

SEXUAL MURDERERS OF CHILDREN

Mainly because of the apparently low frequency of this type of crime, empirical research on sexual homicide and sexual murderers of children has been limited. As mentioned previously, most of the research conducted on sexual homicide and sexual murderers has focused on those individuals who killed an adult woman. Because of the difficulties related to collecting information on these offenders, very few empirical studies have looked specifically at sexual homicide and sexual murderers of children.

Sexual Homicide of Children

The selection of a child victim could be related to the fact that children represent easy, weak, vulnerable, or available targets. In some instances, the motivation underlying may be that some individuals present a deviant sexual preference and are sexually aroused and gratified by the suffering and the killing of a child victim (Lanning, 1994). This type of pedophile molests children with the express desire to physically harm the victims. Typically, the victim is a stranger and may be stalked or abducted from places where children tend to gather (e.g., playgrounds, schools, shopping centers). Such a crime is often premeditated and ritualized, and the victim's body may be mutilated. Moreover, these offenders report few contacts with children outside of their offense, rank low on social competences, and are more likely to be classified as sadists (Holmes, 1991; Holmes & Holmes, 2002).

Comparative Studies of Sexual Murderers of Children

To date, only three comparative studies involving sexual murderers of children have been conducted. Results from these studies revealed that sexual murderers had significantly higher scores on the Psychopathy Checklist Revised (PCL-R) instrument;

had a greater incidence of psychosis, personality disorders (antisocial), paraphilias (sadism), and addictions; and showed more deviant phallometric responses to depictions of sexual assaults of children and adults. Results also suggested that sexual murderers significantly more often victimized strangers and had been charged or convicted in the past of violent nonsexual and sexual offenses as compared to a nonhomicidal group of sex offenders (Firestone, Bradford, Greenberg, & Larose, 1998; Firestone, Bradford, Greenberg, & Nunes, 2000).

According to Firestone et al. (1998), "there is a limited amount of psychological research available on men who commit sexual murders, and no distinction has been made between those who have victimized adults and those who victimized children" (p. 306). Thus, to better understand sexual murderers of children, Beauregard and colleagues (2008) conducted comparisons of these individuals with a group of sexual murderers of women on developmental, pre-crime, crime, and post-crime factors.

Developmental Factors

Beauregard and his colleagues observed that sexual murderers of women report significantly more daydreaming and enuresis (during childhood) as compared to sexual murderers of children. As to the life and sexual history variables, a larger proportion of sexual murderers of women present with frequent alcohol and drug abuse problems as compared to sexual murderers of children. Sexual murderers of children, however, are significantly more inclined to report experience of sexual abuse in childhood and deviant sexual fantasies than sexual murderers of women.

These last two results are congruent with the attachment model of the development of sexual deviance (Ward, Hudson, Marshall, & Siegert, 1995). According to this model, attachment reflects the bond between child and parent that provides the necessary security and confidence for the child to explore his or her world. The presence of negative childhood experiences—such as sexual abuse—may prevent the development of a secure attachment (Cicchetti & Lynch, 1995). The failure to develop a secure attachment can lead to psychosocial deficits such as low self-esteem and lack of skills necessary to establish adequate relationship with peers (Marshall, Hudson, & Hodkinson, 1993). Owing to this difficulty in relating to peers, the individual may seek alternative means of fulfilling emotional and sexual needs in ways that do not challenge these deficits.

Sexual scripts such as child molestation or even sexual homicide may appeal to these offenders because they make no demands on the self-confidence and social skills that these persons lack, and because they may be interpreted as actions that can provide the illusion of intimacy without fear of being rejected (Marshall & Eccles, 1993). Such scripts may be obtained through a social learning process by being exposed to or being a victim of sexual abuse (Laws & Marshall, 1990). Furthermore, the scripts may be used during masturbatory activities, thereby pairing deviant sexual fantasies with orgasm and creating a conditioning process (Abel & Blanchard, 1974).

In summary, negative childhood experiences, especially sexual abuse, may be seen as developmental risk factors leading to the offender's sexual preference for a child (Lussier, Beauregard, Proulx, & Nicole, 2005). It is noteworthy that experience of sexual abuse during childhood and deviant sexual fantasies are also characteristics of sadistic offenders (Gratzer & Bradford, 1995).

The higher prevalence of deviant sexual fantasies in sexual murderers of children may be related to the motivations underlying sexual homicide. It may be hypothesized that sexual murderers of children are mainly sadistic, accounting for the importance of deviant sexual fantasies (Proulx, Blais, & Beauregard, 2006). Sadists retreat from relationships with adults, where they typically have not been successful, and flee into a world of sexually coercive fantasies. Due to the amount of time they dedicate to their fantasies, they become elaborate and form an outlet for their unexpressed emotional states—namely, rage, humiliation, and suffering (Proulx, McKibben, & Lusignan (1996). If the sadist experiences unusually intense stressful events, however, coercive sexual fantasies may prove insufficient as a coping strategy. The nature of this stress could be a generalized conflict, low self-esteem, or a feeling of rejection, as in the case of sexual murderers of children in the sample described here. Actualizing his fantasies through a sadistic sexual offense constitutes another coping strategy, which the sadist resorts to in an effort to deal with his internal distress.

In terms of their criminal careers, sexual murderers of children present with significantly more prior convictions for sexual crimes without contacts (e.g., indecent exposure, voyeurism, obscene phone calls) as compared to sexual murderers of women. Interestingly, this last result is congruent with the findings from previous studies on recidivism. Individuals who commit sexual crimes against children present a higher risk for sexual recidivism (Hanson, Steffy, & Gauthier, 1993). Hence, Hanson (2002) showed that the recidivism rate of rapists dropped gradually with age in adulthood, whereas for child molesters, it remained steady until the late forties. According to Lussier (2005), "it is not surprising then, that their criminal repertoire tends to include a more important proportion of sexual crimes" (p. 283).

Pre-crime Factors

Pre-crime factors include both predisposing and precipitating factors. Precipitating factors (which occur in the 48 hours prior to commission of the crime) are disinhibitors that are conceived as factors that favor sexual crimes. Predisposing factors (which occur one year or more prior to commission of the crime) represent an obstacle repetitively met by the individual in his life or a zone of vulnerability that has triggered the development of one or many ineffective coping strategies, such as avoidance, denial of emotions, or sexual deviance.

Sexual murderers of children report more perception of rejection and a generalized conflict, such as opposition or avoidance conducts toward a real or symbolic group of individuals, in the year prior to the homicide as opposed to the sexual murderers of women. This finding is congruent with Lanning's explanation of sexual homicide of children; he states that in many cases, the use of lethal violence may be due to poor social and interpersonal skills of the offender. Because of this difficulty of interacting with others, especially with adults, those offenders may possibly target children because they are weak, vulnerable, and available (emotional congruence).

This result also concurs with the finding that sexual murderers of children reported more generalized conflict compared to sexual murderers of women. According to Ward and Beech (2005), perceived rejection and generalized conflict "can be viewed in terms of attachment insecurity leading to problems establishing intimate relationships with adults" (p. 55). Some sexual offenders with a *disorganized* attachment style "are likely to use sexual offending as one of several possible strategies of externally based control in response to the intense negative emotional states which are the sequelae of such an attachment style" (p. 55).

In the 48 hours prior to commission of the crime, sexual murderers of women are significantly more prone to use alcohol and target a victim who is under the influence of drugs and/or alcohol. Almost all sexual murderers of children (90%) report prior contacts with the victim as compared to 56% of the sexual murderers of women. Further, a significantly larger proportion of sexual murderers of children use pornography and are unemployed prior to committing the crime as compared to sexual murderers of women.

Crime Factors

The study by Beauregard et al. tested crime factors including crime-phase variables and crime-scene characteristics. Results showed that adult women are more often killed at night as compared to children. However, premeditation of the crime is significantly more frequent for sexual murderers of children as compared to sexual murderers of women.

Regarding crime-scene characteristics, sexual murderers of children use strangulation to kill their victim, and they dismember and conceal the victim's body more often as compared to sexual murderers of women. These modus operandi characteristics, in addition to deviant sexual fantasies, are largely congruent with results from Dietz, Hazelwood, and Warren (1990), who conducted what is probably the most complete study on the issue of the crime phase of sadistic offenders. Their results showed that crimes of sadists were planned (93%), and the victim was usually unknown to the offender (83%). Victims may be tortured (100%), tied up and gagged (87%), and subjected to diverse sexual acts, including sodomy (73%), fellatio (71%), and vaginal intercourse (57%). When victims are killed, the means is often asphyxia (58%), with the corpse subsequently being concealed (65%). Furthermore, hiding the victim's body is an associated feature of sadism.

Finally, some of the significant differences between sexual murderers of children and sexual murderers of women can be better understood by adopting a routine activity approach. For predatory crime, which usually depends on direct physical contact between the offender and the crime target, the routine activity approach emphasizes the importance of the daily activities of offenders and targets. As illustrated by Beauregard et al.'s results, sexual murderers of women are more likely than sexual murderers of children to (1) be characterized by drug and/or alcohol abuse and dependence, (2) be characterized by frequent consumption of alcohol prior to the crime, (3) target a victim who is under the influence of drugs and/or alcohol, and (4) commit the crime at night. Thus, as part of their routine activities, these offenders are more likely to encounter potential victims in places where consumption of drugs or alcohol occurs. Hence, it is possible that these offenders will meet the victim in a bar, at a club, or at a party where a potential victim has been consuming alcohol or drugs, thereby increasing her risk and vulnerability.

Sexual murderers of children present a different pattern. Pedophile murderers were more likely to be unemployed prior to the crime, to have had prior contact with the victim, and to have used pornography prior to the crime as compared to sexual murderers of women. It may be hypothesized that sexual murderers of children spend some of their leisure time at home watching pornography and then go to places where potential victims gather (e.g., playgrounds, schools, and convenience stores) to establish prior contacts with victims (e.g., grooming). They wait for an opportunity, such as the absence of parents or guardians, and then attract the child victim to an isolated location where they commit the assault and homicide. A recent study demonstrated that the type of victim (e.g., child versus adult woman) would influence the hunting field of offenders (i.e., the type of area where offenders hunt for victims), which in turn is influenced by the victim's routine activities (Beauregard, Rossmo, & Proulx, 2007).

Interestingly, sexual murderers of children do not seem to present especially diverse offending patterns. Most demonstrate the core characteristics and predatory behaviors of the sadistic sex offender. Notably, however, situational factors leading to the commission of the crime are specific to this particular type of victim, which clearly suggests that sexual murderers of children are different from those individuals who kill a woman or a man in a sexual context.

CONCLUSION

Sexual homicide is a crime of a relatively rare occurrence, but it is the sexual crime that attracts the most attention from the media and the community, mainly due to apparently random victim selection. Moreover, this particular form of crime is difficult to understand because it links sex, which is normally associated with pleasure and creation, with the painful, destructive act of killing. To make sense of these horrible crimes, it is important to investigate the offending processes of these murderers and the *entire* criminal event, which includes the offender, the victim, and the context of the crime. The results presented in this chapter clearly demonstrate that, depending on the type of victims (women, men, or children), sexual murderers seem to be driven by different criminal motivations and will exhibit different predatory behaviors linked to specific routine activities.

A better understanding of the different types of sexual murderers is beneficial for at least two important actors of the criminal justice system: corrections officials and police. First, knowledge of the different offending processes associated with sexual murderers may help clinicians involved in treatment and in risk evaluation to better identify their offense cycle and factors that may be linked to high-risk situations. Second, the police may use this information in the criminal profiling process. The different offending pathways identified and their related characteristics may help law enforcement better understand what happened at the crime scene and suggest a potential portrait of the offender (Beauregard, 2007).

REFERENCES

Abel, G. G., & Blanchard, E. B. (1974). The role of fantasy in the treatment of sexual deviation. *Archives of General Psychiatry, 30*, 467–475.

Beauregard, E. (2007). The role of profiling in the investigation of sexual homicide. In J. Proulx, E. Beauregard, M. Cusson, & A. Nicole (Eds.), *Sexual murderers: A comparative analysis and new perspectives* (pp. 193–212). Chichester, UK: Wiley.

Beauregard, E., & Proulx, J. (2002). Profiles in the offending process of non-serial sexual murderers. *International Journal of Offender Therapy and Comparative Criminology, 46*, 386–399.

Beauregard, E., & Proulx, J. (2007). A classification of sexual homicide against men. *International Journal of Offender Therapy and Comparative Criminology, 5*, 420–432.

Beauregard, E., Proulx, J., Brien, T., & St-Yves, M. (2005). Colérique et sadique, deux profils de meurtriers sexuels [Anger and sadistic, two profiles of sexual murderers]. In J. Proulx, M. Cusson, E. Beauregard, A. Nicole (Eds.), *Les meurtriers sexuels: Analyse comparative et nouvelles perspectives* (pp. 203–232). Montréal, Québec: Les Presses de l'Université de Montréal.

Beauregard, E., Proulx, J., Stone, M., & Michaud, P. (2008). Sexual murderers of children: Developmental, pre-crime, crime, and post-crime factors. *International Journal of Offender Therapy and Comparative Criminology, 52*, 253–269.

Beauregard, E., Rossmo, D. K., & Proulx, J. (2007). A descriptive model of the hunting process of serial sex offenders: A rational choice approach. *Journal of Family Violence, 22*, 449–463.

Beck, A. T., & Freeman, A. (1990). *Cognitive therapy of personality disorders.* New York: Guilford Press.

Beech, A. D., Fisher, D., & Ward, T. (2005). Sexual murderers' implicit theories. *Journal of Interpersonal Violence, 20*, 1366–1389.

Beech, A. D., Robertson, D., & Clarke, J. (2001). *Towards a sexual murder typology.* Paper presented October 31–November 3, 2001, at the Association for the Treatment of Sexual Abusers, San Antonio, TX.

Bell, M. D., & Vila, R. I. (1996). Homicide in homosexual victims: A study of 67 cases from the Broward County, Florida, medical examiner's office (1982–1992), with special emphasis on overkill. *American Journal of Forensic and Medical Pathology, 17*, 65–69.

Cicchetti, D., & Lynch, M. (1995). Failures in the expectable environment and their impact on individual development: The case of child maltreatment. In D. Cicchetti & D. J. Cohen (Eds.), *Developmental psychopathology, Vol. 2: Risk, disorder, and adaptation* (pp. 32–71). New York: John Wiley.

Clarke, J., & Carter, A. (1999). *Sexual murderers: Their assessment and treatment.* Paper presented October 29–November 1, 1999, at the meeting of the Association for Treatment of Sexual Abusers, Orlando, FL.

Dietz, P. E., Hazelwood, R. R., & Warren, J. (1990). The sexually sadistic criminal and his offenses. *Bulletin of the American Academy of Psychiatry and the Law, 18*, 163–178.

Felson, M. (2002). *Crime and everyday life* (3rd ed.). Thousand Oaks, CA: Sage.

Firestone, P., Bradford, J. M., Greenberg, D. M., & Larose, M. R. (1998). Homicidal sex offenders: Psychological, phallometric, and diagnostic features. *Journal of the American Academy of Psychiatry and the Law, 26*, 537–552.

Firestone, P., Bradford, J. M., Greenberg, D. M., & Nunes, K. L. (2000). Differentiation of homicidal child molesters, nonhomicidal child molesters, and nonoffenders by phallometry. *American Journal of Psychiatry, 157*, 1847–1850.

Firestone, P., Bradford, J. M., Greenberg, D. M., et al. (1998). Homicidal and non-homicidal child molesters: Psychological, phallometric, and criminal features. *Sexual Abuse: Journal of Research and Treatment, 10*, 305–323.

Folino, J. O. (2000). Sexual homicides and their classification according to motivation: A report from Argentina. *International Journal of Offender Therapy and Comparative Criminology, 44*, 740–750.

Geberth, V. (1996). *Practical homicide investigation: Tactics, procedures, and forensic techniques* (3rd ed.). Boca Raton, FL: CRC Press.

Gratzer, T., & Bradford, M. W. (1995). Offender and offense characteristics of sexual sadists: A comparative study. *Journal of Forensic Sciences, 40*, 450–455.

Grubin, D. (1994). Sexual murder. *British Journal of Psychiatry, 165*, 624–629.

Hanson, R. K. (2002). Recidivism and age: Follow-up data from 4673 sexual offenders. *Journal of Interpersonal Violence, 17*, 1046–1062.

Hanson, R. K., Steffy, R. A., & Gauthier, R. (1993). Long-term recidivism of child molesters. *Journal of Consulting and Clinical Psychology, 61*, 646–652.

Holmes, R. M. (1991). *Sex crimes.* Newbury Park, CA: Sage.

Holmes, S. T., & Holmes, R. M. (2002). *Sex crimes: Patterns and behavior* (2nd ed.). Newbury Park, CA: Sage.

Keppel, R. D., & Walter, R. (1999). Profiling killers: A revised classification model for understanding sexual murder. *International Journal of Offender Therapy and Comparative Criminology, 43*, 417–437.

Kocsis, R. N. (1999). Criminal profiling of crime scene behaviors in Australian sexual murders. *Australian Police Journal, 53*, 113–116.

Langevin, R., Ben-Aron, M. H., Wright, P., et al. (1988). The sex killer. *Annals of Sexual Research, 1*, 263–301.

Lanning, K. V. (1994). Sexual homicide of children. *APSAC Advisor, 7*, 40–44.

Laws, D. R., & Marshall, W. L. (1990). A conditioning theory of the etiology and maintenance of deviant sexual preference and behavior. In W. L. Marshall, D. R. Laws, & H. E. Barbaree (Eds.), *Handbook of sexual assault: Issues, theories, and treatment* (pp. 209–229). New York: Plenum.

Lussier, P. (2005). The criminal activity of sexual offenders in adulthood: Revisiting the specialization debate. *Sexual Abuse: Journal of Research and Treatment, 17*, 269–292.

Lussier, P., Beauregard, E., Proulx, J., & Nicole, A. (2005). Developmental factors related to deviant sexual preferences in child molesters. *Journal of Interpersonal Violence, 20*, 999–1017.

Marshall, W. L., & Eccles, A. (1993). Pavlovian conditioning processes in adolescent sex offenders. In H. E. Barbaree, W. L. Marshall, & S. M. Hudson (Eds.), *The juvenile sex offenders* (pp. 118–142). New York: Guilford Press.

Marshall, W. L., Hudson, S. M., & Hodkinson, S. (1993). The importance of attachment bonds in the development of juvenile sex offending. In H. E. Barbaree, W. L. Marshall, & S. M. Hudson (Eds.), *The juvenile sex offenders* (pp. 164–181). New York: Guilford Press.

Meier, R. F., Kennedy, L. W., & Sacco, V. F. (2001). Crime and the criminal event perspective. In R. F. Meier, L. W. Kennedy, & V. F. Sacco (Eds.), *The process and structure of crime: Criminal events and crime analysis* (pp. 1–27). New Brunswick, NJ: Transaction Publishers.

Meloy, R. J. (2000). The nature and dynamics of sexual homicide: An integrative review. *Aggression and Violent Behavior, 5*, 1–22.

Miller, B., & Humphreys, L. (1980). Lifestyles and violence: Homosexual victims of assault and murder. *Qualitative Sociology, 3*, 169–185.

Proulx, J., Blais, E., & Beauregard, E. (2006). Sadistic sexual aggressors. In W. L. Marshall, Y. M. Fernandez, L. E. Marshall, & G. A. Serran (Eds.), *Sexual offender treatment: Controversial issues* (pp. 61–77). Winchester, UK: Wiley.

Proulx, J., McKibben, A., & Lusignan, R. (1996). Relationship between affective components and sexual behaviors in sexual aggressors. *Sexual Abuse: Journal of Research and Treatment, 8*, 279–289.

Ressler, R. K., Burgess, A. W., & Douglas, J. E. (1988). *Sexual homicide: Patterns and motives.* New York: Lexington Books.

Revitch, E., & Schlesinger, L. B. (1981). *Psychopathology of homicide.* Springfield, IL: Charles C. Thomas.

Sagarin, E., & Macnamara, D. E. (1975). The homosexual as a crime victim. *International Journal of Criminology and Penology, 3*, 13–25.

Schlesinger, L. B., & Revitch, E. (1999). Sexual burglaries and sexual homicide: Clinical, forensic, and investigative considerations. *Journal of the American Academy of Psychiatry and the Law, 27*, 227–238.

Tremblay, P., Boucher, E., Ouimet, M., & Biron, L. (1998, January). Rhetoric of overvictimization: The gay village as a case study. *Canadian Journal of Criminology*, 1–20.

Ward, T., & Beech, A. R. (2006). An integrated theory of sexual offending. *Aggression and Violent Behavior, 11*, 44–63.

Ward, T., Hudson, S. M., Marshall, W. L., & Siegert, R. (1995). Attachment style and intimacy deficits in sexual offenders: A theoretical framework. *Sexual Abuse: Journal of Research and Treatment, 7*, 317–335.

Yates, E., Barbaree, H. E., & Marshall, W. L. (1984). Anger and deviant sexual arousal. *Behavioral Therapy, 15*, 287–294.

PART II

Policy and Practice

CHAPTER 15

Focused Deterrence Strategies and the Reduction of Gang and Group-Involved Violence

Anthony A. Braga
Rutgers University and Harvard University

A number of jurisdictions have been experimenting with new problem-oriented frameworks to understand and respond to gun violence among gang-related and criminally active group-involved offenders. These interventions are based on the "pulling levers" focused deterrence strategy, which directs criminal justice and social service attention toward the small number of chronic offenders who are responsible for the bulk of urban gun violence problems (Braga, Kennedy, & Tita, 2002). While the research evidence on the crime prevention value associated with this approach is still developing, the pulling levers strategy has been embraced by the U.S. Department of Justice as an effective approach to crime prevention. In his address to the American Society of Criminology, former National Institute of Justice Director Jeremy Travis (1998) announced that the pulling levers approach has made "enormous theoretical and practical contributions to our thinking about deterrence and the role of the criminal justice system in producing safety."

Pioneered in Boston, Massachusetts, to halt youth violence, the pulling levers framework has been applied in many U.S. cities through federally sponsored violence prevention programs such as the Strategic Alternatives to Community Safety Initiative and Project

Safe Neighborhoods (Coleman, Holton, Olson, Robinson, & Stewart, 1999; Dalton, 2002). In its simplest form, the approach consists of selecting a particular crime problem, such as youth homicide; convening an interagency working group of law enforcement practitioners; conducting research to identify key offenders, groups, and behavior patterns; framing a response to offenders and groups of offenders that uses a varied menu of sanctions ("pulling levers") to stop them from continuing their violent behavior; focusing social services and community resources on targeted offenders and groups to match law enforcement prevention efforts; and directly and repeatedly communicating with offenders to make them understand why they are receiving this special attention (Kennedy, 2006).

The first section of this chapter introduces the development of the pulling levers focused deterrence approach in Boston. The subsequent sections consider the available evaluation evidence on the effectiveness of pulling levers in Boston, speculate on the violence reduction mechanisms that may be associated with the Boston approach, and examine experiences with pulling levers focused deterrence strategies in other jurisdictions. The chapter concludes by reviewing the key elements that make up these promising gang and group-involved violence reduction strategies.

THE BOSTON GUN PROJECT AND OPERATION CEASEFIRE

The Boston Gun Project was a problem-oriented policing enterprise expressly aimed at taking on a serious, large-scale crime problem—homicide victimization among young people in Boston. Like many large cities in the United States, Boston experienced a large sudden increase in youth homicide between the late 1980s and early 1990s. The Boston Gun Project began in early 1995 and implemented what is now known as the "Operation Ceasefire" intervention, which began in the late spring of 1996 (Kennedy, Piehl, & Braga, 1996). The working group of law enforcement personnel, youth workers, and Harvard University researchers diagnosed the youth violence problem in Boston as one of patterned, largely vendetta-like ("beef") hostility among a small population of chronic offenders, and particularly among those involved in some 61 loose, informal, mostly neighborhood-based groups. These 61 gangs included between 1100 and 1300 members, representing fewer than 1% of the city's youth between the ages of 14 and 24. Although small in number, these gangs were responsible for more than 60% of youth homicides in Boston.

The Operation Ceasefire focused deterrence strategy was designed to prevent violence by reaching out directly to gangs, saying explicitly that violence would no longer be tolerated, and backing up that message by "pulling every lever" legally available when violence occurred (Kennedy, 1997). The chronic involvement of gang members in a wide variety of offenses made them, and the gangs they formed, vulnerable to a coordinated criminal justice response. Authorities sought to disrupt street drug activity, focus police attention on low-level street crimes such as trespassing and public drinking, serve outstanding warrants, cultivate confidential informants for medium- and long-term investigations of gang activities, deliver strict probation and parole enforcement, seize drug proceeds and other assets, ensure stiffer plea bargains and sterner prosecutorial attention, request stronger bail terms (and enforce them), and bring potentially severe federal investigative and prosecutorial attention to gang-related drug and gun activity.

Simultaneously, youth workers, probation and parole officers, and later churches and other community groups offered gang members services and other kinds of help. These partners also delivered an explicit message that violence was unacceptable to the community and that "street" justifications for violence were mistaken. The Ceasefire Working Group disseminated this message in formal meetings with gang members (known as "forums" or "call-ins"), through individual police and probation contacts with gang members, through meetings with inmates at secure juvenile facilities in the city, and through gang outreach workers. The deterrence message was not a deal with gang members to stop violence. Rather, it was a promise to gang members that violent behavior would evoke an immediate and intense response. If gangs committed other crimes but refrained from violence, the normal workings of police, prosecutors, and the rest of the criminal justice system dealt with these matters. If gang members hurt people, however, the Working Group concentrated its enforcement actions on their gangs.

The Operation Ceasefire "crackdowns" were not designed to eliminate gangs or stop every aspect of gang activity, but rather to control and deter serious violence (Kennedy, 1997). To do so, the Working Group explained its actions against targeted gangs to other gangs: "This gang did violence, we responded with the following actions, and here is how to prevent anything similar from happening to you." The ongoing Working Group process regularly monitored the city for outbreaks of gang violence and framed any necessary responses in accord with the Ceasefire strategy. As the strategy unfolded, the Working Group continued communication with gangs and gang members to convey its determination to stop violence, to explain its actions to the target population, and to maximize both voluntary compliance and the strategy's deterrent power.

A central hypothesis within the Working Group was the idea that a meaningful period of substantially reduced youth violence might serve as a "firebreak" and result in a relatively long-lasting reduction in future youth violence. The idea was that youth violence in Boston had become a self-sustaining cycle among a relatively small number of youth, with objectively high levels of risk leading to nominally self-protective behaviors such as gun acquisition and use, gang formation, tough "street" behavior, and the like—behavior that then became an additional input into the cycle of violence. If this cycle could be interrupted, it was thought, a new equilibrium at a lower level of risk and violence might be established, perhaps without the need for continued high levels of either deterrent or facilitative intervention (Kennedy et al., 1996).

Evaluation Evidence

A large reduction in the annual number of youth homicides in Boston followed immediately after Operation Ceasefire was implemented in mid-1996. A U.S. Department of Justice (DOJ)–sponsored evaluation of Operation Ceasefire revealed that the intervention was associated with a 63% decrease in the monthly number of Boston youth homicides, a 32% decrease in the monthly number of shots-fired calls, a 25% decrease in the monthly number of gun assaults, and, in one high-risk police district given special attention in the evaluation, a 44% decrease in the monthly number of youth gun assault incidents. The evaluation also suggested that Boston's significant youth homicide reduction associated with Operation Ceasefire was distinct when compared to youth homicide trends in most major U.S. and New England cities (Braga, Kennedy, Waring, & Piehl, 2001).

In subsequent reviews of the evaluation, some researchers have been less certain about the magnitude of the violence reduction effect associated with Operation Ceasefire. Given the complexities of analyzing city-level homicide trend data, Ludwig (2005) and Rosenfeld et al. (2005) urged caution in drawing strong conclusions about this program's effectiveness. Others reviewers, however, have been more supportive of the evaluation evidence on violence reductions associated with the Ceasefire intervention. The National Research Council's Panel on Improving Information and Data on Firearms acknowledged the uncertainty in specifying the size of violence reduction effect associated with Operation Ceasefire by virtue of its quasi-experimental evaluation design (Wilford, Pepper, & Petrie, 2005). Nonetheless, it concluded that the Ceasefire evaluation was "compelling" in associating the intervention with the subsequent decline in youth homicide. In their close review of the Ceasefire evaluation, Morgan and Winship (2007) found the analyses to be of very high methodological quality and to make a strong case for causal assertions.

Explaining the Effectiveness of Operation Ceasefire

The Harvard research team, unfortunately, did not collect the necessary pre-test and post-test data to shed light on the specific mechanisms responsible for the significant violence reductions associated with the Operation Ceasefire intervention. The research team focused on problem analysis and program development and, a priori, did not know which form the intervention would take and who the target audience of that intervention would be. Because the necessary evaluation data are not available, it is necessary to draw on the research literature on gang intervention programs to speculate on the effectiveness of the Operation Ceasefire approach to controlling gang violence (Braga & Kennedy, 2002).

As part of the Office of Juvenile Justice and Delinquency Prevention (OJJDP) National Youth Gang Suppression and Intervention Program, Spergel and Curry (1990, 1993) surveyed 254 law enforcement, school, and community representatives in 45 cities and 6 institutional sites on their gang intervention programs. From these survey data, Spergel and Curry developed a four-category typology of the interventions that these areas used to deal with gang problems: (1) suppression, (2) social intervention, (3) opportunity provision, and (4) community organization. Although Operation Ceasefire was a problem-oriented policing project centered on law enforcement interventions, the other elements of Operation Ceasefire that involved community organization, social intervention, and opportunity provision certainly supported and strengthened the ability of law enforcement to reduce gang violence. Beyond deterring violent behavior, Operation Ceasefire was designed to facilitate desired behaviors among gang members. As Spergel (1995) observed, coordinated strategies that integrate these varied domains are most likely to prove effective in dealing with chronic youth gang problems.

Suppression

The typical law enforcement suppression approach assumes that most street gangs are criminal associations that must be attacked through an efficient gang tracking, identification, and targeted enforcement strategy (Spergel, 1995). The basic premise of this approach is that improved data collection systems and coordination of information across different criminal justice agencies lead to more efficiency and to more gang members being removed from the streets, rapidly prosecuted, and sent to prison for longer sentences. Typical suppression programs included street sweeps in which police officers round up hundreds of suspected gang members, special gang probation and parole caseloads where gang members are subjected to heighten levels of surveillance and more stringent revocation rules, prosecution programs that target gang leaders and serious gang offenders, civil procedures that use gang membership to define arrest for conspiracy or unlawful associations, and school-based law enforcement programs that use surveillance and buy–bust operations (Klein, 1993).

These suppression approaches are loosely based on deterrence theory. Law enforcement agencies attempt to influence the behavior of gang members or eliminate gangs entirely by dramatically increasing the certainty, severity, and swiftness of criminal justice sanctions. Unfortunately, gangs and gang problems usually persist even in the wake of these intensive operations. Malcolm Klein (1993) suggests that law enforcement agencies do not generally have the capacity to "eliminate" all gangs in a gang-troubled jurisdiction, nor do they have the capacity to respond in a powerful way to all gang offending in such jurisdictions. Pledges to do so, though common, are simply not credible to gang members. Klein also observes that the emphasis on selective enforcement by deterrence-based gang suppression programs may increase the cohesiveness of gang members, who often perceive such actions as unwarranted harassment, rather than as cause to withdraw from gang activity. Therefore, suppression programs may have the perverse effect of strengthening gang solidarity.

Beyond the certainty, severity, and swiftness of sanctions, the effective operation of deterrence depends on the communication of punishment threats to the public. As Zimring and Hawkins (1973) observed, "The deterrence threat may best be viewed as a form of advertising" (p. 142). The Operation Ceasefire Working Group recognized that, for the strategy to be successful, it was crucial to deliver a *credible* deterrence message to Boston gangs. Therefore, its intervention targeted only those gangs who were engaged in violent behavior rather than wasting resources on those who were not. Spergel (1995) suggests that problem-solving approaches to gang problems based on more limited goals such as gang violence reduction rather than gang destruction are more likely to be effective in controlling gang problems. Operation Ceasefire did not attempt to eliminate all gangs or eliminate all gang offending in Boston. Despite the large reductions in youth violence, Boston still has gangs and Boston gangs still commit crimes. Nevertheless, Boston gangs now do not commit violent acts as frequently as they did in the past.

The Ceasefire focused deterrence approach attempted to prevent gang violence by making gang members believe that consequences would follow on the heels of violence and gun use, and encouraging them to choose to change their behavior. A key element of the strategy was the delivery of a direct and explicit "retail deterrence" message to a relatively small target audience regarding which kinds of behavior would provoke a special response and what that response would be. In addition to any increases in certainty, severity, and swiftness of sanctions associated with acts of violence, the Operation Ceasefire strategy sought to gain deterrence through the *advertising* of the law enforcement strategy and the personalized nature of its application. It was crucial that gang youth understood the new regime that the city was imposing. Beyond the particular gangs subjected to the intervention, the deterrence message was applied to a relatively small audience (all gang-involved youth in Boston) rather than diffused across a general audience (all youth in Boston), and it operated by making explicit cause-and-effect connections between the behavior of the target population and the behavior of the authorities. Knowledge of what happened to others in the target population was intended to prevent further acts of violence by gangs in Boston.

In the communication of the deterrence message, the Working Group also wanted to develop a common piece of shared moral ground with gang members. The Group wanted the gang members to understand that most victims of gang violence were gang members, that the strategy was designed to protect both gang members and the community in which they lived, and that the Working Group had gang members' best interests in mind even if the gang members' own actions required resorting to coercion in an effort to protect them. The Working Group also hoped that the process of communicating on a face-to-face basis with gangs and gang members would undercut any feelings of anonymity and invulnerability they might have, and that a clear demonstration of interagency solidarity would enhance offenders' sense that something new and powerful was happening.

Social Intervention and Opportunity Provision

Social intervention programs encompass both social services agency–based programs and detached "streetworker" programs; opportunity provision strategies attempt to offer gang members legitimate opportunities and means to success that are at least as appealing as the available illegitimate options (Curry & Decker, 1998; Klein, 1995; Spergel, 1995).

Boston streetworkers were key members of the Operation Ceasefire Working Group and, along with the Department of Youth Services (DYS) case workers, probation officers, and parole officers in the group, added a much needed social intervention and opportunity provision dimension to the Ceasefire strategy. The mayor of the city of Boston established the Boston Community Centers' Streetworkers social services program in 1991. The streetworkers were charged with seeking out at-risk youth in Boston's neighborhoods and providing them with services such as job skills training, substance abuse counseling, and special education.

Many Boston streetworkers are themselves former gang members. Gang researchers have suggested that meaningful gang crime prevention programs should recruit gang members to participate in the program as staff and consultants (Bursik & Grasmick, 1993; Hagedorn, 1988).

Beyond their important roles as social service providers, streetworkers attempted to prevent outbreaks of violence by mediating disputes between gangs. Streetworkers also ran programs intended to keep gang-involved youth safely occupied and to bring them into contact with one another in ways that might breed tolerance, including a Peace League of gang-on-gang basketball games held at neutral, controlled sites.

Through the use of these resources, the Ceasefire Working Group was able to pair criminal justice sanctions, or the promise of sanctions, with help and with services. When the risk to drug-dealing gang members increases, legitimate work becomes more attractive. In turn, when legitimate work is more widely available, raising risks will be more effective in reducing violence. The expansion of social services and opportunities was intended to increase the Ceasefire's strategy's preventive power by offering gang members any assistance they might want: protection from their enemies, drug treatment, access to education and job training programs, and the like.

Community Organization

Community organization strategies to cope with gang problems include attempts to create community solidarity, networking, education, and involvement (Spergel & Curry, 1993).

The Ten Point Coalition of activist black clergy played an important role in organizing Boston communities suffering from gang violence (Winship & Berrien, 1999). This organization was formed in 1992 after gang members invaded the Morningstar Baptist Church, where a slain rival gang member was being memorialized, and attacked mourners with knives and guns. In the wake of that watershed moment, the Ten Point Coalition decided to respond to violence in its community by reaching out to drug-involved and gang-involved youth and by organizing within Boston's black community.

Over time, the Ten Point clergy members came to work closely with the Boston Community Centers' streetworkers program to provide at-risk youth with opportunities. Although the Ten Point Coalition was initially very critical of the Boston law enforcement community, participants eventually forged a strong working relationship. Ten Point clergy and others involved in this faith-based organization accompanied police officers on home visits to the families of troubled youth and also acted as advocates for youth in the criminal justice system. These home visits and street work by the clergy were later incorporated into Operation Ceasefire's portfolio of interventions. Ten Point clergy also provided a strong moral voice at the gang forums in the presentation of Operation Ceasefire's antiviolence message.

Although its members did not become involved in Operation Ceasefire until after the strategy had been designed and implemented, the Ten Point Coalition played a crucial role in framing a discussion in Boston that made it much easier to speak directly about the nature of youth violence in Boston. Members of the Ceasefire Working Group could speak with relative safety about the painful realities of minority male offending and victimization, "gangs," and chronic offenders. The Ten Point clergy also made it possible for Boston's minority community to have an ongoing conversation with Boston's law enforcement agencies on legitimate and illegitimate means to control crime in the community.

Although the clergy supported Operation Ceasefire's tight focus on violent youth, they condemned any indiscriminate, highly aggressive law enforcement sweeps that might put nonviolent minority youth at risk of being swept into the criminal justice system. Before the Ten Point Coalition developed its role as an intermediary, Boston's black community viewed past activities of law enforcement agencies to monitor violent youth as illegitimate and with knee-jerk suspicion. As noted by Winship and Berrien (1999), the Ten Point Coalition evolved into an institution that provides an umbrella of legitimacy for the police to work under. With the Ten Point organization's approval of and involvement in Operation Ceasefire, the community supported the approach as a legitimate youth violence prevention campaign.

EXPERIENCES AND EVALUATION EVIDENCE FROM OTHER JURISDICTIONS

At first blush, the effectiveness of the Operation Ceasefire intervention in preventing violence might seem unique to Boston. Operation Ceasefire was constructed largely from the assets and capacities available in Boston at the time and deliberately tailored to the city's specific gang violence problem. Operational capacities of criminal justice agencies in other cities will be different, of course, and gang violence problems in other cities will have important characteristics that distinguish them from the situation in Boston. Nevertheless, the basic working group problem-oriented policing process and the "pulling levers" approach to deterring chronic offenders are transferable to violence problems in other jurisdictions.

To date, a number of cities have experimented with these analytic frameworks and have reported noteworthy crime control gains. This section highlights gang violence reduction efforts in Los Angeles, California; Indianapolis, Indiana; Stockton, California; and Lowell, Massachusetts. Research evidence on some promising applications of focused deterrence strategies to individual offenders and overt drug markets is also presented. Consistent with the problem-oriented policing approach (Eck & Spelman, 1987; Goldstein, 1990), these cities have tailored the approach to fit their unique violence problems and operating environments.

Operation Ceasefire in Los Angeles

In March 1998, the U.S. National Institute of Justice (NIJ) funded the RAND Corporation to develop and test strategies for reducing gun violence among youth in Los Angeles. In part, the goal of this effort was to determine which parts of the Boston Gun Project might be replicable in Los Angeles. In designing the replication, RAND drew a clear distinction between the process governing the design and implementation of the strategy (data-driven policy development; problem-solving working groups) and the elements and design (pulling levers, collective accountability, retailing the message) of the Boston model. Processes, in theory, can be sustained and adaptive, and as such can be utilized to address dynamic problems. By singling out process as an important component, the RAND team hoped to clarify the point that process can affect program effectiveness independently of the program elements or the merits of the actual design (Tita, Riley, & Greenwood, 2003).

The Los Angeles replication was unique in several important ways. First, the implementation was not citywide, but rather was carried out only within a single neighborhood (Boyle Heights) within a single Los Angeles Police Department (LAPD) Division (Hollenbeck). The project site, Boyle Heights, had a population that was relatively homogenous. More than 80% of the residents were Latinos of Mexican origin. The same was true for the gangs operating in this area, many of which were formed prior to World War II. These gangs were clearly "traditional" gangs, with memberships exceeding 100 members or more. The gangs were strongly territorial, encompassed age-graded substructures, and were intergenerational in nature (Maxson & Klein, 1995).

Unlike in some other cities where gang and group-involved violence is a rather recent phenomenon, Los Angeles represented an attempt to reduce gun violence in a "chronic gang city" with a long history of gang violence, and an equally long history of gang reduction strategies. The research team had to first convince members of the local criminal justice community and the at-large community that the approach espoused differed in important ways from these previous efforts to combat gangs. In fact, it did: The RAND project was not about "doing something about gangs," but rather about "doing something about gun violence" in a community where gang members were responsible for an overwhelming proportion of gun violence. The independent analysis of homicide files confirmed the perception held by police and community alike that gangs were highly over-represented in homicidal acts. From 1995 to 1998, 50% of all homicides in the area had a clear gang motivation. Another 25% of the homicides could be coded as "gang related" because they involved a gang member as victim or offender, but were motivated by reasons other than gang rivalries.

Given the social organization of violence in Boyle Heights, the multidisciplinary working group fully embraced the pulling levers focused deterrence strategy that had been developed in Boston. A high-profile gang shooting that resulted in a double homicide in Boyle Heights triggered the implementation of the Operation Ceasefire intervention in October 2000. The processes of retailing the message were formally adopted, although message delivery was mostly accomplished through personal contact rather than in a group setting. Police, probation, community advocates, street gang workers, a local hospital, and local clergy all passed along the message of collective accountability for gangs whose members continued to commit gang violence.

Unfortunately, as Tita, Riley, Ridgeway, and colleagues (2003) report, the Los Angeles pulling levers intervention was not fully implemented as planned. The implementation of the Ceasefire program in the Boyle Heights was negatively affected by the well-known Ramparts LAPD police corruption scandal and a lack of ownership of the intervention by the participating agencies.

Despite the implementation difficulties, the RAND research team was able to complete an impact evaluation of the violence prevention effects of the pulling levers focused deterrence strategy that was implemented. Using a variety of methods, the investigators determined that consistent, noteworthy short-term reductions in violent crime, gang crime, and gun crime were associated with the Ceasefire program. In addition to their analyses of the main effects of the intervention, RAND researchers examined the effects of the intervention on neighboring areas and gangs. Their analyses suggested a strong diffusion of violence prevention benefits emanating from the targeted areas and targeted gangs (Tita et al., 2003).

Indianapolis Violence Reduction Partnership

The Indianapolis Violence Reduction Partnership (IVRP) working group consisted of Indiana University researchers and federal, state, and local law enforcement agencies (McGarrell & Chermak, 2003). During the problem analysis phase, the researchers examined 258 homicides from 1997 and the first 8 months of 1998 and found that a majority of homicide victims (63%) and offenders (75%) had criminal and/or juvenile records. Those individuals with a prior record often had a substantial number of arrests. The working group members performed the same structured, qualitative data-gathering exercises used in Boston to gain insight into the nature of homicide incidents. The qualitative exercise revealed that 59% of the incidents involved "groups of known chronic offenders" and 53% involved drug-related motives such as settling business and turf disputes. The terminology "groups of known chronic offenders" was initially used in Indianapolis because, at that point in time, there was not a consensual definition of "gang" and the reality of much gang activity in Indianapolis was of a relatively loose structure.

The working group developed two sets of overlapping strategies. First, the most violent chronic offenders in Indianapolis were identified and targeted for heightened attention, leading to greater arrest, prosecution, and incarceration rates. Second, the working group adopted the pulling levers approach to reduce violent behavior by gangs and groups of known chronic offenders.

The IVRP strategy implemented by the Indianapolis working group closely resembled the Boston version of pulling levers. The communications strategy, however, differed in an important way. The deterrence and social services message was delivered in meetings with high-risk probationers and parolees organized by neighborhoods. Similarly, home visits by probation and parolees were generally organized by neighborhood. As the project progressed, when a homicide or series of homicides involved certain groups or gangs, the working group attempted to target meetings, enforcement activities, and home visits on the involved groups or gangs (McGarrell & Chermak, 2003).

A DOJ-funded evaluation revealed that the IVRP strategy was associated with a 34% reduction in Indianapolis homicides. The evaluation further revealed that the homicide reduction in Indianapolis differed from the homicide trends observed in six comparable Midwestern cities during the same time period (McGarrell, Chermak, Wilson, & Corsaro, 2006).

Operation Peacekeeper in Stockton

Beginning in mid-1997, criminal justice agencies in Stockton began experimenting with the pulling levers approach to address a sudden increase in youth homicide. The Stockton Police Department and other local, state, and federal law enforcement agencies believed that most of the youth violence problem was driven by gang conflicts and that the pulling levers approached used in Boston might be effective in reducing Stockton's gang violence problem. The strategy was implemented by the Stockton Police Department's Gang Street Enforcement Team and grew into what is now known as "Operation Peacekeeper" as more agencies joined the partnership (Wakeling, 2003).

The Peacekeeper intervention was managed by a working group of line-level criminal justice practitioners; social service providers also participated in the working group process as appropriate. When street gang violence erupted or when it came to the attention of a working group member that gang violence was imminent, the working group followed the Boston model by sending a direct message that gang violence would not be tolerated, pulling all available enforcement levers to prevent violence, engaging in communications, and providing social services and opportunities to gang members who wanted them.

To better document the nature of homicide and serious violence in Stockton, the working group retained Harvard University researchers to conduct ongoing problem analyses (Braga, 2005). Their results revealed that many offenders and victims involved in homicide incidents had noteworthy criminal histories and prior criminal justice system involvement. Gang-related conflicts were identified as the motive in 41% of homicides between 1997 and 1999.

The research analysis also revealed that the California city is home to 44 active gangs with a total known membership of 2100 individuals. Most conflicts among these gangs fall into three broad categories: Asian gang beefs, Hispanic gang beefs, and African American gang beefs. Within each broad set of ethnic antagonisms, particular gangs may form alliances with other gangs. Conflicts among Asian gangs involve clusters of different gangs comprising mostly Laotian and Cambodian youth. Conflicts among Hispanic gangs mainly involve a very violent rivalry between Norteño gangs from Northern California and Sureño gangs from Southern California. African American gangs tend to form fewer alliances and are divided along the well-known Blood and Crip lines.

An evaluation of Operation Peacekeeper found that this strategy was associated with a 42% decrease in gun homicide incidents. The same investigation also compared Stockton gun homicide trends to gun homicide trends in eight other midsized California cities, reporting that the large reduction in gun homicides in Stockton was not echoed in the comparison cities (Braga, 2008).

Project Safe Neighborhoods in Lowell

Supported by funds from the DOJ-sponsored Project Safe Neighborhoods initiative, an interagency task force implemented a pulling levers focused deterrence strategy to prevent gun violence among Hispanic and Asian gangs in Lowell, Massachusetts, in 2002 (Braga, McDevitt, & Pierce, 2006). While Lowell authorities felt very confident about their ability to prevent violence among Hispanic gangs by pursuing a general focused deterrence strategy, they were much less confident about their ability to prevent Asian gang violence by applying the same set of criminal justice levers to Asian gang members. As Malcolm Klein (1995) suggests, Asian gangs have some key differences from typical black, Hispanic, and white street gangs. Notably, they are more organized, have identifiable leaders, and are far more secretive. They also tend to be far less territorial and less openly visible, so their street presence is low compared to other ethnic gangs. Relationships between law enforcement agencies and the Asian community are often characterized by mistrust and a lack of communication (Chin, 1996). As such, it is often difficult for the police to develop information on the participants in violent acts to hold offenders accountable for their actions.

During the intervention time period, the Lowell Police Department (LPD) had little reliable intelligence about Asian gangs in the city. The LPD had attempted to recruit informants in the past but most of these efforts had been unsuccessful. With the increased focus on Asian gang violence, the LPD increased its efforts to develop intelligence about the structure of the city's Asian gangs and particularly the relationship between Asian gang violence and ongoing gambling that was being run by local Asian businesses. Asian street gangs are sometimes connected to adult criminal organizations and assist older criminals in extortion activities and protecting illegal gambling enterprises (Chin, 1996). In many East Asian cultures, rituals and protocols guiding social interactions are well defined and reinforced through a variety of highly developed feelings of obligation, many of which are hierarchical in nature (Zhang, 2002). This factor facilitates some control over the behavior of younger Asian gang members by elders in the gang.

In Lowell, Cambodian and Laotian gangs consisted of youth whose street activities were influenced by "elders" of the gang. Elders were generally long-time gang members in their thirties and forties who no longer engaged in illegal activities on the street or participated in street-level violence with rival youth. Rather, these older gang members were heavily involved in running illegal gambling dens and informal casinos that were operated out of cafes, video stores, and warehouses located in the poor Asian neighborhoods of Lowell. The elders used young street gang members to protect their business interests and to collect any unpaid gambling debts. Illegal gaming was a very lucrative business that was much more important to the elders than any ongoing beefs the youth in their gang had with other youth (Braga et al., 2006). Relative to acquiring information on individuals responsible for gun crimes in Asian communities, it was much easier to detect the presence of gambling operations through surveillance or a simple visit to the suspected business establishment.

The importance of illegal gaming to influential members of Asian street gangs provided a potentially potent lever for law enforcement to exploit in preventing violence. The authorities in Lowell believed that they could systematically prevent street violence among gangs by targeting the gambling interests of older members. When a street gang was violent, the LPD targeted the gambling businesses run by the older members of the gang. The enforcement activities ranged from serving a search warrant on the business that housed the illegal enterprise and making arrests to simply placing a patrol car in front of the suspected gambling location to deter gamblers from entering. The LPD coupled these tactics with the delivery of a clear message: "When the young gang members associated with you act violently, we will shut down your gambling business. When violence erupts, no one makes money" (Braga et al., 2006). Between October 2002 and June 2003, during the height of the focused attention on Asian gangs, the LPD executed some 30 search warrants on illegal gambling dens that resulted in more than 100 gambling-related arrests.

An impact evaluation found that the Lowell pulling levers strategy was associated with a 43% decrease in the monthly number of gun homicide and gun aggravated assault incidents. A comparative analysis of gun homicide and gun aggravated assault trends in Lowell relative to other major Massachusetts cities also supported the contention that a unique program effect was associated with the pulling levers intervention (Braga, Pierce, McDevitt, Bond, & Cronin, 2008).

While this approach to preventing violence among Asian street gangs represents an innovation in policing, it is not an entirely new idea. The social control exerted by older Asian criminals over their younger counterparts is well documented in the literature. For example, in his study of Chinese gangs in New York City, Chin (1996) suggests that gang leaders exert influence over subordinate gang members to end violent confrontations so they can focus their energies on illegal enterprises that make money. The prospect of controlling street violence by cracking down on the interests of organized crime is also familiar to law enforcement. In his classic study of an Italian street gang in Boston's North End, Whyte (1943) described the activities of beat officers in dealing with outbreaks of violence by cracking down on the gambling rackets run by organized crime in the neighborhood. Nevertheless, the systematic application of this approach, coupled with a communications campaign, represents an innovative way to deal with Asian street gang violence.

Applying Focused Deterrence to Individual Offenders

A variation of the Boston model was applied by Papachristos et al. (2007) in Chicago. Gun- and gang-involved parolees returning to selected highly dangerous Chicago neighborhoods went through "call-ins" where they were informed of their vulnerability as felons to federal firearms laws, with stiff mandatory minimum sentences for violating these laws; offered social services; and addressed by community members and ex-offenders. A rigorous evaluation showed a neighborhood-level homicide reduction impact of 37%. Individual-level effects were remarkable. For offenders with prior gun offenses but no evidence of gang involvement, for example, attendance at a call-in reduced the recidivism rate (i.e., return to prison) at around five years from release from approximately 50% to 10% (Fagan, Meares, Papachristos, & Wallace, 2008).

Applying Focused Deterrence to Overt Drug Markets

There is less experience in applying the focused deterrence approach to other crime and disorder problems. In High Point, North Carolina, a focused deterrence strategy was aimed at eliminating public forms of drug dealing such as street markets and crack houses by warning dealers, buyers, and their families that enforcement was imminent. With individual "overt" drug markets cited as the unit of work, the project employed a joint police–community partnership to identify individual offenders; notify them of the consequences of continued dealing; provide supportive services through a community-based resource coordinator; and convey an uncompromising community norm against drug dealing. This application of focused deterrence is generally referred to as the drug market intervention (DMI) strategy.

The DMI seeks to shut down overt drug markets entirely. Enforcement powers are used strategically and sparingly, with arrest and prosecution being enforced only against violent offenders and when nonviolent offenders have resisted all efforts to get them to desist and to provide them with help. Through the use of "banked" cases, the strategy makes the promise of law enforcement sanctions against dealers extremely direct and credible, so that dealers are in no doubt concerning the consequences of offending and have good reason to change their behavior. The strategy also brings powerful informal social control to bear on dealers from immediate family and community figures. The strategy organizes and focuses services, help, and support for dealers so that those who are willing to change their lives have what they need to do so. Each operation also includes a maintenance strategy.

A preliminary assessment of the High Point DMI found noteworthy reductions in drug and violent crime in the city's West End neighborhood (Frabutt, Gathings, Hunt, & Loggins, 2004). A more rigorous evaluation of the High Point DMI is currently being conducted.

An evaluation of a similar DMI strategy in Rockford, Illinois, found noteworthy crime prevention gains associated with the approach (Corsaro, Brunson, & McGarrell, 2011). Study findings suggest that the Rockford strategy was associated with a statistically significant and substantive reduction in crime, drug, and nuisance offenses in the target neighborhood.

DISCUSSION

The available research evidence suggests that these new focused deterrence approaches to prevent gang and group-involved violence have generated promising results. Certain core elements of this approach seem to be the key ingredients in their apparent success. These key elements are delineated here (Braga et al., 2002).

Recognition of the Concentration of Urban Violence Problems Among Groups of Chronic, Often but not Always Gang-Involved, Offenders

Research has demonstrated that the character of criminal and disorderly youth gangs and groups varies widely both within cities and across cities (Curry, Ball, & Fox, 1994). The diverse findings on the nature of criminally active groups and gangs in the jurisdictions described in this chapter certainly support this assertion. The research also suggests that the terminology used to describe the types of groups involved in urban violence matters less than their behavior. Gangs, their nature, and their behavior remain central questions for communities, police, and scholars.

At the same time, where violence prevention and public safety are concerned, the gang question is not the central one. The more important observation is that urban violence problems are, in large measure, concentrated among groups of chronic offenders; thus the dynamics between and within these groups have major implications for crime rates (Kennedy, 2001). This is an old observation in criminology, and one that is well known among line law enforcement personnel, prosecutors, probation and parole officers, and other authorities. The new crime prevention strategies offer a way of responding to this reality without setting the usually unattainable goals of eliminating chronic offending or eliminating criminal gangs and groups.

Recognition of Dynamic or Self-Reinforcing Positive Feedback Mechanisms Among Group and Gang Violence

The research findings indicate that groups of chronic offenders are locked in a self-sustaining dynamic of violence often driven by fear, "respect" issues, and vendettas. The promising reductions observed in the cities employing these strategic crime prevention frameworks suggest that the "firebreak hypothesis" may be right. If this cycle of violence among these groups can be interrupted, perhaps a new equilibrium at a lower level of risk and violence can be established. This relationship may be one explanation for the rather dramatic impacts apparently associated with what are, in fact, relatively modest interventions.

The Utility of the Pulling Levers Approach

The pulling levers strategy at the heart of these new focused deterrence approaches was designed to influence the behavior, and the environment, of the groups of chronic offenders who were identified as the core of the cities' violence problems. The pulling levers approach attempts to prevent gang and group-involved violence by making these groups believe that dire consequences will follow from violence and gun use and choose to change their behavior. A key element of the strategy is the delivery of a direct and explicit "retail deterrence" message to a relatively small target audience regarding which kinds of behavior will provoke a special response and what that response will be.

Drawing on Practitioner Knowledge to Understand Violence Problems

The experiences, observations, local knowledge, and historical perspectives of police officers, streetworkers, and others with routine contact with offenders, communities, and criminal networks represent an underutilized resource for describing, understanding, and crafting interventions aimed at crime problems (Kennedy, Braga, & Piehl, 1997). The semi-structured qualitative research performed by the academics in these initiatives essentially refined and specified existing practitioner knowledge. Combining official data sources with street-level qualitative information helps to paint a dynamic, real-life picture of the violence problem.

An Interagency Working Group with a Locus of Responsibility

Criminal justice agencies work largely independently of one another, often at cross-purposes, often without coordination, and often in an atmosphere of distrust and dislike (Kennedy, 2001). Different elements operating within agencies may be plagued by the same problems. The ability of the cities profiled in this chapter to deliver a meaningful violence prevention intervention was created by convening an interagency working group of line-level personnel with decision-making power who could assemble a wide range of incentives and disincentives for the target audience. It was also important to place a locus of responsibility for reducing violence within the group. Prior to the creation of the working groups, no one in these cities was responsible for developing and implementing an overall strategy for reducing violence.

Researcher Involvement in an Action-Oriented Enterprise

The activities of the research partners in these initiatives departed from the traditional research and evaluation roles usually played by academics (Sherman, 1991). The integrated researcher–practitioner partnerships in the working group setting more closely resembled policy analysis exercises that blend research, policy design, action, and evaluation (Kennedy, 2009; Kennedy & Moore, 1995). Researchers have been important assets in all of the projects described in this chapter, providing what is essentially "real-time" social science aimed at refining the working group's understanding of the problem, creating information products for both strategic and tactical use, testing (often in a very elementary, but important, fashion) candidate intervention ideas, and maintaining a focus on clear outcomes and the evaluation of performance.

They have begun to produce accounts of both basic findings and intervention designs and implementation processes, which will be helpful to other jurisdictions moving forward. In addition, in several sites, researchers played important roles in organizing the projects.

CONCLUSION

The cumulative experience described in this chapter appears supportive, at this preliminary stage, of the proposition that the basic Boston approach has now been replicated, with promising results, in a number of disparate sites. This extension suggests that there was nothing particularly unique about either the implementation or the impact of Operation Ceasefire in Boston. Further, it suggests that the fundamental pulling levers focused deterrence framework behind Ceasefire can be successfully applied in other jurisdictions, with other sets of partners, with different particular activities, and in the context of different basic types of gangs and groups. Further operational experience and more refined evaluation techniques will tell us more about these questions, as experience and analysis continue to accumulate. At the moment, there appears to be reason for continued optimism that serious violence by gangs and other groups is open to direct and powerful prevention.

REFERENCES

Braga, A. A. (2005). Analyzing homicide problems: Practical approaches to developing a policy-relevant description of serious urban violence. *Security Journal, 18*, 17–32.

Braga, A. A. (2008). Pulling levers focused deterrence strategies and the prevention of gun homicide. *Journal of Criminal Justice, 36*, 332–343.

Braga, A. A., & Kennedy, D. M. (2002). Reducing gang violence in Boston. In W. Reed & S. H. Decker (Eds.), *Responding to gangs: Research and evaluation* (pp. 265-288). Washington, DC: National Institute of Justice, U.S. Department of Justice.

Braga, A. A., Kennedy, D. M., & Tita, G. (2002). New approaches to the strategic prevention of gang and group-involved violence. In C. R. Huff (Ed.), *Gangs in America* (pp. 271-286) (3rd ed.). Thousand Oaks, CA: Sage.

Braga, A. A., Kennedy, D. M., Waring, E. J., & Piehl, A. M. (2001). Problem-oriented policing, deterrence, and youth violence: An evaluation of Boston's Operation Ceasefire. *Journal of Research in Crime and Delinquency, 38*, 195–225.

Braga, A. A., McDevitt, J., & Pierce, G. L. (2006). Understanding and preventing gang violence: Problem analysis and response development in Lowell, Massachusetts. *Police Quarterly, 9*, 20–46.

Braga, A. A., Pierce, G. L., McDevitt, J., Bond, B. J., & Cronin, S. (2008). The strategic prevention of gun violence among gang-involved offenders. *Justice Quarterly, 25*, 132–162.

Bursik, R. J., & Grasmick, H. G. (1993). *Neighborhoods and crime: The dimensions of effective community control.* New York: Lexington Books.

Chin, K. (1996). *Chinatown gangs: Extortion, enterprise, and ethnicity*. New York: Oxford University Press.

Coleman, V., Holton, W. C., Olson, K., Robinson, S., & Stewart, J. (1999, October). Using knowledge and teamwork to reduce crime. *National Institute of Justice Journal*, 16–23.

Corsaro, N., Brunson, R. K., & McGarrell, E. (In press). Problem-oriented policing and open-air drug markets: Examining the Rockford pulling levers strategy. *Crime & Delinquency*.

Curry, G. D., Ball, R. A., & Fox, R. J. (1994). *Gang crime and law enforcement recordkeeping*. Research in Brief. Washington, DC: National Institute of Justice, U.S. Department of Justice.

Curry, G. D., & Decker, S. H. (1998). *Confronting gangs: Crime and community*. Los Angeles: Roxbury Press.

Dalton, E. (2002). Targeted crime reduction efforts in ten communities: Lessons for the Project Safe Neighborhoods Initiative. *U.S. Attorney's Bulletin, 50*, 16–25.

Eck, J. E., & Spelman, W. (1987). *Problem-solving: Problem-oriented policing in Newport News*. Washington, DC: Police Executive Research Forum.

Fagan, J., Meares, T., Papachristos, A. V., & Wallace, D. (2008, November). *Desistance and legitimacy: Effect heterogeneity in a field experiment with high-risk offenders*. Presented at the annual meeting of the American Society of Criminology, St. Louis, MO.

Frabutt, J., Gathings, M. J., Hunt, E. D., & Loggins, T. J. (2004). *High Point West End Initiative: Project description, log, and preliminary impact analysis*. Greensboro, NC: Center for Youth, Family, and Community Partnerships.

Goldstein, H. (1990). *Problem-oriented policing*. Philadelphia: Temple University Press.

Hagedorn, J. (1988). *People and folks: Gangs, crime, and the underclass in a Rustbelt city*. Chicago: Lakeview Press.

Kennedy, D. M. (1997). Pulling levers: Chronic offenders, high-crime settings, and a theory of prevention. *Valparaiso University Law Review, 31*, 449–484.

Kennedy, D. M. (2001). A tale of one city: Reflections on the Boston Gun Project. In G. Katzmann (Ed.), *Managing youth violence* (pp. 14-33). Washington, DC: Brookings Institution Press.

Kennedy, D. M. (2006). Old wine in new bottles: Policing and the lessons of pulling levers. In D. L. Weisburd & A. A. Braga (Eds.), *Police innovation: Contrasting perspectives* (pp. 155-170). New York: Cambridge University Press.

Kennedy, D. M. (2009). Drugs, race, and common ground: Reflections on the High Point Intervention. *National Institute of Justice Journal, 262*, 12–17.

Kennedy, D. M., Braga, A. A., & Piehl, A. M. (1997). The (un)known universe: Mapping gangs and gang violence in Boston. In D. L. Weisburd & J. T. McEwen (Eds.), *Crime mapping and crime prevention* (pp. 219-262). Monsey, NY: Criminal Justice Press.

Kennedy, D. M., & Moore, M. H. (1995). Underwriting the risky investment in community policing: What social science should be doing to evaluate community policing. *Justice System Journal, 17*, 271–290.

Kennedy, D. M., Piehl, A. M., & Braga, A, A. (1996). Youth violence in Boston: Gun markets, serious youth offenders, and a use-reduction strategy. *Law and Contemporary Problems, 59*, 147–196.

Klein, M. (1993). Attempting gang control by suppression: The misuse of deterrence principles. *Studies on Crime and Crime Prevention, 2*, 88–111.

Klein, M. (1995). *The American street gang: Its nature, prevalence, and control*. New York: Oxford University Press.

Ludwig, J. (2005). Better gun enforcement, less crime. *Crime and Public Policy, 4*, 677– 716.

Maxson, C. L., & Klein, M. W. (1995). Investigating gang structures. *Journal of Gang Research, 3*, 33–38.

McGarrell, E. F., & Chermak, S. (2003). Problem solving to reduce gang and drug- related violence in Indianapolis. In S. H. Decker (Ed.), *Policing gangs and youth violence* (pp. 77-101). Belmont, CA: Wadsworth.

McGarrell, E. F., Chermak, S., Wilson, J., & Corsaro, N. (2006). Reducing homicide through a "lever-pulling" strategy. *Justice Quarterly, 23,* 214–229.

Morgan, S., & Winship, C. (2007). *Counterfactuals and causal models.* New York: Cambridge University Press.

Papachristos, A., Meares, T., & Fagan, J. (2007). Attention felons: Evaluating Project Safe Neighborhoods in Chicago. *Journal of Empirical Legal Studies, 4,* 223–272.

Rosenfeld, R., Fornango, R., & Baumer, E. (2005). Did Ceasefire, Compstat, and Exile reduce homicide? *Crime and Public Policy, 4,* 419–450.

Sherman, L. (1991). Herman Goldstein: Problem-oriented policing. *Journal of Criminal Law and Criminology, 82,* 693–702.

Spergel, I. A. (1995). *The youth gang problem: A community approach.* New York: Oxford University Press.

Spergel, I. A., & Curry, G. D. (1990). Strategies and perceived agency effectiveness in dealing with the youth gang problem. In C. R. Huff (Ed.), *Gangs in America* (pp. 254–265). Newbury Park, CA: Sage.

Spergel, I. A., & Curry, G. D. (1993). The National Youth Gang Survey: A research and development process. In A. Goldstein & C. R. Huff (Eds.), *Gang intervention handbook* (pp. 359–400). Champaign/Urbana, IL: Research Press.

Tita, G. E., Riley, K. J., & Greenwood, P. W. (2003). From Boston to Boyle Heights: The process and prospects of a "pulling levers" strategy in a Los Angeles barrio. In S. H. Decker (Ed.), *Policing gangs and youth violence* (pp. 102-130). Belmont, CA: Wadsworth.

Tita, G., Riley, K.J., Ridgeway, G., Grammich, C., Abrahamse, A., & Greenwood, P. (2003). *Reducing gun violence: Results from an intervention in East Los Angeles.* Santa Monica, CA: RAND Corporation.

Travis, J. (1998, November 12). *Crime, justice, and public policy.* Plenary presentation to the American Society of Criminology, Washington, DC. Retrieved January 21, 2011, from http://www.ojp.usdoj.gov/nij/speeches/asc.htm

Wakeling, S. (2003). *Ending gang homicide: Deterrence can work.* Perspectives on Violence Prevention, No. 1. Sacramento, CA: California Attorney General's Office/California Health and Human Services Agency.

Wellford, C., Pepper, J., & Petrie, C. (Eds.). (2005). *Firearms and violence: A critical review. Committee to Improve Research Information and Data on Firearms. Committee on Law and Justice, Division of Behavioral and Social Sciences and Education.* Washington, DC: National Academies Press.

Whyte, W. F. (1943). *Street corner society: The social structure of an Italian slum.* Chicago: University of Chicago Press.

Winship, C., & Berrien, J. (1999, Summer). Boston cops and black churches. *Public Interest,* 52–68.

Zhang, S. (2002). Chinese gangs: Familial and cultural dynamics. In C. R. Huff (Ed.), *Gangs in America* (pp. 219-236) (3rd ed.). Thousand Oaks, CA: Sage.

Zimring, F., & Hawkins, G. (1973). *Deterrence: The legal threat in crime control.* Chicago: University of Chicago Press.

CHAPTER 16

Arrested Today, Out Tomorrow: Patrol Officers' Perceptions of a Broken Justice System

Jonathan W. Caudill
California State University, Chico

Chad R. Trulson
University of North Texas

Policing violence is nothing new, and the realities of policing violence have been well established. Frequently, patrol officers attempt to help complainants and witnesses who are uninterested in the services the police can immediately provide. From exaggerated initial reports of violence to unwilling witnesses, patrol officers have historically been at a disadvantage when responding to violent crime calls. The consequences of these disadvantages have led some officers to suggest that a faulty judicial system acts as a turnstile by failing to enforce preventive detention for violent crime suspects. The study described in this chapter sought to explore this phenomenon by observing the court records of pretrial detention. The results indicated that officers' perceptions of a faulty judicial system were inaccurate. In addition, an ethnographic approach was employed to focus on the frustrations of policing violence, including the police department's organizational management style.

Since the crime peak in 1993, the United States has seen dramatic decreases in crime overall—specifically, a more than 40% decline in violent crime. While the reasons for such striking decreases have divided scholars (Conklin, 2003; Fagan, Zimring, & Kim, 1998), police agencies and administrators, such as William Bratton, formerly of the New York City Police Department, have been quick to discount scientific rigor and pat themselves on the back, holding their programs up as the source of this phenomenon. Even though this approach bodes well politically when crime declines, it makes law enforcement accountable when crime increases or even remains stagnant. The logic goes as follows: If you have the wherewithal to reduce crime but rates are not declining, then you are failing in your job. The reality, of course, is that crime is a function of society, not a function of the police. Indeed, the police have a reactionary approach to crime and, therefore, are a function of crime within their crime fighting strategies. Although crime is a function of society and not the police, Americans have been lulled into a false sense of police security.

The consequences of this façade are far reaching, but have specifically demanding effects on law enforcement. If, as police administrators claimed, law enforcement agencies were capable of significant crime reductions, then they must be held accountable when such reductions fail to materialize. Thus, when crime fails to decrease at the desired rates, police agencies were called on the carpet by city managers, the media, and the public in general.

This chapter reviews violent crime from the patrol officer's perspective, explores officers' tendency to blame a faulty judicial system for the actions of violent suspects, and deciphers the reasons for patrol officers' frustrations within the criminal justice system. To set the stage, we briefly discuss violence from a sociological perspective and in terms of patrol officers' exposures to it.

THE ENVIRONMENT OF VIOLENCE

Sociologists have explored the environmental factors associated with crime and discovered that, in general, communities with poor residential stability and low-income residents experience disproportionately more violence (Ruback & Thompson, 2001; Shaw & McKay, 1969). Within these communities, there is a strong sense of individual honor, distrust of law enforcement, expectations of retaliation and vigilantism for personal affronts, and an overall tolerance for violence (Anderson, 1999; Horowitz, 1987; Jacobs & Wright, 2006; Virgil, 2003). Violence is accepted as a part of the landscape in these communities, and reliance on the police is perceived as weak and ill advised.

Even though certain situational characteristics increase the likelihood of calling the police, this decision is much more complex in these violence-infected communities (Hart & Rennison, 2003; Hindelang & Gottfredson, 1976). The decision to invoke the police typically involves several steps, but a paramount concern is whether the responding officer can fix the problem (Greenberg & Ruback, 1985). Unfortunately, residents of these communities view the police as a temporary fix to their long-term problems at best and, in many cases, as just plain incompetent. Furthermore, residents view the official sanctions for violence (e.g. community supervision or even incarceration) to be less consequential than retaliation—from their perspective, any period of incarceration actually serves to protect their attacker from them. Finally, residents view retaliation, or justified violence, as a legitimate mechanism to restore a sense of order within their community. As this discussion suggests, violence serves a function in these communities (Duck, 2009).

Ultimately, members of these communities have a higher tolerance for violence, especially justified violence, and greater distrust of officials than average communities. These characteristics reduce their willingness to rely on and cooperate with law enforcement. It is, therefore, unsurprising that approximately 50% of all violent crime goes unreported to the police (Moore, 1980; Wilson, 1980). The lack of reporting to and cooperation with them place patrol officers at a distinct disadvantage when policing violence.

THE COMPLICATIONS OF POLICING VIOLENCE

Pronounced distrust and exacerbation of adversarial relationships between violence-ridden communities and the police who patrol them not only restrict cooperation between witnesses and the police, but also have practical implications for the officers who work in these areas. When citizens become uncooperative with police, officers are left to their own limited interpretation and accounts of the incidents and experience a sense of helplessness. For instance, police have reported believing that violence is the by-product of gangs, gun smuggling, and drug trafficking, but readily express the view that they are ill equipped to reduce violence solely through police actions (Ezeonu, 2010; Wilson, 1980). This conflicting approach to policing violence may merely be the product of limited officer exposure to violence in their everyday activities.

Contrary to the commonly held belief, an encounter with violence is a rarity for the average patrol officer. In reality, the majority of patrol work involves situations that do not involve violence or, for that matter, even serious crime. Officers spend a disproportionate amount of their time responding to "medical emergencies, family quarrels, auto accidents, barking dogs, [and] minor traffic violations" (Wilson, 1978, p. 203). To put an officer's exposure to violence in perspective, one scholar reported that officers categorized only 18% of their calls for service as needing an "urgent response," with that percentage including violent and all other types of crimes (Reiss, 1971, p. 6). In another study of the police, researchers were able to use only approximately one-third of calls for service in their study; among those calls, only one of every ten situations involved anything more than a verbal altercation or posturing gestures (Bayley & Garofalo, 1989). On actual calls of violence, the perpetrator was gone before the police arrived at the scene in approximately half of situations, and many serious crimes (e.g., robberies) were reduced to cold trails by the time officers arrived (Smith, 1987). Finally, the overwhelming majority of calls for service with a suspect still at the scene were domestic violence incidents, which is substantially different than the fear-invoking stranger violence. Collectively, the reality of police work precludes regular interactions with the violent offenders who are responsible for what police consider serious crimes, such as robberies, aggravated assaults, and murders.

In the rare event that officers get the opportunity to perform real police work—that is, handle an individual suspected of immediate violence—certain factors influence their decisions. Consistent with the discussion on tolerance of violence and tattered relationships between the police and citizens in urban areas, police officers are generally more likely to make an arrest, versus mediate conflict, in lower-socioeconomic communities (Smith, 1987). These findings suggest that the police–citizen interactions in these environments are contaminated with poor communication, further exacerbating their already tenuous relationship.

The lack of communication between witnesses and the police without doubt strains their relationships, but the official processing of these cases may complicate it as well. Because witnesses are unlikely to cooperate with the police when they perceive the incident to be justified or when they lack confidence in the police, officers are frequently left to create their own interpretations of the event. In attempts to make sense of the violence they failed to understand, police officers and prosecutors have applied the illicit drug market and gang affiliation nexuses to violent crimes, even when the crimes involve neither drug deals nor gang affiliations. For the police and prosecutors, these inaccurate recounts are utilitarian, in the sense that they enable authorities to close a case by arrest and/or conviction. Thus, however inaccurate they are, these descriptions satisfy common sense logic to those *outside* the communities. While they may help officers and prosecutors to clear the cases, however, the accounts bewilder community residents who were privy to the incident (Duck, 2009). This official approach of making sense of violent crime not only labels communities and those that live there as undesired, but also isolates residents from patrol officers. This process serves to compound the extant mistrust of the police, thereby diminishing officers' abilities to intervene in violence.

ONE CITY'S APPROACH

The aforementioned factors, uncooperative citizens, ever elusive violent suspects, reoccurring disappointment of trumped-up calls for service, poor officer–citizen communication, and official misunderstanding of violent encounters create occupational frustrations for patrol officers. Although research has suggested that proactive policing strategies fail to produce crime-reducing outcomes, patrol officers (and police administrators for that matter) continue to hold onto the idea that directed patrols in high-crime areas, during high-crime times, will reduce crime (Kelling, Pate, Dieckman, & Brown, 1974).

In Fort Portland (a pseudonym for the large Southwestern city where we conducted the research described here; the Fort Portland Police Department maintained a sworn staff of more than 2500 officers), for instance, police ranks from top to bottom repeatedly emphasized the need for proactive policing. Fort Portland used a three-pronged approach to proactive policing that included a centralized saturation patrol unit that identified and policed several locations across the city each day, district-level officers who were instructed to avoid answering calls for service and to "hunt" for crime, and patrol officer–initiated contacts when a lull in dispatched calls occurred. Although these policing strategies ran counter to the available research and even failed to produce crime reductions in Fort Portland, officers defended these approaches through anecdotal stories of suspect apprehension. This continued allocation of police resources to proactive policing strategies that failed to produce suitable outcomes (crime reduction) merely served to increase officer frustrations and engender a self-preservation strategy of displacing blame for crime.

Although a systematic decline in violent crime has occurred in recent years, police remain puzzled by violence and continue to equivocate about which, if any, steps they can take to solve their violent crime problem. While national reports have revealed that less than half of all violent crime is reported to the police, with police ultimately clearing less than half of those cases reported (Snyder & Sickmund, 2006; Truman & Rand, 2010), Fort Portland officials focused on policing strategies to combat violence. Regardless of the national-level data, commanders in this city struggled with crime rates on a weekly basis, always pushing for deeper reductions in crime. When crime decreased, commanders attributed the decline to their "crime fighting" strategies; when they experienced an increase in crime, however, they sought to lay blame beyond their doorstep. From violent offender task forces to proactive patrols in low-income areas, commanders were unable to isolate crime rate fluctuations. This inability to predict outbreaks of violence simply led to greater frustrations.

Police officials and patrol officers alike frequently placed the blame for their inability to reduce Fort Portland's crime on the local prosecutor and court systems. Officers of varying ranks indicted the "liberal D.A." (district attorney) who "was too busy being a politician to prosecute criminals" in both private and official capacities. One Sergeant complained, "We have a bunch of Democratic judges here and a D.A. that doesn't do [expletive]! He [the D.A.] doesn't want to put bad guys in jail." While the district attorney was, indeed, involved in several high-profile situations, this perception of a faulty legal system was not unique to Fort Portland or only the adult court system. Other studies have discovered similar situations where police officials blamed the judicial system for perceived soft treatment of criminals (Ezeonu, 2010).

Fort Portland officers extended this accusation to the juvenile justice system as well. "They just slap 'em on the wrist and send them home to mommy and daddy," exclaimed one officer when asked about juvenile detention.

Collectively, this attitude suggests that police officers exhausted their resources apprehending suspects of violent crimes and that the courts, in turn, placed suspects "back on the streets before [an officer's] shift is over." These statements and the general sentiment of the Fort Portland police department suggested that pretrial detention was akin to a turnstile, where suspects were logged in and then released quickly back to the streets. From a policy perspective, this situation raised two important questions: Do violent crime suspects experience preventive detention? If so, are they detained for longer periods than nonviolent crime suspects?

JUSTIFIED BLAME?

To explore the possibility of a faulty system as suggested by Fort Portland police officers, we pulled the records of a representative sample of offenders from the corresponding juvenile probation department. We decided to use juvenile records for two reasons: (1) juvenile records are typically more comprehensive and contain richer information and (2) juveniles were not afforded the right to bail in this jurisdiction, so we were able to isolate any financial disparity introduced by the bail system. Under state mandate, the juvenile court was obligated to release a juvenile unless the court was able to establish at least one of the following conditions:

- The juvenile was likely to be removed from the jurisdiction.
- The juvenile lacked suitable supervision in the home.
- The juvenile was unlikely to return for court hearings.
- The juvenile was a danger to self or the public.

- The juvenile had a previous adjudication for which the punishment included incarceration.
- The juvenile was suspected of a crime that involved a firearm.

While several of the detention requirements allude to extra-legal factors, the mandate also allowed for offense severity to be a detention justification. Considering the alternative, we used the most appropriate data available.

Sample Descriptives

Table 16.1 presents an overall description of the suspects referred to the juvenile court for an offense in Fort Portland. Of the 8205 juveniles, slightly less than three-fourths (72%, 5907) were males and more than 80% of the sample were categorized as minorities (42% African American, 39% Hispanic, and 2% another race). Familial dynamics introduced three distinct groups as well, with slightly less than half (45%) of the sample residing with a single parent, less than one-third of the sample (30%) living in a traditional family environment, and one-fourth of the sample (25%) living in some alternative form of home environment (e.g., living with a relative or group home). Additionally, the majority of the sample group did not report gang affiliation or experiencing emotional, physical, or sexual abuse (only 8% reported gang affiliation or abuse). As these statistics suggest, the average juvenile was male, was a minority, and lived in a nontraditional family environment.

Beyond the extra-legal distinctions, the sample was divided by legal factors. Of the total sample, slightly less than half of the juveniles were suspected of either a property crime or a violent crime (24% and 21%, respectively), slightly more than one-third (36%) for violating a court order, and slightly less than one-fifth (19%) for other offenses, including a drug offense. Finally, and the focus of this exploration, more than half of the sample members (59%) were detained for some period of time after being arrested; on average, a detained juvenile remained in detention for slightly less than three weeks (19.18 days). The state mandated that the juvenile court afford all detained juveniles a detention hearing within two working days after the juvenile arrived in detention or, if referred during the weekend, on the first working day after arrival. After the initial detention hearing, juveniles returned to the juvenile court every 10 days for a detention review.

If, as officers indicated, the court system failed to deal with suspects correctly, we would expect to see little distinction between juveniles with varying numbers of previous referrals. In short, the number of days a suspect was detained would not be contingent on the number of previous referrals. **Figure 16.1** depicts the percentage

TABLE 16.1 Sample Descriptives

	Total Sample N = 8,205	
	f	%
Variables		
Male	5,907	72
Age ($\bar{x}$, sd)	15.17	0.78
Ethnicity		
African-American	3,458	42
Caucasian	1,444	18
Hispanic	3,176	39
Other Race	127	02
Family Dynamics		
Single Parent	3,664	45
Two-Parent	2,496	30
Extended Family	579	07
Non-Family	1,223	15
Other Living Arrangement	54	01
Gang Affiliate	665	08
Abused	673	08
Legal		
Violent Offense	1,746	21
Property Offense	1,975	24
Drug Offense	858	10
Violation of Court Order	2,962	36
Failure to Attend School	129	02
Other Offense	524	06
Incapacitation		
Detained	4,804	59
Days Detained ($\bar{x}$, sd)	19.18	27.99

Note: Some percentages may not sum to 100 due to rounding error or missing data.

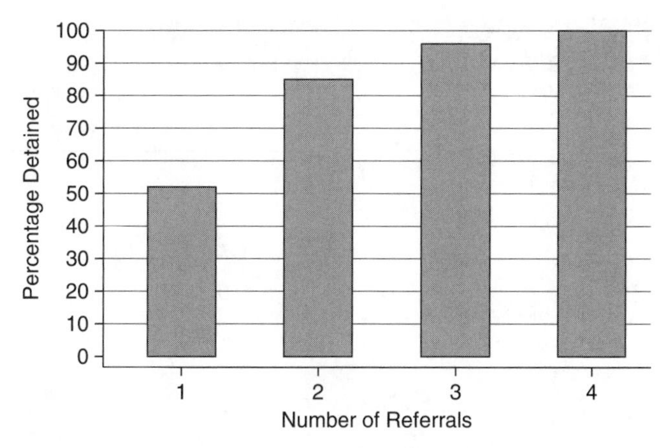

FIGURE 16.1 Percentage detained per referral.

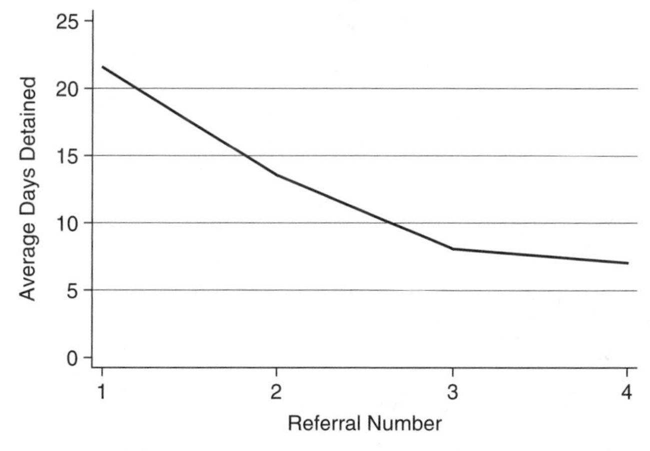

FIGURE 16.2 Average days detained by the number of referrals.

of detained juveniles by referral to the juvenile court. As indicated in the figure, more than half (52%) of juveniles with no previous referral were detained, with this percentage significantly increasing (analysis not shown in tabular form) with each new referral. Eighty-five percent of juveniles with one previous referral were detained, 96% of juveniles with two previous referrals were detained, and *every* juvenile who had three previous referrals experienced detention.

While the likelihood of detention significantly increased as the number of referrals increased, **Figure 16.2** suggests that juveniles with more previous referrals actually spent significantly less (analysis not shown in tabular form) time in detention than first-time juvenile suspects. Overall, these data indicate that while the likelihood of pretrial detention may have increased, those at greatest risk of detention (i.e., more

TABLE 16.2 Sample by Misdemeanor and Felony Offense

	Misdemeanor n = 6,174		Felony n = 2,031	
	f	%	*f*	%
Variables				
Male**	4,144	67	1,763	87
Age ($\bar{x}$, sd)	15.18	0.01	15.15	0.02
Ethnicity				
African-American*	2,555	41	903	44
Caucasian**	1,042	17	402	20
Hispanic**	2,475	40	701	35
Other Race	102	02	25	01
Family Dynamics				
Single Parent**	2,876	47	788	39
Two-Parent**	1,931	31	565	28
Extended Family**	402	07	177	09
Other Living Arrangements	38	01	16	01
Gang Affiliate**	438	07	227	11
Abused	501	08	172	08
Legal				
Violent Offense**	932	15	814	40
Property Offense**	1,112	18	863	42
Drug Offense**	530	09	328	16
Incapacitation				
Detained**	3,189	52	1,615	80
Day Detained ($\bar{x}$, sd)**	15.21	0.42	27.01	0.84

*Significant at $p < 0.05$
**Significant at $p < 0.01$
Note: Violations of Court Orders and Failure to Attend School were excluded as they were only misdemeanors in this jurisdiction.

previous referrals) spent less time there. However, even those juveniles with three previous referrals averaged a week in detention.

Comparisons

To further explore the Fort Portland officers' perceptions of a faulty legal system, we observed the difference between misdemeanor and felony suspects, as presented in

Table 16.2. The overwhelming majority (75%, or 6174 juveniles) were suspected of a misdemeanor offense, while 2031 juveniles were suspected of a felony offense. Although misdemeanants shared some similar characteristics with felonious suspects (e.g., age and abuse), overwhelming differences were noted between these two groups. In terms of extra-legal factors, felony suspects were significantly more likely to be males, African American or Caucasian, living with an extended family member, and a gang affiliate, but significantly less likely to be Hispanic, and from a single-parent or two-parent home. In terms of legal factors, felony suspects were also significantly different than their misdemeanant counterparts. Specifically, felony suspects were significantly more likely to be arrested for violent, property, or drug offenses. Moreover, and pertinent to the focus of this study, felonious suspects were significantly more likely to be detained than misdemeanants (80% versus 52%, respectively) and, when detained, felonious suspects remained detained for significantly longer periods than misdemeanants (27 days versus 15 days, respectively). Thus, when their crimes were detected, felony suspects remained in detention for almost twice as long as the misdemeanor suspects.

Table 16.3 compares property crime suspects and violent crime suspects. Violent crime suspects were similar to property crime suspects with respect to several characteristics. From a probability perspective, violent crime suspects were just as likely to be male; Hispanic; from a single-parent, two-parent, or other living arrangements home; or a gang affiliate as property crime suspects. Furthermore, violent crime suspects were similar in age to property crime suspects. Similarities aside, violent crime suspects were distinguishable from property crime suspects in several aspects. First, violent crime suspects were significantly more likely to be African American and significantly less likely to be Caucasian or of the "other race" category. Second, violent crime suspects were significantly more likely to reside with an extended family member and significantly more likely to have experienced abuse. Finally (and the focus of this study), violent crime suspects were significantly more likely to be detained than property crime suspects (77% versus 65%, respectively). When they were detained, violent crime suspects remained in detention for a significantly longer period than property crime suspects (25 days versus 16 days, respectively). These findings suggest that violent crime suspects spend, on average, slightly less than a month in pretrial detention—substantially longer than property crime suspects.

PERCEIVED REALITY

Comparisons of detention outcomes for misdemeanor and felony offenses, for property and violent offenses, and by the number of previous referrals produced several distinguishing relationships. In relation to Fort Portland police officers' perceptions,

TABLE 16.3 Sample by Property or Violent Offense

	Total n = 3,721		Property n = 1,975		Violent n = 1,746	
	f	%	*f*	%	*f*	%
Variables						
Male	2,871	77	1,528	77	1,343	77
Age ($\bar{x}$, sd)	15.15	0.01	15.16	0.02	15.15	0.02
Ethnicity						
African-American**	1,743	47	837	42	906	52
Caucasian**	821	22	489	25	332	19
Hispanic	1,091	29	601	30	490	28
Other Race**	66	02	48	02	18	01
Family Dynamics						
Single Parent	1,691	45	903	46	788	45
Two-Parent	1,052	28	563	29	489	28
Extended Family**	294	08	133	07	161	09
Other Living Arrangements	36	01	17	01	19	01
Gang Affiliate	328	09	183	09	145	08
Abused**	365	10	134	07	231	13
Incapacitation						
Detained**	2,625	71	1,284	65	1,341	77
Day Detained ($\bar{x}$, sd)**	20.33	0.92	15.52	0.65	24.94	0.92

Note: All non-property and non-violent offenses were excluded from these analyses.
*Significant at $p < 0.05$
**Significant at $p < 0.01$

the data suggested that the juvenile probation department utilized detention more frequently and for longer periods than believed. More than half of in-custody referred suspects were detained beyond the initial entry into the detention facility. Among suspects who were detained, they averaged slightly less than three weeks in detention before being released. While those suspects who had more previous referrals were more likely to experience detention, as previous referrals increased, the amount of time in detention decreased. Given that the decision to detain was largely influenced by the offense severity (i.e., misdemeanor versus felony offenses), felony suspects were significantly more likely to be detained, and to be detained for longer periods, than misdemeanor suspects. Specific to violent offense suspects, the data suggested that they were significantly more likely to be detained and remained in detention for a significantly longer period. Overall, the results produced a consistent theme: Not only was detention used for suspects arrested by police officers, but the offense magnitude also influenced the detention decision and length of detention.

The only inconsistent finding was that as the number of previous referrals increased, the number of days detained decreased. On its face, this result appears disturbing and especially discouraging for patrol officers who work to reduce crime on their beat. Police officers reported, and their various policing strategies supported, a stance that habitual offenders were to blame for the majority of crimes. If habitual offenders were disproportionately responsible for crimes and the judicial system did not exercise preventive detention, then it is possible that officers' perceptions of a faulty legal system were a reality.

While such an outcome might potentially be a product of an ill-equipped legal system, several other factors should be considered before making such a condemning accusation. First, we were unable to discern whether subsequent referrals were original or connected to some continued probation supervision. In other words, we were unable to identify those suspects who were on community supervision at the time of the current referral. If the suspect was on probation at the time of the new referral, then changing supervision orders would have required less time and resources than an original referral.

Second, referrals for those under supervision could have included violations of supervision orders, with suspects being considered delinquent only because of the prior court-ordered supervision (e.g., associating with a known gang member while on probation). This pattern would suggest that these violations were less severe than the original referral and, therefore, produced less time in detention.

Third, although the number of days detained significantly decreased as the number of previous referrals increased, suspects with three previous referrals still remained in detention for approximately a week before being released. This is hardly the "in and out" process described by officers.

Finally, while the number of days in detention decreased as the number of previous referrals increased, the percentage of suspects detained significantly increased as the number of previous referrals increased. By the fourth referral, all suspects were detained. Despite the fact that the number of days in detention decreased as the number of previous referrals increased, several explanatory caveats are possible.

If, as suggested here, violent offenders do experience preventive detention but police officers perceive an alternative reality, it becomes necessary to determine why such a discrepancy occurs. In other words, why do officers perceive the judicial system to be "broken"? To answer this question, it is essential to look beyond the reality of detention decisions and length of detention stays and explore violence from the officers' perceptive—that is, what they experience on the streets.

As noted earlier in this chapter, subcultural acceptance of violence and distrust and a lack of confidence in the police tend to reduce reporting to and cooperating with the police in violent crime situations. In addition, the majority of cases reported to the police go unsolved in this situation, and research has suggested that detectives and prosecutors focus more on case solution than situation resolution. What we have yet to explore is how officers' environments influence their perceptions. The following discussion focuses on the organizational demands in this system as well as patrol officers' exposure to violent crime suspects.

The Fort Portland police department used the Compstat model of policing and conducted a weekly meeting for the entire department. The Compstat policing model focuses on mission clarification, internal accountability, geographical organization of command, organizational flexibility, data-driven problem identification and assessment, and innovative problem solving to reduce crime (Weisburd, Mastrofski, Willis, & Greenspan, 2006). In preparation for these meetings, upper- and mid-level management conducted similar Compstat meetings at their respective levels. This process produced at least three different meetings that covered the same topic each week. Almost without exception, Compstat meetings followed a rigid format of crime statistics consisting of year-to-date, month-to-date, and percent change metrics; strategies, including what worked and what didn't work; and anecdotal evidentiary support. With accountability at the foundation of the Compstat model, mid-level police managers were confronted with crime statistics and, more importantly, were expected to provide reasons for the fluctuations. This process was extended to the lower-level Compstat meetings as well.

While the effectiveness of Compstat is beyond the goal of this discussion, there were several unintended consequences of this practice that influenced how officers dealt with crime. For starters, just the design encouraged mid-level managers to attribute decreases in crime to patrol strategies: "We have the second worst areas for robberies, but we're making arrests, so we've had some successes" and "Officers are getting into the swing of looking for [offenders]." This posturing continued even if the managers lacked empirical support for their assertions. Conversely, police managers were quick to blame environmental factors when crime increased: "It's gettin' warm, people are gettin' out [in the community]," "They're in [jail] and they're right back out and we're chasing them again," and, in response to a 278% increase in crime, "Crime statistics can be manipulated, what can I say?" Finally, when a particular patrol bureau was the focus of inquiry, it was common for the bureau chief to provide a laundry list of auxiliary support units—for example, "We're going to get mounted [patrol], bike [patrol], and walking patrols with *Operation Stop and Contact* [a pseudonym for an operation that involved making contact with suspected criminals in communities]; I think we'll be good". From a utility perspective, the Compstat approach failed to provide consistently decreasing crime rates, but did place additional pressures on line-level officers.

Beyond the additional pressures placed on patrol officers to reduce crime rates on their beats, they were inundated with special operations. In less than six months, four (two for violent offenders and two for illegal narcotics) separate department-wide special operations pulled officers from their assigned beats to reduce serious crime. Although a significant relationship between illegal narcotics enforcement and serious crime has yet to be established, command staff stated that the two drug offender apprehension operations would reduce violent crime. In addition to command staff being unable to empirically establish these claims, the operations placed increased strain on patrol by pulling officers from their respective beats. Not only were beat officers reassigned to these sorts of department-wide operations, but each patrol bureau was responsible for additional special operations. One such operation targeted a gang-infested apartment complex with more than 100 officers, air support, and inter-agency collaboration. This operation produced only five felony arrests, approximately ten misdemeanor arrests, a small amount of marijuana and cocaine, and one handgun—hardly a bountiful reward given the cost.

While the premise of the Compstat policing model may seem logical, it focused officers' attention on measures that did not consistently produce lower crime rates. This forced activity and unreliable results served to further frustrate officers when they achieved an adequate activity level, only to see increased crime rates.

With organizational directives pressing for efficient policing, patrol officers were sent out to capture perpetrators and combat crime. In reality, however, patrol officers spent very little time working calls that were remotely associated with violent suspects. This finding was not especially novel, as previous scholars have highlighted the relatively quiet daily life of patrol work (Reiss, 1971). While patrol work did not produce the excitement portrayed in mass media, it offered opportunities to encounter felony and, more to the point, violent suspects. Unfortunately (given the organizational pressures to capture these perpetrators), the lack of an incident, a suspect, or a willing complainant almost always foiled officers' opportunities. When officers did come in contact with violent suspects, they were typically previously convicted violent offenders and the reason for contact was of a less serious nature. The following account is representative of these contacts with violent offenders.

We were dispatched to assist another unit with mutual combatants in the street and blocking traffic. When we arrived, the two bloody white males who had been fighting were in handcuffs on the sidewalk. The suspects claimed to be friends and did not want to press charges against each other. After questioning the suspects, the responding officers discovered that both suspects were homeless and one had been incarcerated for a felony offense until his release the day before. The ex-con continued to cuss at the officers and act belligerent, which resulted in the officers booking him in jail for being intoxicated in public. His co-combatant was told to go away.

The ex-con escalated the situation by being disrespectful and being what previous research has referred to as an "asshole" (Van Maanen, 1978, p. 221), which resulted in his arrest for a public order offense while the other suspect was released at the scene. More frequently, however, officers arrested suspects based on a warrant issued for a previous investigation, not due to any immediate danger.

This lack of arrest for an immediate danger should in no way be misconstrued as suggesting that officers were unwilling to arrest violent suspects. On the contrary, officers generally jumped at the opportunity to engage violent suspects and valued these opportunities. Almost without exception, however, code calls (those in which officers were authorized to use lights and sirens and that typically involved an "in progress" crime) produced no actual offense, a missing suspect, or an unwilling complainant. On two occasions, officers were dispatched to violent crimes in progress (gunfire at a nightclub and a sexual assault in progress) in which, respectively, there was no gunfire and the sexual assault complainant accused a 19-year-old male, her daughter's boyfriend, of sexually assaulting her 17-year-old daughter. While the sexual assault accusation merited investigation, the complainant had been aware of the sexual relationship between her daughter and the adult male for several months and the alleged sexual assault had occurred several days prior. The suspect, as might be anticipated, could not be located.

On those occasions when officers were able to determine that an immediate danger existed, officers were rarely able to make contact with the suspect during the event. Although the local bus station was a nexus of criminal activities, officers were typically unable to catch suspected violent criminals in the midst of their activities. From the suspect who asked citizens for money and then threatened to assault them if they refused to the victim of robbery with serious bodily injury, officers were unable to locate the suspects upon arrival. This phenomenon was not unique to highly transient locations such as the bus station, either. Of all the robbery or shooting calls encountered during the course of the research, none produced an immediate suspect. Even more frustrating for the responding officers were unwilling complainants. In responding to two different calls where individuals had been shot (one inside a house and the other just outside the door of this apartment), officers encountered complainants (and witnesses) who were unwilling to provide descriptions of the suspect.

The combination of organizational pressures to apprehend these violent suspects and the environmental restraints (lack of reporting, untimely reporting, false reporting, lack of a suspect, and unwilling complainants) merely served to increase patrol officers' frustrations. These frustrations evolved into cynicism directed toward the system—specifically, those aspects beyond those over which officers had control. Ultimately, officers experienced significant pressures to apprehend criminals in unfavorable situations; when they did apprehend someone, it was easy for them to associate the frequently elusive suspects with the one they actually caught. In other words, officers associated the magnitude of a suspect's crime with the frequency of the ones who got away. From this mentality and their proximity to the criminal event, they often discounted the need for a trial to establish guilt and perceived anything less than an immediate, long-term incarceration to be an insufficient outcome.

CONCLUSION

The primary goal of this chapter was to evaluate the judicial system's response to violent offenders as it relates to patrol officers' perceptions. While the results of the study described here refuted officers' perceptions of a lenient judicial system, a further exploration of officers' experiences with violence was the cornerstone to understanding the associated discrepancy between their perceptions of a faulty system and reality. By exploring violence from a patrol officer's perceptive, it became possible to identify the structural factors that influenced patrol officers' experiences with and approaches to violence.

From the outset, patrol officers were at a disadvantage concerning violent crime. Approximately half of all violent crime was unreported to the police. When such crime was reported, officers experienced an overwhelmingly lack of cooperation from complainants and witnesses alike. Sociologists have framed this environment as one where machismo is coveted, vigilantism trumps reporting crimes to officials, violence is tolerated, and distrust of and lack of faith in the police is frequent (Anderson, 1999; Jacobs & Wright, 2006); thus it is unsurprising that patrol officers reported frustrations with violence on their beats.

The frustration that arose from providing services to persons who did not want help was compounded by the organizational pressures that officers experienced under the Compstat policing model. Patrol officers were constantly pressured to improve performance measures (e.g., response times) with the expectation that these measures were somehow correlated to violent crime. When the performance measures improved but crime remained constant, officers invoked survival mechanisms to explain away the blame. Unfortunately, the judicial system—a system generally misunderstood or at least loathed in police circles—was an easy target.

The findings of our study suggested that the judicial system was not at fault, however, as it distinguished between offense levels and typologies. Given that violent suspects spent almost a month in pretrial detention, the results did not support officers' perceptions of a faulty judicial system.

If we were charged with finding fault, there is no question that the police management strategy used by the Fort Portland Police Department would be considered suspect. This is not to say that the police department was incapable of apprehending violent suspects. On the contrary, there was always a steady stream of detectives filing cases with the district attorney. Nevertheless, the Compstat process focused on patrol strategies, not investigative expertise. Ultimately, patrol officers were held accountable for crimes that typically involved missing suspects and unwilling complainants.

REFERENCES

Anderson, E. (1999). *Code of the street: Decency, violence, and the moral life of the inner city.* New York: Norton.

Bayley, D. H., & Garofalo, J. (1989). The management of violence by police patrol officers. *Criminology, 27,* 1–25.

Conklin, J. E. (2003). *Why crime rates fell.* New York: Allyn & Bacon.

Duck, W. (2009). "Senseless" violence: Making sense of murder. *Ethnography, 10,* 417–434.

Ezeonu, I. (2010). Gun violence in Toronto: Perspectives from the police. *Howard Journal of Criminal Justice, 49,* 147–165.

Fagan, J., Zimring, F. E., & Kim, J. (1998). Declining homicide in New York City: A tale of two trends. *Journal of Criminal Law and Criminology, 88,* 1277–1324.

Greenberg, M. S., & Ruback, R. B. (1985). A model of crime-victim decision making. *Victimology, 10,* 600–616.

Hart, R. C., & Rennison, C. (2003). *Reporting crime to the police, 1992–2000.* Bureau of Justice Statistics (Special Report). Washington, DC: U.S. Department of Justice.

Hindelang, M. J., & Gottfredson, M. (1976). The victim's decision not to invoke the criminal justice process. In W. F. McDonald (Ed.), *Criminal justice and the victim* (pp. 57-78). Beverly Hills, CA: Sage.

Horowitz, R. (1987). Community tolerance of gang violence. *Social Problems, 34,* 437– 450.

Jacobs, B. A., & Wright, R. (2006). *Street justice: Retaliation in the criminal underworld.* New York: Cambridge University Press.

Kelling, G., Pate, T., Dieckman, D., & Brown, C. (1974). *The Kansas City preventative patrol experiment: A summary report.* Washington, DC: Police Foundation.

Moore, M. H. (1980). The police and weapons offenses. *Annals of the American Academy of Political and Social Science, 452,* 22–32.

Reiss, A. J. Jr. (1971). *The police and the public.* New Haven, CT: Yale University Press.

Ruback, R. B., & Thompson, M. P. (2001). *Social and psychological consequences of violent victimization.* Thousand Oaks, CA: Sage.

Shaw, C. R., & McKay, H. D. (1969). *Juvenile delinquency and urban areas.* Chicago, IL: University of Chicago.
Smith, D. A. (1987). Police response to interpersonal violence: Defining the parameters of legal control. *Social Forces, 65*, 767–782.
Snyder, H. N., & Sickmund, M. (2006). *Juvenile offenders and victims: 2006 national report.* Washington, DC: U.S. Department of Justice, Office of Justice Programs, Office of Juvenile Justice and Delinquency Prevention.
Truman, J. L., & Rand, M. R. (2010). *Criminal victimization, 2009.* Washington, DC: U.S. Department of Justice.
Van Maanen, J. (1978). The asshole. In P. K. Manning & J. Van Maanen (Eds.), *Policing: A view from the street* (pp. 221–238). Santa Monica, CA: Goodyear.
Virgil, J. D. (2003). Urban violence and street gangs. *Annual Review of Anthropology, 32*, 225– 242.
Weisburd, D., Mastrofski, S.D., Willis, J. J., & Greenspan, R. (2006). Changing everything so that everything can remain the same: Compstat and American policing. In D. Weisburd & A. A. Braga (Eds.), *Police innovation: Contrasting perspectives* (pp. 284-304). New York: Cambridge University Press.
Wilson, J. Q. (1978). The police and crime. In P. K. Manning & J. Van Maanen (Eds.), *Policing: A view from the street* (pp. 3-25). Santa Monica, CA: Goodyear.
Wilson, J. Q. (1980). What can the police do about violence? *Annals of the American Academy of Political and Social Science, 452*, 13–21.

CHAPTER 17

Sex Offenders on the Internet: Cyber-Struggles for the Protection of Children

Frank Kardasz
Arizona Internet Crimes Against Children Task Force and the Phoenix, Arizona Police Department

The Internet provides resources that facilitate communication, education, commerce, and, unfortunately, crime. Research from the University of Southern California indicates that 78% of Americans, including children, visit cyberspace (University of Southern California, Annenberg School, 2005). Other researchers who have specifically examined children's use of the Internet found that young people are frequently being exposed to sexually-related material and are also encountering predators who solicit them for sex (Wolak, Finkelhor, & Mitchell, 2005a). Some criminals use cyberspace to traffic images depicting the sexual abuse of children. Some offenders use the Internet for the purpose of luring and enticing minors towards sex. The quiet collision of young people and sex offenders on the Internet has resulted in a cyber-struggle for the protection of children. This chapter explores topics related to the sex offenders who use the Internet to victimize youth.

There are many sex offenders, both registered and unregistered, in every U.S. state. The Family Watchdog group (2006) reports that there are 344,362 registered sex offenders in the United States, many of whom offended against minors. According to the U.S. Justice

Department, two-thirds of the victims of reported sexual assault are minors, and one out of every three victims of a sex offense is younger than age 12 (Allen, 2006). Among the most disturbing sex crimes being facilitated via the Internet are those that involve child and teen victims. In 2002, President George W. Bush said, "In the hands of incredibly wicked people, the Internet is a tool that lures children into real danger." (U.S. Department of Justice, 2006). One study found that 34% of children using the Internet were exposed to unwanted sexually explicit material. The same study also revealed that one in seven youngsters received unwanted sexual solicitations and that 4% received aggressive solicitations involving a stranger who wanted to meet in person (Wolak, Mitchell, & Finkelhor, 2006).

It is difficult to calculate the true number of Internet crimes against children. Unlike spectacular and conspicuous crimes involving crashes, explosions, fires, shootings, and public bloodletting, sex crimes against children are committed in quiet and private places. Internet predators may use chat rooms, web-cams, and social networking sites to find and befriend victims. Offenders may use cyberspace to slowly groom, psychologically control, humiliate, or intimidate their young victims—first into compliance, and then into lifelong silence. As a consequence, sometimes the crimes against children are never reported.

Offenders often use Internet resources in deviously creative ways. They may construct online identities portraying themselves as amiable adult mentors. They may pose as other children of the same age as their intended victims. They may anonymously browse web pages for many hours until they find a preferred target and then work to gather enough information to facilitate an offense.

Moreover, the cyber-deviants who lure children and those who traffic child pornography sometimes find hobbies and occupations that facilitate their crimes. Some are professional photographers, computer technicians, or information technology specialists. Others find occupations in positions of trust where they can constantly be near children. Internet sexual criminals may be found in occupations including teachers, priests, day-care workers, police officers, doctors, or nurses.

There is no single profile of an Internet sex offender. These individuals come from all walks of life and work in a variety of occupations. Most offenders are males, but occasionally females are also involved in Internet crimes against children.

CHILDREN AS THE MARGINALIZED IDEAL VICTIMS

Offenders worldwide are targeting children for pornography as the international trade in unlawful images expands across the globe. Sex offenders everywhere consider children to be the perfect victims for several reasons. In some places, including Russia, children have become a commodity. There are five reasons for this phenomenon have been suggested: (1) children are plentiful and easily accessed; (2), child pornography is easy and inexpensive to produce; (3), there is a huge consumer market for child pornography; (4) it, it is enormously profitable; and (5) trafficking in online pornography carries far less risk than working with traditional illegal commodities such as drugs, guns, and tobacco (Allen, 2006).

Helpless child victims cannot summon the assistance of law enforcement. Indeed, some victims of child pornography are too young to know how to use a telephone. The evidence of a child's victimization is typically invisible to the general public, and the crimes are often unreported. Because crimes against children are not publicly apparent, law enforcement agencies may marginalize the problems and give the crimes lower priority than other offenses. Consequently, there are fewer law enforcement resources devoted to the problems of children. Most police departments have many more traffic cops than investigators devoted to crimes against children investigators.

Attacks against children are mostly committed in private locations. These offenses do not create the conspicuous noise of gunfire or a car crash. They are not accompanied by the noticeable smell of smoke from a fire and cannot be seen from the street like graffiti or broken windows, so there is these crimes attract little public attention drawn to the crimes. Offenders know that children can be easily intimidated into silence and often cannot communicate well enough to be understood by authorities. For offenders, children are ideal victims.

THE SCOPE OF THE PROBLEM

The luring and enticement of minors via the Internet is a widespread problem that is difficult to measure. Crimes involving minors who are lured or enticed are probably under-reported. Researchers who conducted telephone interviews with teens were discouraged to find that very few youths ever told authorities about episodes involving online misconduct. In more than half the cases, youths did not tell their parents or any others about solicitations from possible Internet sexual predators (Wolak et al., 2006). Thus the true extent of the Internet luring and enticement problem remains unknown.

Internet crimes involving pornographic images of the sexual abuse of children are also widespread and probably too numerous to be measured. A recent Congressional study identified several key factors that have contributed to the proliferation of child pornography on the Internet. First, and perhaps most problematic according to the study, is the sheer number of child abuse images on the Internet. United States law enforcement sources estimated that there are approximately 3.5 million known child pornography images online (U.S. House of Representatives, Committee on Energy and Commerce, 2007). The exact number of child pornography websites is also difficult to determine. In 2001, the National Center for Missing and Exploited Children's (NCEMC) CyberTipline received more than 24,400 reports of child pornography. Five years later, that number had climbed to more than 340,000 (NCEMC, 2006). The Internet has fueled a tremendous and immeasurable increase in the amount of child pornography being produced, trafficked, and possessed worldwide.

INTERNET LURING AND ENTICEMENT OF MINORS

Curious, unsuspecting youngsters visit the Internet each day seeking friendship and information, but sometimes instead encountering sexual deviance and predators. The law enforcement investigators who assume online undercover identities for the purpose of apprehending predators well know that there are many offenders lurking in chat rooms, appearing on web cams, and placing false profiles on social networking sites. Internet social networking sites are places in cyberspace where subscribers may post personal information and share it with others. Both children and adults use social networking sites to communicate and to make friends. A recent study estimated that 55% of young people have established online profiles in one or more of the dozens of social networking sites (Pew Internet and American Life Project, 2007).

Most social networking sites are free and permit users to register without providing information about their true identity and whereabouts. Perhaps not surprisingly given the anonymity possible on such sites, many sites are well suited for molesters, who can pose as harmless mentors while disguising their true intent. There have been many incidents reported in which registered sex offenders who have created online profiles portraying themselves as inoffensive individuals seeking romance without any reference to their malevolent pasts (Kardasz, 2007). Although proactive undercover "sting" investigations often lead to the apprehension of Internet sexual predators, few of the actual teens who are victimized by such predators ever report the crimes. Victimized teens are often too embarrassed to report their experiences and fearful of their parents' wrath for disobeying rules against communicating with strangers online.

Sometimes a teen returns home after secretly meeting an Internet stranger without his or her parents ever discovering the illicit tryst. Recently, an Arizona Internet Crimes Against Children (ICAC) Task Force undercover officer posing online as a young girl was contacted by a man who requested a meeting for sex. When the man went to a location where he believed that he would be meeting a minor for sex, he was arrested. Investigators learned that the offender had previously met two girls whom he had victimized and to whom he had given sexually transmitted diseases. In their shame, the girls had never notified their parents of the crimes. The girls' distraught parents only learned of the offenses when detectives informed them of the suspect's confessions (Phoenix, Arizona Police Department, 2002).

In some cases, a child's natural curiosity leads them to Internet places where they do not belong, and with unintended results. Beginning at the age of 13, a California boy was repeatedly victimized by offenders who met him via the Internet after first seeing his image on a web cam. The boy suffered sexual abuses at the hands of the men who had first contacted him online. Some of the boy's Internet acquaintances had even assisted him in operating commercial pornography websites featuring sexual images and videos of himself (Eichenwald, 2005).

In many cases, the children who are lured by sexual predators will never come forward due to fear or a misplaced sense of guilt. A few of them are forever silenced by Internet sexual predators who lured them to meetings via the Internet, sexually victimized them, and then killed them.

Child prostitution is also being facilitated via the Internet. Pimps use message boards and social networking sites to find customers seeking to engage in paid sex acts with minors. For example, Cook County, Illinois police arrested three adults who used Craigslist, a free Internet advertising site, to offer the sexual services of girls as young as 14 years old. The illegal prostitution business resulted in profits of tens of thousands of dollars for the pimps. Undercover officers investigated and solved the case by responding to postings on Craigslist (Gutierrez, 2007).

In the past, child molesters were characterized as often lurking near schoolyards. Folklore held that child molesters frequented schoolyards because that is where the children were. The Internet is the new proverbial schoolyard. Cyberspace provides a ready hunting-ground for those who seek children.

IMAGES OF THE SEXUAL ABUSE OF MINORS

The Internet facilitates a variety of crimes involving images and videos depicting the sexual abuse of minors. According to Wolak, Finkelhor, and Mitchell (2005b), the Internet supports the child pornography market by making images and videos easily accessible. The Internet also contributes to the problem in other ways. Researchers believe that the online child pornography market may motivate offenders to produce images for trade and that the potential for child pornography distribution exists whenever an image is created. Because images can be easily scanned or uploaded to the Internet, so that production and trafficking can be easily accomplished. In the past, film had to go through third parties for development before it could be distributed. Today however, images can be produced without risk of the developer revealing the offenders' identity. The production of child pornography is also facilitated by the fact that images can be captured by hidden cameras and victims can be secretly filmed without their knowledge (Wolak, Finkelhor, & Mitchell, 2005b).

In 1986, a U.S. Attorney General's Commission found "substantial evidence" that child pornography is often used as a tool for molesting children. The report stated that pedophiles sometimes show their intended victims pictures of children engaging in sexual activity in order to persuade them that it is not wrong because other children are doing it. The Commission also noted, "child pornography is extraordinarily harmful both to the children involved and to society" and recommended that combating all forms of child pornography should be "a governmental priority of the greatest urgency." (National Law Center for Children and Families, 2006a).

The use of the Internet has contributed to an immeasurable enormous increase in the trafficking of images of the sexual abuse of children. Digital images and videos depicting the victimization of children are easily traded by Internet users who also derive sexual gratification from viewing child sexual abuse. Some videos are accompanied by the audio sounds of the victim crying or begging the offender to stop. New images are being created and reproduced each day, even as the stories accompanying the images become more and more disturbing. In 2002, an alert computer information technology professional working for an electronics-manufacturing firm in Arizona found child pornography on the computer of an engineer employed by the firm. Investigators learned that the engineer had viewed live web cam molestations being performed at another location by a man whose wife operated an in-home day care facility. Both men were subsequently arrested (National Law Center for Children and Families, 2006b).

Because cyberspace has no borders, it permits the rapid global distribution of child pornography. Nefarious entrepreneurs have turned the illegal trade into big business while partnering with others around the world. In April 2000, Thomas and Janice Reedy of Fort Worth, Texas, were indicted for commercial distribution of child pornography. Investigators from the U.S. Postal Inspection Service and the Dallas ICAC Task Force learned that the Reedys' illegal business had 70,000 customers worldwide. In one month, the business earned more than $1 million dollars, which was shared with Internet webmasters in Russia and Indonesia. The business employed a dozen U.S. workers and trafficked child pornography to subscribers across the globe (U.S. Department of Justice, 2006). The global connectivity of the Internet demands coordinated efforts between nations of the world so that the problem of child pornography can be stopped. The monetary profits from the trafficking of unlawful images must be curtailed so that children can be protected from the threat of those who would exploit them.

CHILD PORNOGRAPHY: GATEWAY TO DANGER

Some apologists for possessors of child pornography argue that possessors are simply harmless observers of images. Although it may be true that not all possessors can be assumed to be "hands-on" or "contact" offenders, recent research has indicated that child pornography possession is a strong indicator of pedophilia (Blanchard, Cantor, & Seto, 2006).

Some of the possessors of child pornography who are also pedophiles struggle psychologically with their tendencies. A 27-year-old Phoenix man was arrested at a residence where he lived with his wife, mother, and brother. The arrest was for the crime of trafficking images of child pornography via Internet file-sharing systems. No "contact" offenses involving actual children were uncovered. The man admitted to possessing child pornography and expressed deep remorse. He lamented about the inner psychological struggles he had experienced because of his intense sexual attraction to children. According to this offender, he had once attempted suicide by hanging because of his self-loathing. The man said that he had to avoid being near his brother's children because of his attraction to them. His pattern of child pornography possession included a repeating cycle. He would download images from the Internet, view them, and then delete them because of his shame. He would later reacquire additional images by again downloading them and then again deleting them. This pattern continued repeatedly because the offender could not overcome his obsession with the images that fed his sexual fantasies (Phoenix, Arizona Police Department, 2006).

Some pedophiles can suppress their urges to offend against children; others cannot. A study of 1713 child pornography possessors arrested in Internet-related crimes showed that 40% of those arrested for child pornography were "dual offenders" who both sexually victimized children and possessed child pornography, with both crimes discovered in the same investigation. In some cases, the pornography drove the offender to violent crimes (Wolak, Finkelhor, & Mitchell, 2005b). Although some theorists argue that there is no definite proof that a link exists between possessors and "hands-on" offenders, a study of 54 prisoners incarcerated for possession of child pornography showed that 79% had molested significant numbers of children without ever being detected. On average, each offender had sexually victimized more than 26 children, and overall the group admitted to more than 1400 contact sexual crimes. Of those contact sex offenses, only 53 were detected or known at the time the offender was sentenced for the child pornography offense (Hernandez, 2000).

Dr. Anna Salter, a psychologist who treats sex offenders, testified before a Congressional Subcommittee that was investigating Internet child pornography. Dr. Salter confirmed that a "considerable percentage" of individuals convicted for child pornography crimes had also committed contact offenses. She also suggested that the number of contact offenders may be underestimated, as one report found that only 3% of individuals who have committed contact offenses are ever caught. In addition to finding a link between possessing child pornography and committing contact offenses, Dr. Salter testified that she believes there is a link exists between viewing online pornography and committing contact offenses. With regard to reducing the number of contact offenses against children, she noted that reducing online child pornography would also reduce contact crimes, especially among those offenders who are emboldened to act on their sexual desires or urges after viewing child pornography. According to Dr. Salter, "child pornography increases the arousal to kids and is throwing gasoline on the fire." (U.S. House of Representatives, Committee on Energy and Commerce, 2007).

In some cases, the possessor of child pornography is not a contact molester. In other cases, actual molestation accompanies the suspect's collection of pornography. In either circumstance, the possibility that a child molester is collecting child pornography or the possibility that a child pornography collector is molesting children should always be aggressively investigated. Collecting child pornography should be viewed as significant criminal behavior by itself (Lanning, 2001).

DATA COLLECTION EFFORTS

The true scope of the Internet child pornography problem is difficult to measure. Among the dilemmas in trying to determine the actual number of offenses are the systemic failings of the national crime reporting methods that should capture the data but do not. Clearly, data collection methods for Internet crimes against children need improvement. Under the Uniform Crime Reporting (UCR) definitions, for example, there are neither specific crime analysis categories for either child pornography nor for the crime of luring minors via the Internet. Such crimes might be tabulated under the catch-all category of forcible rape, but only when an actual victim is subsequently identified. Thus offenses against children are truly marginalized under the UCR processes.

Newly devised national data collection methods are not much better. One research study suggested that the highly-touted National Incident-Based Reporting System (NIBRS) does not accurately portray the scope of the child pornography problem. The collection methods and processes meant to provide an accurate picture of crimes were deemed unreliable by these researchers and unable to delineate juvenile victims' exact connection in pornography crimes—either as subjects in images or as victims in sex crimes in which pornography was used. The researchers called for increased training of law enforcement officials so that information about pornography involving juveniles can be accurately and uniformly recorded in NIBRS. Nationally, the true number of Internet crimes against children is undoubtedly under-reported (Finkelhor & Ormrod, 2004).

Other agencies have collected data that indicates a burgeoning problem involving Internet crimes against children. Between 1996 and 2005, the number of child exploitation cases investigated by the FBI increased 2026%, from 113 investigations to 2402 reports. Both the Immigration and Customs Enforcement Service and the U.S. Postal Service have each also investigated thousands of cases. In 2005, the 46 Internet Crimes Against Children Task Forces nationwide in the United States conducted thousands of investigations, resulting in more than 1600 arrests and 6000 computer forensics examinations (U.S. Department of Justice, 2006).

One organization that collects and reports accurate statistics is the National Center for Missing and Exploited Children (NCMEC). Since 1984, it has compiled data on reported crimes against children. Today, it also operates a Cyber Tipline that permits Internet users to report misconduct either online or via telephone. Since it was established in March 1998, and as of January 2007, the Cyber Tipline had received 444,084 complaints of child pornography, child prostitution, sex tourism, online enticement, and other child-related Internet crimes. By comparison, in 2006, the 46 ICAC Task Forces throughout the United States received 17,346 cyber tips. The number of offenses reported to NCMEC is intriguing considering that the organization is not as well known to most citizens as their local police department and that only a fraction of the actual number of Internet crimes against children are likely reported to NCMEC.

NCMEC also maintains a database of the child victims of pornography whose identities are known to investigators. The Child Victim Identification Program began in 2002. As of January 2007, over more than 6.3 million images had been reviewed with more than 880 victims being identified. These numbers probably represent only a fraction of the actual number of images being trafficked via the Internet and only a small number of the child victims of sex abuse (Rabun, 2007).

As these data indicate, the scope of the Internet child pornography problem is beyond measure. Accurate accounting of the number of Internet crimes against children is needed.

THE MUTED COMMUNITY-ORIENTED POLICING RESPONSE

Traditional community-oriented policing approaches are ineffectual in evaluating the child pornography problem. Internet sex crimes against minors place substantial burdens on law enforcement for the following reasons:

- The crimes are widespread, occurring throughout the criminal justice system;
- They are multi-jurisdictional, so that they require extensive collaboration among law enforcement agencies;
- They involve constantly changing technology; and
- They require specialized investigation methods.

Former FBI Agent Ken Lanning has studied the problem of child molestation extensively. He described the unique systemic challenges of enforcing crimes against children:

> *Law-enforcement investigators must deal with the fact that the identification, investigation, and prosecution of child molesters may not be welcomed by their communities—especially if the molester is a prominent person. Individuals may protest, and community organizations may rally to the support of the offender and even attack the victims. City officials may apply pressure to halt or cover up the investigation. Many law-enforcement supervisors, prosecutors, judges, and juries cannot or do not want to deal with the details of deviant sexual behavior. They will do almost anything to avoid these cases (Lanning, 2001).*

Traditional approaches based on community-oriented policing theories are not applicable in the area of Internet crimes against children. In fact, many Internet crimes against children are only discovered as the result of undercover proactive investigations. Many more crimes are not reported and will never be reflected in crime statistics. Because few police agencies have investigators who work online in undercover assignments, the number of offenses reported will naturally be small. Anyone using an analysis based on the number of crimes reported may reach the wrongful conclusion that because there are few reports, no problem exists.

Thus an administrator who follows the community-oriented policing services guidebook for child pornography by examining the recommended statistics will undoubtedly notice the low number of reported crimes. Based on these data, officials will assume no problem exists and, therefore, will not devote any police resources to the issue. The approach suggested in the guidebook provides an easy way for police administrators to dismiss the problem and to deny that it exists.

Some proponents of community-oriented policing recommend surveys of the citizenry as a way to measure problems and evaluate police effectiveness. Such surveys are intended to gauge police performance and to identify community needs. The results from such surveys are often used to guide the allocation of resources towards those issues identified by respondents as needing attention. Sadly, child victims of Internet sexual predators and child pornographers cannot or do not respond to surveys. If they were capable of completing customer satisfaction surveys, child victims would likely rate police services as nonexistent. Unfortunately, citizen surveys are another poor way to measure crimes against children.

As this discussion suggests, traditional community-oriented policing approaches are not designed to accurately evaluate Internet crimes against children. Administrators should consider re-evaluating their responses to the crimes of child pornography and the luring of minors in cyberspace.

LAW ENFORCEMENT SPENDING

The Internet crime problem requires increased resources for law enforcement services, training, and equipment. All too often, however, resources for law enforcement functions of any kind are scarce. The competition for the limited available funds occurs in environments where agencies contend with one another for a larger slice of the budget. Knowledgeable advocates for children continually hope that more government dollars will be devoted to the enforcement of Internet crimes against children, but are often disappointed. In testimony to Congress, Grier Weeks, Executive Director of the National Association to Protect Children, said:

> *The federal government must get serious. We are losing this war [against child pornography], and we are not supporting our troops on the front lines. Recent estimates of the size of the exploding global criminal market in child pornography are in the multi-billion-dollar range. Yet, by no objective measure can we claim to be serious or prepared as a nation about stopping what is being done to these children. The FBI's Innocent Images National Initiative is funded at $10 million annually. By comparison, the Department of Housing and Urban Development just announced it was awarding more money than the entire Innocent Images budget to build 86 elderly apartment units in Connecticut and almost 7 times*

their budget on the homeless in Ohio. The administration has proposed 20 times the entire Innocent Images budget for abstinence-only education programs through the Department of Health and Human Services. The Department of Justice's Internet Crimes Against Children (ICAC) Task Force program received about $14.5 million in fiscal year 2006. That is less than one-fifth the amount proposed for a new initiative to help former prisoners reintegrate into society. Last year's budget included $211 million for the Department of the Interior for "high-priority brush removal" and related projects. $14.5 million doesn't clear much brush (Weeks, 2006).

In recent years, federal resources have gravitated towards anti-terrorism efforts, drug enforcement, and border control, all of which have legitimate nationwide importance. Local resources are typically devoted to curtailing homicides, sex assaults, gangs, drugs, burglaries, property crimes, and other offenses of great and legitimate local importance. Consequently, those who fight Internet crimes against children often find that their operations are underfunded and understaffed. Advocates for children agree that more funding for the enforcement of cyber-crime is needed. As the vital nature of these issues becomes more apparent to the public and to lawmakers, it is hoped that funding will increase.

WHERE IS THE INTERNET CRIME PROBLEM?

Internet crimes involving sex offenses are not confined to individual homes. Libraries, schools, businesses, and wireless free-access points are all part of the Internet crime domain. Many public libraries have resisted placing the restrictions and filters on computers that would make it tougher for offenders to commit crimes. In turn, many incidents involving pornography at library computers have occurred across the United States.

For example, in 2004, an Arizona man, while on parole for child molestation, repeatedly used unrestricted and unfiltered computers and printers at the City of Phoenix public library to download and copy child pornography. His crimes were only discovered during a routine search of his residence by parole officers who were following up on his earlier child molestation conviction (Johnson, n.d.).

In another case, a homeless man in Phoenix was apprehended in 2005 while online at a library computer attempting to sell child pornography to an online undercover investigator from the Arizona ICAC Task Force. The offender was observed using open-access City of Phoenix public library computers that did not require identification or a library card to operate. Investigators found computer storage media in his possession containing dozens of illegal images (Phoenix, Arizona Police Department, 2005a).

Private businesses and large corporations are not immune to the threat of Internet crime. Wherever there is Internet connectivity, predators may try to exploit the service for the purpose of finding victims. In 2005, an information technology specialist employed by an Arizona health care provider reported that an employee, a prominent neuropsychologist, had used his office workplace computer to contact young men for sex. The misconduct was reported to the Arizona ICAC Task Force and a proactive investigation began. A detective posing online as a minor was subsequently solicited for sex by the man; the man, in turn, was arrested. Follow-up investigators then learned of a Phoenix boy who had been molested by the man. The neuropsychologist was later sentenced to 17 years in prison (Phoenix, Arizona Police Department, 2005b).

Cyber-crimes against children occur from homes, businesses, and public places. The increasingly widespread unrestricted wireless access to computers means that criminals are, in turn, increasingly elusive. To apprehend these offenders, law enforcement must combine advanced technology with old-fashioned police work in order to apprehend the offenders.

Internet crimes cannot occur without an Internet service provider (ISP) allowing the offender access to the Internet. In most cases, the ISP is an unwitting facilitator to the offense. Other unwitting parties who benefit from Internet services may include those retailers that advertise on websites for the purpose of drawing buyers to their products. Some businesses have been surprised to learn that their advertising banners have appeared on web pages featuring child pornography.

Most ISPs charge a fee for their services, and individual subscribers often pay these charges with a credit card. The subscription and payment process provides a path for law enforcement to use subpoenas or search warrants to trace back to a subscriber by following the money trail. The subpoena and search warrant response process can be time-consuming for both law enforcement and the ISPs. Delays in the response process can, in turn, stall law enforcement investigative efforts.

In 1998, a federal law (U.S. Code, Title 42, Chapter 132) was passed requiring ISPs to report child pornography to the National Center for Missing and Exploited Children (NCMEC). By 2002, thousands of reports were flooding into NCMEC from the ISPs that chose to comply with the law. Those reports were subsequently sent to federal, state, and local agencies for investigation. The large number of reports quickly overwhelmed the small staffs of those few agencies that employed investigators who had the technical expertise needed to investigate Internet crimes. Investigators also began to complain that ISPs were sometimes failing to respond in a timely manner to subpoenas or search warrants requesting subscriber information. Investigators noted that, in some cases, ISPs retained no information whatsoever, leaving investigations at a dead end.

The need for ISP's to retain data and to subsequently respond with alacrity to the legal process generated by law enforcement is critical to successful investigation of Internet crimes. Lack of response to law enforcement subpoenas and search warrants can have dire consequences for victims and stall or end law enforcement investigations before an offender can be identified. Internet service providers, credit card companies, social networking sites, gaming sites, providers of chat rooms, e-mail services, and those who advertise in cyberspace are all among the facilitators who can become caught in the middle of the Internet crime problem. Notably, ISPs may tacitly assent to Internet crime while profiting from subscribers and advertisers. Providers should logically bear some of the responsibility for correcting the problems. In the same way that automobile manufacturers begrudgingly gave way, after thousands of roadway deaths, to regulations mandating vehicle safety, ISPs must commit to improved Internet safety before the annual number of Internet crimes matches the annual number of vehicular accidents.

For ISPs, preserving information and providing it to law enforcement in response to legal process is an unwanted and unprofitable chore. As the tragedies associated with some ISPs' reluctance to preserve and provide information gain increased attention, public pressure and legislative action may dictate that ISPs work harder to help law enforcement officers identify the suspects associated with Internet crimes. Eventually, the reluctant ISPs will be unable to turn a blind eye to the crimes and might be forced to become partners in justice instead of facilitators of injustice.

SEX OFFENDERS AND THE INTERNET

Offenders against children typically befriend their young victims through a process that has come to be called "grooming." Grooming may be defined as the way in which an adult predator patronizes and gains the trust of a child or teen. It sometimes involves a long series of carefully planned steps that may occur over an extended period of time; alternatively, it may proceed quite rapidly. When grooming a child, an Internet predator may exchange chat messages with the child and then request a phone number in order to engage in a personal voice conversation. The offender might then gradually escalate to sending money or gifts to the child. These acts not only advance the bonding process, but also help the offender to gain additional information such as a phone number and address.

Luring victim Katie Tarbox, who met and was victimized by an Internet predator when she was 13 years old, described the grooming process:

> *He took the time to understand me, and that is why these predators are so successful. He talked about the things I was interested in, however mundane they may have been. More importantly, he told me the things that I needed to hear as a vulnerable teenager: That I was intelligent, mature, and beautiful. So he bolstered my confidence, which won my trust as well (CourtTV.com, 2004).*

That many teens are curious about their sexuality and seeking their own sense of identity helps the predators who wish to victimize them. Young people often share private personal information with strangers on the Internet because they feel a sense of security there. They may establish relationships based upon interests they have in common with Internet strangers.

Some offenders appear to be upstanding members of the community. Many have spent their lives trying to gain the trust of others for the purpose of being entrusted with access to children. Often the neighbors, friends, and colleagues of offenders are shocked at their misdeeds and come to the defense of the offender. Former FBI agent Ken Lanning describes the tactics that offenders sometimes use:

> *At sentencing some offenders play the "sick and sympathy" game in which the offender expresses deep regret and attempts to show he is a pillar of the community, a devoted family man, a military veteran, a churchgoer/leader, nonviolent, without prior arrests, and a victim of abuse with many personal problems. They get the courts to feel sorry for them by claiming they are hard-working "nice guys" or decorated career military men who have been humiliated and lost everything. In view of the fact that many people still believe in the myth that child molesters are evil weirdos or social misfits, this tactic can unfortunately be effective, especially at sentencing (Lanning, 2001).*

The offenders who commit Internet crimes sometimes use the latest technologies to conceal their identities. Wireless networking now provides a means for offenders to try to hide themselves. Investigators must sometimes combine their technical knowledge with old-fashioned police work to apprehend Internet criminals. For example, in August of 2005, a 13-year-old Wisconsin boy was reported missing by his mother. Milwaukee detectives checked his computer and traced his last Internet contact to an address in Phoenix. Arizona ICAC Task Force detectives went to the address and learned that the resident there was unwittingly providing unencrypted wireless Internet access to the surrounding neighborhood. Detectives conducted surveillance and began diligently canvassing the area for the boy until they found him in the company of a wanted sex offender who had a history of abuse. The offender had surreptitiously used his unwitting neighbors' wireless Internet service to lure the boy. Investigators learned that the boy had been molested by the offender (Herman, 2005).

Law enforcement investigators must stay abreast of the latest techniques and equipment in use by the computer industry and subsequently by sex offenders. Training is the only way for investigators to keep up-to-date with the many ways in which offenders will misuse technology for criminal purposes.

Although many offenders against children are not outwardly aggressive and do not have violent criminal records of assaults against other adults, law enforcement officers should not let down their guard. There are a disproportionate number of offenders who contemplate, attempt, or complete the act of suicide after being identified. There have also been cases of offenders with vendettas against the police who have attacked officers. Some offenders may be dangerous. In 2004, a Broward County, Florida, Sheriff's Deputy was assisting in the service of a search warrant at the home of an offender who possessed child pornography. Using a high-powered rifle, the offender killed one deputy and wounded another before being apprehended (Aaronson, 2005). Investigators must remain vigilant to the possibility that Internet offenders may be dangerous. Many of the crimes against children are felony offenses, and officers should consider using appropriate safety tactics during confrontations with offenders.

REFERENCES

Aaronson, T. (2005). Deputy down: One year later, questions remain about Deputy Todd Fatta's tragic death. *New Times, Broward—Palm Beach, Florida.* Retrieved January 13, 2011, from http://www.browardpalmbeach.com/Issues/2005-08-25/news/feature_3.html

Allen, E. (2006). Closing speech: Project Safe Childhood Conference. Washington, D.C. Retrieved January 13, 2011, from http://www.missingkids.com/missingkids/servlet/PublicHomeServlet?LanguageCountry=en_US

Blanchard, R., Cantor, J. M., & Seto, M. C. (2006). Child pornography offenses are a valid diagnostic indicator of pedophilia. *Journal of Abnormal Psychology, 115*, 610–615.

CourtTV.com. (2004). On line discussion with Katie Tarbox. Retrieved January 13, 2011, from http://www.courttv.com/talk/chat_transcripts/2004/0728tarbox.html

Eichenwald, K. (2005). Through his webcam, a boy joins a sordid online world. *The New York Times.com.* Retrieved January 13, 2011, from http://www.nytimes.com/2005/12/19/national/19kids.ready.html?ex=1292648400&en=aea51b3919b2361a&ei=5090

Family Watchdog, LLC. (2006). Registered sex offenders counts by state. Family Watchdog Web site. Retrieved January 13, 2011, from http://www.familywatchdog.us/OffenderCountByState.asp

Finkelhor, D., & Ormrod, R. (2004). *Child pornography: Patterns from NIBRS.* Washington, DC: U.S. Department of Justice, Office of Justice Programs, Office of Juvenile Justice and Delinquency Prevention.

Gutierrez, T. (2007). *Online probe uncovers underage prostitution.* ABC7chicago.cnews.com Retrieved January 13, 2011, from http://abclocal.go.com/wls/story?section=local&id=4928094

Hermann, W. (2005). *Parents can watch out for Net predators.* AZCentral.com. Retrieved January 13, 2011, from http://azcentral.com/families/articles/0824netpredator24.html

Hernandez, A. E. (2000). *Self-reported contact sexual offenses by participants in the Federal Bureau of Prisons's sex offender treatment program: Implications for internet sex offenders.* Paper presented at the 19th Annual Research and Treatment Conference of the Association for the Treatment of Sexual Abusers.

Johnson, H. (n.d.). *Man used library computers to download child pornography. Arizona Republic.* Retrieved January 13, 2011, from http://www.azcentral.com/arizonarepublic/local/articles/0813childsex13.html

Kardasz, F. (2007). *Internet social networking sites.* Retrieved January 13, 2011, from http://www.kardasz.org/Case_Studies.html

Lanning, K. V. (2001). *Child molesters: A behavioral analysis* (4th ed.). National Center for Missing and Exploited Children, p. 86. Retrieved January 13, 2011, from http://www.missingkids.com/en_US/publications/NC70.pdf

National Center for Missing and Exploited Children (NCEMC). (2006). *Financial and Internet industries to combat Internet child pornography: Child pornography fact sheet.* NCMEC Web site. Retrieved January 13, 2011, from http://www.ncmec.org/.missingkids/servlet/NewsEventServlet?LanguageCountry=en_US&PageId=2314

National Law Center for Children and Families. (2006a). *Child pornography investigation and prosecution manual.* Alexandria, VA: Author.

National Law Center for Children and Families. (2006b). *The NLC manual on child pornography law: Cases and analysis.* National Law Center for Children and Families. Alexandria, VA: Author, pp. 42–43.

PewEW, Internet & and American Life Project. (2007). *Social networking web sites and teens: An overview.* Reports: Family, friends & and community. Retrieved January 13, 2011, from http://www.pewinternet.org/PPF/r/198/report_display.asp

Phoenix, Arizona Police Department. (2002). Phoenix Police Report no. 2002-2233604. Retrieved from the Phoenix, Arizona Police Department Records and Identification Bureau: http://phoenix.gov/POLICE/pub1.html

Phoenix, Arizona Police Department. (2003). Phoenix Police Report no. 2003-31701542. Retrieved from the Phoenix, Arizona Police Department Records and Identification Bureau: http://phoenix.gov/POLICE/pub1.html

Phoenix, Arizona Police Department. (2005a). Phoenix Police Report no. 2005-52427690. Retrieved from the Phoenix, Arizona Police Department Records and Identification Bureau: http://phoenix.gov/POLICE/pub1.html

Phoenix, Arizona Police Department. (2005b). Phoenix Police Report no. 2005-137156001. Retrieved from the Phoenix, Arizona Police Department Records and Identification Bureau: http://phoenix.gov/POLICE/pub1.html.

Phoenix, Arizona Police Department. (2006). Phoenix Police Report no. 2006-62346540. Retrieved from the Phoenix, Arizona Police Department Records and Identification Bureau: http://phoenix.gov/POLICE/pub1.html

Rabun, J. (2007). CVIP weekly activity report #223. E-mail to ICAC Task Force List serve.

University of Southern California, U.S.C. Annenberg School. (2005). Fifth study of the Internet by the Digital Future Project. Center for the Digital Future. University of Southern California, Annenberg

School. Retrieved January 13, 2011, from http://www.digitalcenter.org/pdf/Center-for-the-Digital-Future-2005-Highlights.pdf

U.S. Department of Justice. (2006). *Project Safe Childhood: Protecting children from online exploitation and abuse*. Retrieved January 13, 2011, from http://www.projectsafechildhood.gov/guide.htm.

U.S. House of Representatives, Committee on Energy and Commerce. (2007). *Sexual exploitation of children on the Internet: Bipartisan staff report for the use of the Committee on Energy and Commerce, 109th Congress.*

Weeks, G. (2006). Sexual exploitation of children over the Internet: What parents, kids and Congress need to know about child predators. U.S. House of Representatives, Committee on Energy and Commerce, Subcommittee on Oversight and Investigations. Serial No. 109-126. Retrieved January 13, 2011, from www.access.gpo.gov/congress/house.

Wolak, J., Finkelhor, D., & Mitchell, K. J. (2005a). *Child pornography possessors arrested in Internet-related crimes: Findings from the National Juvenile Online Victimization Study.* National Center for Missing and Exploited Children. Retrieved January 13, 2011, from www.missingkids.com/en_US/publications/NC144.pdf

Wolak, J., Finkelhor, D., & Mitchell, K. J. (2005b). The varieties of child pornography production. In E. Quayle & M., E., Taylor, M. (Eds.), *Viewing child pornography on the Internet: Understanding the offense, managing the offender, helping the victims* (pp. 31–48). Dorset, UK: Russell House Publishing. Retrieved January 13, 2011, from http://www.unh.edu/ccrc/pdf/jvq/CV100.pdf

Wolak, J., Mitchell, K., & Finkelhor, D. (2006). Online victimization of youth: Five years later. *National Center for Missing and Exploited Children Bulletin.* Retrieved January 13, 2011, from http://www.missingkids.com/en_US/publications/NC167.pdf

CHAPTER 18

Prosecuting Criminal Predators

Denise Timmins
Iowa Attorney General's Office

I am an attorney. It is a profession of which I am proud to be a part. I am even more proud that I am a prosecutor. I know that none of the jokes and stereotypes generally associated with lawyers really apply to me. The negative images people have about attorneys are of ambulance chasers and slick corporate lawyers who profit from the misery of others. Prosecutors have a higher calling: We help make society a safer place. Because of our work, criminals go to prison, and we hope that they stay there for as long as the law allows. Prosecutors are defenders of truth and justice.

Strangely, though, when I tell someone for the first time that I am a prosecutor, I usually receive one of two responses, neither of which is positive. The inquirer may smile politely and move as quickly as possible to another subject that does not involve the "dirty" details of my job. Too often, however, people want to have a prolonged discussion involving the intimate details of any recent case in which I have been involved. To someone outside of the criminal justice system, it all sounds like a good episode of *Law and Order*. We, as prosecutors, know it is reality.

Crime evokes interesting responses in people. No one will deny its existence. No one can. The media, both in news and entertainment programming, always seems ready to exploit human suffering. With television shows dramatizing and glorifying violent acts, and with news programs broadcasting increasingly graphic images of real violence, even the most unwilling viewer has to recognize the prevalence of violence in our society. However, even as most people acknowledge crime's existence, they will do everything possible to distance themselves from it. They tell themselves, "That doesn't happen here," or insist that they would never allow themselves to be in a situation like that. This denial allows them to shelter themselves from the reality of violence, replacing it with a comforting belief that those horrors exist elsewhere.

A prosecutor can never enjoy that comfort. Every day, prosecutors deal with the reality that comes from the gruesome details of another person's pain and suffering. Every day, prosecutors work to bring criminals to justice, and hopefully to provide closure and healing to their victims. And every day, prosecutors deal with the worst of the worst: predators.

CRIMINAL PREDATORS

"Predator" is a term that is often misused. In today's media, the term is applied to almost any heinous act committed by an individual. My experience as a prosecutor has proved that definition to be too broad. There are limitless reasons for criminal acts, and usually those reasons are known only to the criminals who commit them. We do know, however, that a true predator is part of a small but extremely dangerous group of criminals. The live-in boyfriend who abuses his girlfriend or molests her young children is a danger to that specific pool of victims, but he is not likely to go outside of the home in search of other victims. The bookkeeper who embezzles thousands of dollars from her employer most likely becomes a criminal due to her employment circumstances and the temptation of easy money. It is not likely that she researched the classified ads and applied for a job where she would have access and control over a business's bank accounts. Most criminals are like the boyfriend and the bookkeeper. The common denominator is opportunity and easy access.

This statement is not intended to minimize the seriousness of these crimes or lessen the responsibility of their perpetrators, but it does differentiate these offenders from predators. Predators are not driven by easy opportunities. Just as a true hunter would derive no pride from shooting a tethered deer, so predators do not find satisfaction in attacking an easy mark. They seek out their victims. They research and study them to know as much as they can. Their lives revolve around finding the next victim, and they use any means necessary to give themselves the advantage over their prey. Unlike an ordinary criminal, a predator's crime is his life, his identity. (The vast majority of predators are men.) Predators thrive on the thrill of the hunt, and their successes only encourage them to continue.

In my experience, the two most common types of predatory criminals are sexual predators and financial predators. These two types of predators are surprisingly similar. While comparing sexual and financial crimes may seem like comparing apples to oranges, a closer inspection leads to an understanding of the parallel motivations for these two categories of predators.

The biggest difference between the two is the crime itself. Obviously, sexual and financial crimes are glaringly different, and society's attitudes toward sex and money make the differences even more extreme. It seems obscene to even suggest the two have anything in common. Yet the similarities in the methods of the predators, and the effects of their crimes on their victims, become quite clear to the prosecutors who deal with these cases. For these prosecutors, these similarities simply cannot be overlooked. An important element in the prosecution of both sexual and financial predators is understanding how the predatory mind works and the process the predator uses to stalk his victim and finally strike his mark. Just as the overall approach to dealing with financial crimes is different from dealing with sexual crimes, so is the approach to dealing with regular financial and sexual criminals different from dealing with their predatory counterparts.

The first similarity between sexual and financial predators relates to how they select their victims. Like their counterparts in the animal kingdom, human predators prey on the weak. Unlike animals, human predators are not motivated by a hunger for the food necessary for physical survival. Animals take only what they need; human predators have no such limitations. Their victims are limited solely by availability. Thus, while the animal predator seeks to fulfill only its basic needs, the human predator hunts more for self-gratification, the sense of power that comes from overcoming and defeating another victim.

The prime targets for the predatory criminal are children, the elderly, and people outside the societal norm, such as prostitutes or illegal drug users, who do not have the same defenses as the majority of society. These victims have weaknesses due to their age, their mental capacity, or their status; these weaknesses make them more desirable to the predator because of the advantages he has over such a victim. A predator may choose a child who is too young to verbalize the abuse he has inflicted on him or her. A child can more easily be intimidated into silence with threats toward the child's own safety, or that of the child's family, friends, or pets. An elderly woman who has grown to trust a financial predator may take a longer time to discover that her bank accounts and retirement funds have slowly been drained, or, if she suffers from some form of dementia, she may never comprehend what happened to the financial security she had hoped to pass on to her children and grandchildren. The predator who sexually assaults a prostitute counts on the fact that very few, if any, will believe her if she reports the rape. He knows that if he attacks an addict whose children were taken away due to neglect, no one is likely to listen to her or care. Victim selection is extremely important to predators, and they choose carefully before striking.

Despite the differences in their actual crimes, the thought processes of sexual and financial predators are strikingly similar. First and foremost, there is a frightening disconnection from reality that allows them to justify their actions, no matter how heinous those crimes would seem to a regular person. Many predatory criminals truly believe that they are not doing anything wrong or immoral. Even though they understand that their actions are illegal, they believe that the law that is supposed to protect their victims is misguided or oppressive. A nonpredatory person will often tell himself that he has done nothing wrong or find some sort of justification for his wrongful actions in order to live with the knowledge of what he has done. It is possible that predators do the same, but it seems just as likely that they truly believe in the righteousness of their behavior. At this stage in my career, I cannot say for certain which is true. Perhaps no one can, including the predators themselves, since their perception of what constitutes normal human behavior is so distorted.

The predators' lack of remorse is staggering, almost unbelievable. Sexual predators will claim that molested children actually enjoy sexual contact, and that such offenses are a natural human behavior that has been wrongly outlawed by a prudish and oppressive society. They will claim that they were helping their adult rape victims live out fantasies, and that they had full consent for everything they did. Financial predators will blame their investments on sham corporations, on bad judgment, or state that they were fully authorized by their victims to do everything they did. Rarely do they take responsibility for their actions or acknowledge that their behavior was illegal or wrong. Even more rarely do they acknowledge or take responsibility for the physical, emotional, or financial harm they inflict on their victims.

Coupled with the predator's lack of remorse is an inflated sense of entitlement. Predators are almost childlike in their perception of ownership. In their minds, their desire for something justifies the means to obtain it. If a predator wants sex, he should be able to have it with whomever he wants. If something horrible has happened to the predator at some point in his life, then someone else should suffer the same level of violence and degradation. A predator who needs a new car or wants to take a trip to some exotic location finds a new victim in order to pay for their pleasure with someone else's money.

This sense of entitlement is most obvious in financial predators. They invest a great deal of time and effort in grooming their victims. They "take care" of them for so long before they strike that they feel completely justified in taking their victims' money. The financial predator believes that the victim owes him for everything that he has done. The stolen money amounts to a self-imposed service fee for befriending an elderly victim. It is a payment due for all of the time it took to learn enough about the victim to be able to empty the individual's bank account or retirement fund. To financial predators, it is not theft or embezzlement, but rather is simply taking what they have coming to them, what should already be theirs because they deserve it.

Predators rarely accept blame for their actions. On those infrequent occasions when I have seen a predator whom I am prosecuting take responsibility for his crime, it was because he believed doing so would work to his benefit. This apparent remorse usually happens at a sentencing hearing, when the predator is trying to convince the court to impose a light sentence. Otherwise, the actions of predators are claimed to be the fault of other people or of outside influences that are beyond their control. Blaming the victim is commonplace. Sexual predators, for example, will often speak of "promiscuous children." They may also blame the victims' parents, claiming, "if the parents had taken better care of their children, I would never have had access to them." The typical excuse for financial predators is "it was just business." They also blame their victims for not paying better attention to the fine print or being gullible enough to believe that any investment could be a "sure thing."

Drugs, alcohol, and other addictions are also common excuses for either type of predator. They may claim that they would not normally molest, rape, or assault someone, or that they understand that stealing money is wrong, but they just could not control their behavior. If they had not been drinking, on drugs, or gambling, they never would have committed those crimes, they insist. Prosecutors quickly learn how to defeat these arguments because we hear them so often. Alcohol and drugs can never be recognized as the *cause* of criminal activity. As prosecutors, we must recognize these excuses as just another item in an endless list of ways the predator will try to present himself as the victim, rather than as the criminal.

Many people outside of law enforcement would be surprised to learn that the crimes of sexual and financial predators have similar effects on their respective victims. Some may even be offended by the idea of comparing a rapist and a white-collar criminal in such a manner. While it is obvious that the victims of sexual assault endure a different sort of trauma than the victims of financial crimes, the core feelings and emotions endured by both sets of victims are surprisingly similar. Shame, anger, embarrassment, and fear are common reactions reported by both types of predatory victims. It is common knowledge that far too many rapes go unreported because the victim feels ashamed of what has happened. In the same way, many victims of financial predators choose not to report their crimes because they are embarrassed that they placed their trust in someone who, in retrospect, was obviously not trustworthy. They do not want anyone to think they were fools or suckers, and they absorb the financial loss rather than give anyone that opportunity.

Rape victims will typically feel anger that they were violated in such a personal way, that another person took something that most people consider intimate and private and forced it upon them. The financial victim has also had something personal and private taken from them, something that most people share only with a trusted few, and will feel the same kind of anger, despite the fact that they have been violated in a completely different manner. A rape victim will feel embarrassed about having to tell a police officer, a doctor, or a judge and jury all of the degrading details of the assault in order to convict the sexual predator. A financial victim is likewise reluctant to share the details leading up to her loss, perhaps because every detail is an opportunity for anyone who can hear to think the individual was foolish to trust the predator in the first place. Elderly persons who are still clinging to independence may not want to report such crimes because of worries that people might think they are no longer capable of handling their own finances, as well as other areas of their private life.

Finally, both sets of victims feel fear. Victims of sexual predators fear that, due to the nature of sexual crimes and society's attitude toward sex in general, they could be looked upon as promiscuous or immoral, despite the fact that they had no choice in what happened to them. Financial victims fear that their families might step in and take away their personal and financial freedom, that they could be placed in a nursing home, or have their checkbook and credit cards taken away. In addition, some fears are common to both victim pools. They fear that the predator will retaliate if they report the crime. They fear that family members and friends will treat them differently or act differently around them. They fear that they will be partially blamed for the crime. They fear that they will never feel safe and secure, that they will never trust anyone again.

LEGAL PROCEDURES

Charging

A case always comes to a prosecutor after the fact. At this point, the crime has been committed and the damage has been done. Prosecutors cannot change what happened to the victim. We can only help them by holding the offender accountable for his actions.

The legal process begins with a file on the prosecutor's desk. This file contains most, if not all, of the relevant information necessary to prosecute the case. At this point, prosecutors decide who will be charged and with what crime. Interpreting the information in the file is like putting together a puzzle. Each fact is just one piece. Some of those pieces are more important than others, but each has a place in determining what shape the puzzle takes. How well all the pieces fit together helps determine the level of crime that will be charged. A missing piece could reduce a kidnapping case to a charge of simple assault. As prosecutors, we organize, compartmentalize, and dissect every piece of information in order to determine what charges we will be able to bring to court.

The charging stage is important when prosecuting a predator. Both financial and sexual predators come into the legal process with the mindset that they have done nothing wrong. Most predators who get caught have prior experience in getting away with their crimes either because they were never caught before, or, for those who have been arrested or even convicted, because the consequences for their crimes were too slight.

One reason for this predators' disdain for the legal process is that too often prosecutors will take into consideration things that should have no effect on the level of the initial criminal charge. A prosecutor may worry that the young victim of a child molester might have trouble holding up under cross-examination, so he might charge the predator with a lesser crime in hopes of pleading out the case before it reaches trial. In a financial case involving an elderly patient with Alzheimer's disease, the prosecutor may decline to file the case because the victim is in poor health and unable to testify, despite the fact that there is a well-documented paper trail of evidence that led to the predator's arrest in the first place.

Obviously a prosecutor needs hard facts in order to charge a crime. With that said, second-guessing the proper criminal charge because of how a witness might act gives the predator an unnecessary and unfortunate advantage. In a case with a well-documented investigation, the need for a live witness is a myth. Certainly, having a good witness on the stand can help sway a jury's feelings about a case, but a good case is based on facts, not emotions. The prosecutor's responsibility is to charge the case as the facts support it, and then let the other concerns sort themselves out. Certainly, there are cases that sometimes have to be dismissed due to concerns with witnesses or evidence, but a strong investigation almost always reveals facts that can be presented in some way.

Only one person benefits from a prosecutor's undercharging or declining to charge a case, and that person certainly is not the victim: It is the predator himself. This is a person who has already absolved himself of any blame or responsibility for the offense, and the failure to fully and properly charge him for his crime only confirms that belief. Once again, the predator will walk away from his crime with little or no consequence. Even worse, he is now even more convinced that no one can or will do anything to prevent him from striking other victims. That belief will empower him to seek out more victims and to continue with his destructive behavior.

Discovery

The pieces to the prosecutor's puzzle start to fit together during the discovery process. During this stage in the legal procedure, motions are filed, hearings are held, and depositions are conducted. The discovery process allows both the prosecution and the defense to get a feel for the case and to begin formulating a strategy for the best way to present their cases to the jury. For the predator, this point is where the games really begin. Predators are masters of manipulation. They have to be: Whether it is a sexual predator luring a child into his home or a financial predator hoping to clean out an elderly person's pension fund, the predator must know the best ways to make another person believe and trust him, when in reality he knows that he is really acting only out of concern for his own self-interests.

This art of manipulation does not end when a predator enters the judicial system. If anything, it is intensified, because the stakes become so much higher. Until an arrest, the predator who is unable to manipulate his prey into doing what he wants can simply move on to a new mark. Once that predator has been charged with a crime, however, failing to convince the right people of his innocence means a loss of freedom, and with it, a loss of opportunities to attack more victims while he is incarcerated. For the predator, the courts are a new arena with new players against whom he must pit his manipulative skills.

These skills are most obvious with financial predators. The financial predator's lifeblood is his ability to charm and manipulate others. It is how he thrives in business. Because this individual has always presented himself to his victims and to society as an honest, hard-working, educated man in a business suit, he is quite comfortable in a room full of attorneys. When he enters into negotiations or faces prosecution, the financial predator is at the top of his game, because he truly believes that he is the smartest person in the room. He will attempt to establish from the beginning that he is the one who is in charge of the situation. Once he feels he has established control, he will bring out all of the reasons why he is not the one responsible for what happened to his victim, and he will appear quite charming while he attempts to minimize the effects of his crime or insists that he has been wrongly accused.

The financial predator's charm can make convincing the victim that the predator is actually a criminal the most difficult aspect of the discovery stage. Often, the victim does not want to believe that the predator is not the person he appears to be. While the victims of sex crimes understand immediately that a crime has occurred and that they were targets, financial crime victims often have to be convinced of that fact. Financial predators choose their prey wisely. Most often, their targets are elderly people. These offenders take time to form close relationships. They take their victims to church or to the grocery store. They visit them on a regular basis when no one else does. More than one victim has told me that although they would like to get their money back, they do not want anything done to the predator who took it. For these people, the predator is the only person who cared about them.

The relationship that the financial predator creates with his victim is parasitic, but to the victim it seems symbiotic. The predator creates a situation where the victim comes to rely on him for basic physical and emotional needs. This dependence makes it difficult for the victim to sit in the same room with the predator during a deposition. Many victims do not want to tell a prosecutor about what happened to them, because they understand that assisting the prosecution means that this person that they trusted and befriended really did something horrible to them. This betrayal of trust is simply more than many victims can handle emotionally.

Sexual predators are equally skilled in the art of manipulation as their financial counterparts, but they tend to use different techniques. Because they are usually less educated and less charming than financial predators, for them the discovery process becomes a new opportunity to assault their victims. They expect their attorneys to embarrass and humiliate their victims by asking intimate, personal, and often irrelevant questions. While their attorneys grill their victims, they take the opportunity to apply what is often their best form of manipulation: intimidation. They cannot speak to the victims during the questioning, but they will attempt to stare their victims down, or some will continually tap a pencil or find some other way to constantly remind their victims that they are in the room, and that they are listening to every word of every answer the victim gives. They will use whatever body language they can to nonverbally let the victims know that providing evidence against them is a dangerous thing to do, that there will be retribution against the victim once these legal issues have been resolved.

Imagine sitting in a room with the very person who violently raped you, while his attorney questions everything you did before, during, and after the assault. Keeping self-control in this situation is a daunting task for anyone, especially given that sexual predators are just as savvy in their choices of victims as financial predators are. As stated earlier, many sexual assault victims are chosen because of their status in society. Children, for example, are easy prey for predators. They are easy to control because they can be easily threatened and intimidated. Out of fear for their own safety or for the safety of their loved ones, children do not always tell anyone what was done to them, and when they do tell, they are not always believed.

The same is true for adult victims of sexual assault. Who would believe a stripper or a prostitute who claims she was raped? How reliable can an addict be when she tells police that she was held at knifepoint and assaulted? Sexual predators understand how society judges those outside of the mainstream. As a consequence, they often choose victims who are not accepted, not well-liked, or not likely to be believed, knowing that doing so increases their chances of walking away from their crimes without fear of consequences.

Because of the predator's inherent ability to manipulate the system and intimidate or charm his victims, the prosecutor must do everything possible to protect the victim during the discovery process. Prosecutors have a duty to make the victim feel safe and involved while participating in the judicial process. We must help victims understand that, despite their fears, at the end of the process comes healing and the opportunity to move beyond what happened to them.

Many tools are available to help prosecutors achieve this goal. Whether the case involves a financial or sexual predator, the prosecutor should ask for and receive no-contact orders for the victims. Of course, a no-contact order is not some magical device that protects a victim from any wrongdoing by the predator. It is just a piece of paper. However, it presents the first opportunity to give the victims a sense of safety and empowerment and to also assert further control over the predator. It also provides recourse to the prosecutor if the defendant decides to violate the order by contacting a victim. If the predator has been released prior to the pre-trial, that violation gives the prosecutor an easy opportunity to get the predator off the streets and into a jail cell where he belongs. Swift action following a violation also lets the victims know that law enforcement will do everything possible to protect them as the judicial process moves forward.

It is also important for the prosecutor to limit the line of questioning that is allowed during depositions of the victim prior to the deposition itself. This is especially true for victims of sexual predators. In almost all states, the victim's prior sexual history is not admissible during trial. Thus the defense attorney should not be allowed to ask about it during a deposition. For the defense, the only purpose of such questions is to embarrass and intimidate a witness in hopes of scaring them out of testifying during the trial. It is exactly what the sexual predator wants, and should be stopped before it is ever allowed to start.

Prosecutors should also refer victims to any services that might be available to them. Prosecutors are only a small part of the healing process for victims, and it is important in every case that the prosecutor works with other agencies or community groups to provide a well-rounded support system for the victim. Financial victims can be referred to local agencies or nonprofit organizations that are willing to help them get their finances back in order. Sexual assault victims should be referred to local shelters or to counseling services that can help them begin to deal with their trauma. These services cannot change what happened to the victims, but they do give them ways to cope. They help victims live their lives normally again, which initially seems impossible for many victims.

Voir Dire

Trial is the part of my job that I love the most. Athletes spend countless hours conditioning their bodies and practicing their skills to prepare for competition. Musicians rehearse a piece of music over and over before performing it in front of an audience. The same is true for the prosecutors. We spend months preparing for a case. Long hours are spent with witnesses, preparing them for what they could be asked while on the stand. We work late into the night developing questions for witnesses and preparing analyzing evidence to present to a judge and jury. And as much as we try to avoid it, prosecutors become emotionally vested in every case. Well before I go to trial, I know for a fact that the predator is guilty, and that he needs to be put in prison to protect his victims and the rest of society from the dangers he presents. As the trial begins, it is time for me to take what I already know and convince twelve other people to believe it as well.

Everything starts with jury selection. Out of a room full of strangers, the prosecutor is expected to find the 12 people who would be the most fair-minded for the trial. Jury selection is not an opportunity to directly present the case to the jurors—that comes later. It is the first chance the prosecutor has to prepare the jurors for what will be presented to them during the trial. Jury selection is the time to present the problems in my case and to see which potential jurors are willing to accept the prosecutor's line of thought. For example, in a financial case, many defense attorneys will base their strategy on the fact that the victim willingly gave her money to the defendant in full knowledge of his intended uses of it and the risks involved with the "investment opportunity" proposed by the predator. As a prosecutor, I can tailor my questions to potential jurors to address that defense before the defense attorney has the opportunity to present it in detail. The most common defense in sexual assault cases is that the victim consented to the sexual contact. Potential jurors must be questioned to discover what they think defines consent and how they may view whatever contact the victim may have had with the predator. The issue of consent is absolutely vital in selecting jurors for most sexual assault cases.

Much has been written about jury selection and the best strategies for choosing a favorable jury. My experience as a prosecutor says to go with your gut. Jurors may appear to have everything going for them in terms of background and how they answer your questions, but if there is something that simply does not feel right about someone, strike that candidate from the jury pool. Your gut is always right in jury selection.

Trial

After the jury has been selected, prosecutors get to tell their side of the story during opening arguments. Television courtroom dramas make it seem like cases are won and lost during closing arguments, but criminal attorneys on both sides of the aisle know the opposite to be true. A strong opening is vital. If I tell the story wrong, I may never get the jury back on my side. The opening statement is the prosecutor's opportunity to present the facts, to show the jury an overview of the evidence, and to make human the victims of the case. Of course, the defense will follow by tearing apart every detail of the prosecutor's opening argument, but if the story is told right, the jurors can see through rhetoric, or at least be able to have an open mind as the trial evolves.

When presenting the facts of the case through witnesses and evidence, the prosecutor must be well-organized and persuasive. Jurors are like anyone else: They need to have the evidence presented to them in a manner that allows them to process it as well as possible. Jurors also need to understand the story from the very beginning. The victim is the most important person in the prosecutor's story, and he or she needs to be on the witness stand at the beginning of the trial, regardless of where the individual fits into the actual timeline of events. Police officers who responded to the crime, doctors and nurses who treated the assault victim, and detectives and accountants who investigated financial records are all important to the case, but they cannot personalize the crime for the jury. Only the victim can put a face on it. Only the victim can let the jury know that the crime was committed against a real person. In the victim, jurors can see the actual effects of the crime on another human being, and as the victim's testimony unfolds, they can begin to identify with the victim, or see similarities between the victim and people they know from their own lives outside of the courtroom.

Unfortunately, the victim is not always available to testify due to death, age, illness, or some other reason. If that is the case, the prosecutor needs to begin with a witness who is close to the victim. Even without the victim present, it is vital to humanize that person for the jury. In the abstract, it is far too easy for the defense to present the victim as consenting, unreliable, or even the actual cause of the crime. That feat becomes much more difficult once the prosecution has made the victim real for the jury.

Cross-examination of the predator can be the most challenging aspect of a trial. A defendant always has the choice of whether or not to take the witness stand, and most predators choose to do so. Their sense of entitlement and confidence in their skills of manipulation lead them to believe that they can talk their way out of anything. They are practically compelled by their own arrogance to take the stand, thinking that once the jury hears their side of the story, it is impossible that they would end up with anything except for a verdict of "not guilty."

Some predators are more skilled at testifying than others, but in the end, their stories simply cannot stand up against solid evidence. The sexual predator who claims that his victim consented to his sexual contact has a hard time explaining how that is possible when the knife he used to threaten her or pictures of the injuries he caused her are entered into evidence. A financial predator's claim that his victim knowingly and voluntarily gave the predator control of his finances unravels quickly when the jury learns that the victim is 90 years old and has nothing to live on after his bank account and pension fund were emptied.

Despite the patent falsity of their testimony, I am constantly amazed at the sincerity that predators are able to project from the witness stand. They are that good at lying. Fortunately, a trial is not a simple matter of two people having their say before the case goes to the jury. Through witness testimony and indisputable evidence such as DNA or a well-documented paper trail, a prosecutor can chip away at the predator's story piece by piece. A good story is helpful, but it ultimately fails when confronted with solid evidence.

During closing arguments, the prosecutor has to go back and reassemble the pieces to form a completed puzzle for the jury. Evidence comes into a trial piecemeal. There is so much information for the jury to process, and many important pieces of evidence will have been forgotten by the time the trial comes to a close. Other evidence gets presented, and jurors become confused about what its purpose is in relation to the overall case. In the closing arguments, the prosecutor shows the jury the big picture of the case, and explains how all of those pieces of evidence fit into that big picture. As a prosecutor, I retell the jury what really happened. I re-emphasize which evidence really matters. I remind them why they should find my evidence to be more believable and more reliable than the defense's. It is during closing where the prosecutor applies the facts of the case to the law. The law says that these particular acts are a crime. In the end, the jury must understand that these acts were committed by the defendant, recognize how and why he did it, and accept that it is their responsibility to hold him accountable for his actions.

Verdict

After all the hours of preparation, all of the arguments, all of the time and money spent on bringing a predator to justice, everything rests on the final decision of

12 people. Everything comes down to one or two words as the verdict is read: guilty or not guilty. A prosecutor has to prepare the victim and other interested parties for either outcome. I always let them know that regardless of the verdict, everyone has done the best they can do. Whether the predator is sent to prison or he walks out of the courtroom a free man, they need to feel proud of themselves for standing up to him, for doing what is right, and for speaking out against him and speaking the truth about what happened. And I always tell them that, no matter how well they testified, no matter how strong the evidence was, and no matter how badly the predator incriminated himself on the stand, there is no such thing as a sure thing once the case is in the hands of the jury.

For the prosecutor, a guilty verdict always brings with it a sense of relief. At that point, I know that justice has been done, and I feel rewarded for all of the hard work and long hours I put into the case. For the victim, however, it does not make everything right. The guilty verdict is only one small part of the healing process, and while it does help bring closure to that part of the victim's life, it does not and cannot magically make the pain and suffering disappear. I hear from so many victims that they expected to feel happy or relieved when they heard the word "guilty," but in reality they felt the same anger or emptiness or whatever it was they felt before the verdict.

Victims endure unthinkable emotional pain. A guilty verdict never takes that pain away, but it does allow them to move forward in the healing process. With the verdict, they can put behind them the ordeal of maneuvering through the judicial process and instead move on to the all-important process of dealing with their emotions and putting their lives back together however they can. This process can take many forms, but it is always easier to proceed with once the trial has ended and the predator has been found guilty.

As a prosecutor, I also experience mixed emotions with every guilty verdict. I receive great satisfaction in knowing that the predator will spend time behind bars, and that it will be a very long time before he is able to harm another person. There is a lot of pride in knowing the part I played to make that outcome possible.

That satisfaction, however, is accompanied by a sense of emptiness for me as well. After the dust has cleared and the excitement of the trial is over, so much of it seems to be such a waste. All of those skills that the predator used to commit his crime and then try to avoid the consequences could have made him a highly productive member of society. Had he made different choices, he could have been successful in so many different areas of life. Instead, he chose a life where he benefited from the pains of others; instead of contributing to society, he will now waste his life and large amounts of taxpayer money in prison. The victim's life has been completely and irrevocably changed by the predator's actions, and there is no going back there, either. As a result, the lives of those around the victim are changed as well. So many lives changed, ruined, or destroyed—all because of the malicious acts of one person.

No matter how many of these cases I prosecute, I doubt that I will ever understand predators and the reasons why they commit the crimes they do. I have no desire to. My sympathies lie with the victims. As far as I am concerned, there is no justification strong enough to excuse these predators from responsibility for their actions. Besides, there is the chilling fact that to truly and fully understand a predator means that you actually are one.

CONCLUSION

Winning and losing are both part of the job for a prosecutor. I hate to lose, but it is something that a good prosecutor learns to deal with. Dealing with losing is vital for any prosecutor, not only to keep his or her sanity, but also to keep trying cases. Difficult cases are easily lost, but they are still worth trying. Some prosecutors may claim that they have never lost a case. There is a reason for that record: they never try the difficult ones, those cases that are high-risk in terms of getting a conviction, but yet are necessary to undertake because the same predator who makes the case so difficult is the one who most needs to be put into prison.

Predators are a disease in our society, and prosecutors are only able to treat this disease after it starts. Fortunately, there are dedicated people in other fields who study and research the root causes of the disease, working to find out what causes predators to do what they do, and searching for ways to prevent people from ever becoming predators. While I hope they have success, until that time, I know that I will always have a job.

CHAPTER 19

Alternative Adjudication Methods for Offenders with Mental Health Problems

Monic P. Behnken
Iowa State University

The Department of Justice reports that more than half of all adult prison and jail inmates suffer from at least one mental health problem. Specifically, 56% of state prisoners (705,600 inmates), 45% of federal prisoners (78,800 inmates), and 64% of local jail inmates (479,900 inmates) reported a recent history or symptoms of a mental health problem (U.S. Department of Justice, Bureau of Justice Statistics, 2006). The vast majority of inmates suffer from manic or depressive symptoms, with a smaller number suffering from psychotic symptoms. Inmates with a mental health problem are more likely than those without a mental health problem to have a prior criminal record, to have substance dependence or abuse, to use drugs in the month prior to arrest, and to be homeless prior to arrest.

Many offenders with mental health problems suffer abysmal lives both inside and outside of prison. Specifically, both male and female inmates who have a mental health problem are at least two times more likely than those without a mental health problem to have experienced prior physical and/or sexual abuse, with 68% of female prisoners with mental health problems being the victim of physical or sexual abuse at one time in their life (U.S. Department of Justice, Bureau of Justice Statistics, 2006). Mentally ill offenders are also more likely to receive longer sentences than their non-mentally ill counterparts. In fact, mentally ill offenders are likely to serve 15 months longer on an incarceration sentence than a non-mentally ill offender. Once in prison, mentally ill offenders find it more difficult to fit into the confinement setting. Mentally ill inmates are more likely to be involved in a fight, assaulted, or raped in incarceration environments (Ditton, 1999; Schaefer & Bloom, 2005). In addition, these inmates are also more likely to have disciplinary problems related to following rules than other inmates. Moreover, incarceration can exacerbate psychiatric symptoms and increase the risk of suicide among mentally ill inmates (Bender, 2004; Tyuse & Linhorst, 2005).

No federal statistics exist specifically pertaining to the prevalence of mentally ill juvenile offenders in state or federal custody. Moreover, it is difficult to reach a consensus about the prevalence of mental illness in this population due to differing symptoms seen in childhood and variances in developmental trajectories. Available prevalence data suggest that the rate of mental illness is between 50% and 100% of the juvenile offender population. Researchers agree that the rate of mental illness seen in the juvenile offender population is at least double that seen in the general adolescent population (Cocozza & Skowyra, 2000; Grisso, Skowyra & Powell, 2006).

These statistics reveal that a number of individuals with at least one mental health problem are finding their way into the correctional system rather than into mental health treatments. According to the Department of Justice, the percentage of mental illness in the adult correctional setting exceeds that of the general U.S. population, which is estimated to be approximately 11%. These figures have led some to label jails as "America's new mental hospitals," reflecting the fact that there are routinely twice as many mentally ill people in America's U.S. jails than there are in public psychiatric hospitals (Watson, Hanrahan, Luchins, & Lurigio, 2001).

Most experts attribute the increase in mentally disordered offenders seen in jails and prisons to the decline in mental health services provided during the deinstitutionalization phase of the 1970s and 1980s and the inability of the current system to provide community-based treatments (Bailey, 2003; Rabasca, 2000). These deficiencies in services have led many critics to suggest that the current judicial climate is criminalizing mental illness. In the area of adjudicating mentally ill offenders, most would agree that the traditional system of offense-based punishments is not working to deter criminal behavior or produce reformed offenders who are ready to become contributing members of society. The predominate method of offense-based punishments looks at the crime committed to determine the applicable sanction (Arrendondo, 2003). This type of sanctioning system pays little attention to the defendant's characteristics when designing punishment.

A theory called therapeutic jurisprudence has been developed in an attempt to help the law provide a therapeutic outcome for those mentally ill persons who become involved in the legal system. With the goal of changing the way offenders with mental health needs are sanctioned in the United States, many jurisdictions are investing resources into specialty courts that utilize therapeutic jurisprudence principles to facilitate interdisciplinary communication and education among decision-making professionals within the criminal justice system.

THERAPEUTIC JURISPRUDENCE

Therapeutic jurisprudence is the study of the role of the law as a therapeutic agent (Barton, 1999; Winick, 1997). Therapeutic jurisprudence, a term coined by law professor David Wexler and outlined by Bruce Winick, was conceived as an interdisciplinary movement to inform law reform efforts by recognizing that the mental health system carries important knowledge, theories, and insight that could positively inform modern developments in the law. This theory utilizes social science principles to examine the law's impact on the mental and physical health of the people it affects, while recognizing that the law is capable of functioning as a therapeutic agent in our society (Winick, 1999, pp. 1068–1069). The therapeutic jurisprudence model is based on the theory that "all things being equal, positive therapeutic effects are desirable and should generally be a proper aim of law, and that antitherapeutic effects are undesirable and should be avoided or minimized." (Winick, 1997, p. 188).

Although therapeutic jurisprudence principles have many applications, they seem to be most at home in the realm of mental health law. Therapeutic jurisprudence is a new addition to other legal schools of thought that focus on the betterment of special criminal populations. Just as feminist jurisprudence focuses on the law's impact on women and as critical race theory focuses on the law's treatment of racial minorities, so therapeutic jurisprudence is a way of examining the law's effects on those persons with mental health problems who become involved in the justice system. It also can be viewed as making the interaction with the legal system more beneficial to the offender and society than is being accomplished now (Winick, 2008).

One argument made against the incorporation of therapeutic jurisprudence principles into the adjudication of mentally disordered offenders suggests that this model is a paternalistic extension of the government. Winick (1997, 1999, 2008), however, asserts that true therapeutic jurisprudence proponents do not believe in paternalism. Instead, therapeutic jurisprudence emphasizes the importance of individual autonomy as an integral part of self-determination, which is therapeutically desired. Therefore, therapeutic jurisprudence rejects the paternalistic requirement to label people as disturbed or incompetent. Therapeutic jurisprudence also emphasizes the right to refuse treatment and a greater ability of the offender to participate in his or her own defense and sentencing/plea bargaining negotiations.

Another argument against the incorporation of therapeutic jurisprudence principles is that attorneys trying to help mentally ill clients may, in fact, coerce them into treatment decisions that they may might not otherwise make (Grudzinskas & Clayfield, 2004). The proper role of an attorney working under the therapeutic jurisprudence model, however, is one of "persuasion, not coercion." (Winick, 1999). Winick argues that the psychological value of choice should remain paramount in the minds of the criminal justice context. Self-determination, he says, is an essential aspect of psychological health. He suggests that people who make individual choices perceived as non-coercive will function more effectively and with greater satisfaction. When people are coerced into a particular decision, Winick argues, they respond with negative psychological reactions and other psychological difficulties. Winick (1999) also urges that the value of choice should be paramount in the design of any rehabilitative plan.

SPECIALTY COURTS

The specialty court model developed from the idea that the way people who were suffering from mental illness and substance abuse addictions were adjudicated in the United States needed to change. The goal was to move away from the *offense*-based sanctioning system and toward a more appropriate *offender*-based sanctioning system. An offender-based sanctioning system considers the developmental stage of the offender; the developmental, psychiatric, and emotional needs of the offender; and community safety when crafting sanctions. This innovative idea resulted in the creation of many specialty court models that target a specific issue of concern about the defendant that is believed to contribute to the offender's criminal behavior; such courts and attempts to address that specific need in the hopes of reducing further criminal conduct.

Under the guidance of therapeutic jurisprudence principles, many forms of alternative adjudication have risen to serve criminal offenders with particular psychiatric needs. Specifically, drug courts, homelessness courts, domestic violence courts, community courts, competency courts, and mental health courts have been developed in the wake of the therapeutic jurisprudence movement (Finkle, Kurth, Cadle, & Mullan, 2009; Mirchandani, 2005, 2006; Steadman, Davidson, & Brown, 2001; Wolff, 2002). Specialized courts, when created under the guiding principles of therapeutic jurisprudence, provide a nonadversarial forum that takes a problem-solving, treatment-focused approach to helping defendants in their programs. The two most prominent examples of specialty courts are drug courts and mental health courts.

Drug Courts

One of the oldest forms of the specialty court model is that of drug courts. These courts began appearing in the United States in the 1990s and have had large success in treating drug-addicted offenders. Drug courts emerged in response to the understanding that traditional police and justice enforcement techniques were ineffective and had little impact on offenders who were currently involved in substance abuse. These courts were designed to address the substance use problems of the many offenders who were commonly arrested on charges of possession and other drug-related offenses. The goal of these courts was to reduce recidivism by addressing the fact that addiction is the likely cause of the criminal behavior and to focus understanding about drug addiction as a disease and compulsion (Lurigio, 2008; Steadman et al., 2001; Tyuse & Linhorst, 2005).

The success of the drug court model is widely credited to the rapid coordination of local officials and the federal government to standardize drug court adjudication processes. National financial resources and technical assistance were brought to bear to create a common model of drug courts (Belenko, 2002). Funds were made available by the U.S. Attorney General through the Violent Crime Control and Law Enforcement Act of 1994, which provided more than $100 million dollars for the development and implementation of drug courts across the nation. To ensure the continuation of drug court development and integrity, the National Association of Drug Court Professionals, funded by the federal Drug Courts Program Office, developed a standardized list of 10 key components for future courts to incorporate in their operation (Tyuse & Linhorst, 2005).

As determined by the National Association of Drug Court Professionals, the 10 key components of these courts are:

1. Drug courts integrate alcohol and other drug treatment services with justice system case processing;
2. A nonadversarial approach is used by legal professionals to promote public safety while protecting participants' due process rights;
3. Eligible participants are identified early and admitted into the drug court;
4. Alcohol, drug, and associated treatment and rehabilitation services are offered to participants;

5. Frequent alcohol and drug testing is used to monitor participants' sobriety;
6. The drug court personnel respond to participant compliance according to set procedures;
7. Judicial interaction is ongoing and essential;
8. Achievement of program goals are monitored and measured to assess effectiveness;
9. Interdisciplinary education is used to promote court operations; and
10. Drug courts establish and maintain partnerships with community agencies to enhance the effectiveness of the process (Hora, 2002).

Another reason for the success of drug courts is that researchers were able to quickly demonstrate the effectiveness of the early assessment, treatment, monitoring, court supervision, and follow-up services drug courts provided to the defendants participating in the programs. Belenko (2002) reviewed the efficacy of drug courts and reported that the participants showed reduced levels of substance use and recidivism. Lurigio (2008) conducted a thorough review of the past 20 years of drug courts in the United States. His findings indicate that the adoption of the drug court model has successfully impacted the lives of offenders, the justice system, and society. Lurigio states that more than 1000 drug courts were in existence in the United States, Guam and Puerto Rico by 2007. In reviewing the larger courts, Lurigio found that drug court participants have fewer arrests and subsequent incarceration rates than offenders who are not drug court participants. He also found that drug court participants are less likely to recidivate after program completion, have a longer time to re-arrest, and have fewer positive tests for illicit drugs than non-participants. Further, drug court participants are more likely to stay clean for longer periods and to remain gainfully employed than offenders who do not complete a drug court program.

Following the success of drug specialty courts, the idea of mental health courts to adjudicate the large number of mentally ill offenders began to develop.

Mental Health Courts

Mental health courts are the newest and second most common type of specialty court. Such courts are set up to divert mentally ill offenders from the traditional justice system and adjudicate them with an emphasis on providing mental health treatment in lieu of incarceration, with the goal of decreasing recidivism. These courts fill the much-needed role of serving vulnerable criminal populations. Research of four prominent adult mental health courts revealed that 75% of the population served is male. Additionally, 25% of the population consists of racial minorities. Between 25% and 45% suffering from a dual diagnosis of mental illness and substance use, 25% of the total population is homeless, and 50% are receiving disability benefits, but not mental health treatment, at the time of arrest. Some sources even credit these courts with playing a role in social change that has not been seen since the times of the civil rights movement (Behnken, Arredondo, & Packman, 2009; Goldkamp & Irons-Guynn, 2000; Mirchandani, 2005; Miller & Perelman, 2009; Sipes, Schmetzer, Stewart, & Bojrab, 1986).

Notably, mental health courts are lauded for the following characteristics:

- Their ability to offer a viable alternative to incarceration by providing access to community resources that are generally underutilized by this population
- Their ability to provide access to community treatment resources and adapt to meet treatment needs specific to the community of the mental health court
- The creation of new and novel relationships among service providers, policy makers, and funding agencies (Griffin, Steadman, & Petrila, 2002; Petrila & Monahan, 2003, Tyuse & Linhorst, 2005).

Bailey (2003) describes the advantages of mental health courts as providing for the humane treatment of mentally ill offenders, lowering the costs of that care when compared to incarceration, and reducing the docket sizes of already overwhelmed courts. U.S. Representative Ted Strickland, the most notable proponent of mental health courts on the federal level, has stated that "these are human beings who are worthy of receiving appropriate interventions and treatment. . . It is cheaper to provide outpatient services than to pay for their confinement in a prison" (Bailey, 2003, p. 55).

General Mental Health Court Structure/FormatŸn Process

Broward County, Florida, is widely attributed with creating the United States' first mental health court in 1997 (Bailey, 2003). However, the appearance of the first mental health court was actually noted in an article by Sipes et al. (1986) as occurring in Marion County, Indiana. This court was temporarily suspended in 1992 and ultimately reconstituted as the PAIR Mental Health Diversion Project in 1996. According to the Survey of Mental Health Courts, as of 2008 there were 184 mental health courts currently functioning in approximately 34 states. As with drug courts, the federal government has provided funds to assist in the promulgation of these courts, in the form of 38 $150,000 grants from the Bureau of Justice Assistance. However, most jurisdictions do not use or rely on federal funds to operate their mental health court (Behnken et al., 2009; Raines & Laws, 2008).

Most mental health courts accept only non-violent offenders with serious mental illness (e.g., bipolar disorder, schizophrenia, depression, dual diagnosis of mental illness and substance use, and mental retardation) who are charged with a misdemeanor or nonviolent felony crimes (Arredondo et al., 2001; Miller & Perelman, 2009; Poythress, Petrila, McGaha, & Boothroyd, 2002; Winick, 2003). Referrals are screened based on a variety of ineligibility criteria (e.g., unwillingness to accept treatment, severity of mental illness, seriousness of offense, etc.). The defendant's competence to stand trial is then determined. After a finding of competence, each of the courts utilizes various adjudication methods. The courts vary between intervening at the pre-sentence stage to divert mentally ill defendants from incarceration and intervening at the post-sentence stage to reduce incarceration or probation time.

General Specialty Court FunctŸning

Mental health courts are problem-solving courts that differ in structure and are influenced by the characteristics of the court's location. Although these courts vary significantly across all domains of function and structure, some similarities can be found among the courts. First, mental health courts have specialty dockets that handle only mentally ill offenders. Second, most of these specialty dockets are likely to limit eligibility for mental health court services to nonviolent misdemeanants. However, some courts accept felony cases, and a few also require victim consent to hear cases of mentally ill defendants charged with assault in these specialty courts. Last, the goal of each court is quick access to community treatment and utilization of these services as an alternative to incarceration (Goldkamp & Irons-Guynn, 2000).

Most specialty courts have a similar pattern of functioning, although that process is carried out in unique patterns based on the needs of each jurisdiction. The processes used by drug courts and mental health courts, therefore, are fairly similar. Because mental health courts are the newest generation of specialty courts, their functioning will be discussed more in depth here, although roughly the same procedures are used by drug courts.

Screening/Assessment

Most mental health courts uniformly limit their case load to misdemeanor cases, require the presence of a mental illness as a necessity for program entrance, and allow participation by developmentally disabled defendants (Arredondo et al., 2001; New York State Office of Mental Health, 2002; Sipes et al., 1986; Trupin & Richards, 2003). The courts, for the most part, use a team approach to treatment, allowing for multiple inputs on acceptance decisions; more rarely, the district attorney makes the final acceptance decision. Most courts also serve a high number of dually diagnosed defendants and are able to identify eligible defendants within 24 to 48 hours of the defendant's arrest (Cosden, Ellens, Schnell, & Yamini-Diouf, 2005).

Case DispositŸn

Usually, criminal charges are not dropped against the defendant simply because the defendant agreed to participate in the mental health court. Courts favor at least one of three approaches to mandating adherence to the community treatment: preadjudication suspension of prosecution, post-plea strategies that suspend the defendant's immediate sentencing, or probation.

- The *pre-plea* adjudication structure allows deferral of prosecution of the defendant's case until successful completion of the mental health court program, with the goal of dismissing the initial charges upon completion.
- In the *post-plea* adjudication structure, the case is adjudicated by the mental health court, but the sentencing is deferred until the defendant completes the program.
- In the *probation*-based adjudication structure, the defendant has been convicted of the crime and sentenced to probation and sometimes a deferred or suspended jail sentence that can be lessened if the defendant participates in the mental health court program.

A little more than half of mental health courts use dismissal of charges upon successful completion of the mental health court program as an incentive to recruit people into the program. The remaining courts give credit for time served and may even retain the conviction on the defendant's record. Most courts acknowledge successful completion of the mental health court program, with the judge making a congratulatory announcement in open court.

Court SupervisŸn

Supervision is a component in 100% of the mental health courts, although reviews of the defendant's progress may be scheduled differently by each court. Dates for reviewing mental health cases range from an as-needed basis depending on progress reports to regular intervals ranging from every three to four weeks (Griffin et al., 2002).

Defendant AccountabÅity

As opposed to drug courts, which mainly rely on threats of jail sanctions, mental health courts require participants to become self-motivated and rarely rely on the use of incarceration as a motivating factor. Even in the very first model of a mental health court created in the 1980s, the message that the defendant was still a part of the community and was expected to be responsible for his or her own actions was explicitly stated. Uniformly, defendants are made aware that they are expected to comply with the community treatment goals as a condition of participation in the program (Arredondo et al., 2001; Griffin et al., 2002; Sipes et al., 1986).

SanctŸns

Mental health courts are hesitant to use incarceration as a sanction with this population to avoid the issue of coercion that is possible when working with vulnerable defendants. In fact, only courts that allow felony defendants report frequently using jail as a sanction for noncompliance with this offender population. Instead, mental health courts typically try to be more creative when designing sanctions for this population. Specifically, sanctions may consist of bringing the defendant back into the court to answer for noncompliance, the use of reprimands or admonishments by the judge or other team members, the enactment of stricter treatment conditions, changing housing situations, and the use of community service. It is believed necessary for these courts to first and foremost function as courts and use the leverage unique to the court system. The reluctance of mental health courts to use jail sanctions is attributed to the low-level offenders involved in the courts, the ambivalence of the legal system toward punishment of people who are clearly mentally ill, and the understanding that sanctions do not work the same way with mentally ill offenders as they may with non-mentally ill offenders (Griffin et al., 2002; Sipes et al., 1986).

UtÅizatŸn of Treatment Services

CoordinatŸn of Services

The duration of mental health court involvement is limited by the particular state's maximum sentence for misdemeanors, which may range anywhere between one to five years. Early reports indicate that while persons are under the jurisdiction of mental health courts, the use of mental health services increases by at least 20% over pre-program usage levels. The use of mental health services by participants in these courts is also significantly higher than that of mentally ill offenders adjudicated in a jurisdiction that does not have a mental health court. Offenders not being adjudicated through a mental health court process are more likely to utilize severe and costly emergency services and require higher levels of inpatient psychiatric care (Boothroyd, Poythress, McGaha, & Petrila, 2003).

The implementation of a mental health court positively impacts the volume of mental health services received by defendants. In one study, the mean number of service units increased by 61% for mental health court participants, while these same units decreased by 18.3% for non-mental health court participants. More strikingly, the mental health court participants (73%) were more likely to use mental health services even when not ordered to by the court, as compared to mentally ill offenders being adjudicated in a non-mental health court (53.3%) (Boothroyd et al., 2003).

CRITICISMS OF MENTAL HEALTH COURTS

Several criticisms have been cited about the proliferation of mental health and other specialty courts. One such criticism is that some subgroups of criminal offenders with mental health problems also have serious substance abuse problems that are resistant to traditional mental health treatments, making it less likely that these individuals will reduce their level of criminal activity after participating in a mental health court. These critics argue that the level of a defendant's substance abuse may be more important than mental illness when treating a criminal population (Cosden et al., 2005).

Second, some concerns have been raised about the federal support and subsequent proliferation of mental health courts because these courts are reshaping the justice system while using a tool that has, as yet, a brief history, an unclear conceptual model, and unproven effectiveness (Steadman et al., 2001; Wolff, 2002). A consistent complaint about the mental health court movement is the lack of agreement about what constitutes a mental health court and how these courts fit into the specialty court rubric, leaving most courts to function idiosyncratically. Such courts have been called a "hybrid of drug court principles and use of existing community-based services for persons with mental illness." (Steadman et al., 2001, p. 457). Some argue that the concept of a mental health court is so open to interpretation that it is virtually meaningless; therefore, any court claiming to serve the needs of the mentally ill offender population can designate itself as a mental health court. Steadman et al. (2001) urge that until sufficient evidence has accumulated to support the utilization of the mental health court model, jurisdictions should wait to implement these courts in other jurisdictions.

The Department of Justice has proposed that the 10 elements of a mental health court are essential to provide structure and guidance to jurisdictions looking to use this specialty court model. The DOJ suggests that:

1. A broad-based group of stakeholders from several different specialties should participate in the planning and creation of the court;
2. The target population for these courts should be selected only after considering public safety and a community's capacity to provide these individuals with care;
3. The identification of those in need should be made quickly, followed by swift linkages to services;
4. The terms of participation should be clearly communicated to participants;
5. The defendant should make a fully informed decision to participate in the court;
6. Participants should have access to treatment services and support;

7. Information should be shared in ways that protect the participant's confidentiality;
8. Court team members should regularly receive training and advice from relevant professionals;
9. Participants should be monitored while in the program; and
10. Data should be collected and analyzed to evaluate the court's impact. (Thompson, Osher, & Tomasini-Joshi, 2008).

A third criticism is that mental health courts can only serve a limited number of defendants. Because many of these courts are set up at the county level, offenders residing in a county without a mental health court are not eligible to participate in the program, even if they meet all other eligibility requirements. Moreover, the lack of resources needed to fund the court's operations and treatment programs is problematic. Many communities lack programs with the sufficient breadth and depth of services to handle the mentally ill offender population (Powell, 2003). This deficit will, therefore, limit the impact that these courts can have on an already vulnerable population.

Mental health courts are also accused of engaging in a process known as "creaming," whereby individuals are hand-picked to participate in such courts because they are the least impacted by their illness, which again leaves the most vulnerable of the total population to fend for themselves in an admittedly abysmal justice system (Tyuse & Linhorst, 2005; Wolff, 2002). The reasons cited for creaming are the drive to increase positive outcomes and the goal of avoiding embarrassing and politically damaging recidivism among this population. This process is an anathema to the ideals of mental health courts, which are charged with identifying offenders whose mental illness makes it more difficult to handle them in the traditional system because of the very nature of the severity of their illness.

Finally, a serious concern exists that the government is using coercion to force people into unwanted mental health treatment, where they will likely be expected to take psychiatric medications and disclose private and painful personal information. Traditionally, the concern about coercion in the mental health context is whether treatment has to be voluntary to produce long-lasting effects in the patient (Monahan, Bonnie, Applebaum, Hyde, Steadman, & Swartz, 2001).

JUVENILE MENTAL HEALTH COURTS

Early results suggest that *adult* mental health court programs reduce overall incarceration days spent in jail and increase the likelihood that the defendant will engage in treatment when compared to defendants who do not participate in mental health courts (Smith, 2002). Behnken et al. (2009) conducted a program evaluation of the nation's first *juvenile* mental health court in San Jose, California. The Court for the Individualized Treatment of Adolescents (CITA) is a specialty court that has adopted the mental health court model to adjudicate juvenile offenders. In a records review of the 64 graduates of the CITA program court, analysis revealed statistically significant reductions in the number of offenses committed by juveniles after entry into the court when compared to the number of offenses the same juveniles had accumulated before entry into the court. These reductions were seen even though the juveniles were in the CITA program for longer than they had been in the traditional adjudication system. This finding suggests that even though the juveniles may be supervised for longer in a juvenile mental health court, they tend to recidivate less when receiving mental health treatment as part of their adjudicative experience.

In the CITA program, participants showed the most significant reductions in offenses such as assault and battery, threats to commit a crime, theft, possession of dangerous weapons, and vandalism. Rates of many of the other offenses, such as arson, carjacking, driving under the influence, forgery, escaping from detention, causing a public disturbance, and indecent exposure, had results that were trending in a positive direction.

CONCLUSION

Offenders with mental health problems have complex psychiatric presentations that require a full understanding of their past traumas, psychiatric medication usage, use of illicit substances, birth history, medical history, school performance, and familial history. The use of the specialty court model represents a policy shift away from the traditional offense-based sanctioning system and toward an offender-based sanctioning system. The shift toward increased use of specialty courts to address offenses committed by the mentally ill population has found support in the literature, which cites reasons for the high cost of caring for mentally ill offenders in custody situations and the recidivism rate of untreated mentally disordered offenders.

Compared to other inmates, those offenders with mental health problems have a higher chance of being incarcerated for a violent offense, are more likely to be actively abusing drugs and alcohol, more often have a prior history of incarceration, and are more than twice as likely as other inmates to be homeless. These factors and the results of these studies presented in this chapter indicate that the policy shift toward embracing the therapeutic jurisprudence model through the use of specialty courts is a useful way to intervene in the lives of these offenders to reduce recidivism. These results of available research conducted to date clearly indicate that promulgating the specialty court model has the potential to reduce recidivism for special population offenders and increase public safety.

REFERENCES

Arredondo, D. E. (2003). Child development, children's mental health and the juvenile justice system: Principles for effective decision-making. *Stanford Law & Policy Review, 14*(1), 13–28.

Arredondo, D. E., Kumli, K., Soto, L., Colin, E., Ornellas, J., Davilla, R., Jr., Edwards, L., & Hyman, E. M. (2001, Fall). Juvenile mental health court: Rationale and protocols. *Juvenile and Family Court Journal*, 1–19.

Bailey, D. S. (2003). Alternatives to incarceration. *Monitor on Psychology, 34*, 54–56.

Barton, T. D. (1999). Therapeutic jurisprudence/preventive law and the lawyering process: Therapeutic jurisprudence, preventive law, and creative problem solving: An essay on harnessing emotion and human connection. *Psychology, Public Policy, and Law, 5*, 921–943.

Behnken, M. P., Arredondo, D. E., & Packman, W. L. (2009). Reduction in recidivism in a juvenile mental health court: A pre- and post-treatment outcome study. *Juvenile and Family Court Journal, 60*, 23–44.

Belenko, S. (2002). Drug courts. In C. G. Leukefeld, F. Tims, & D. Farabee (Eds.), *Treatment of drug offenders: Politics and issues* (pp. 301–318). New York, NY: Springer.

Bender, K. J. (2004). Pending legislation addresses mental health treatment in prisons. *Psychiatric Times, 21*, 8–9.

Boothroyd, R. A., Poythress, N. G., McGaha, A., & Petrila, J. (2003). The Broward mental health court: Process, outcomes, and service utilization. *International Journal of Law & Psychiatry, 26*, 55–71.

Cocozza, J. J., & Skowyra, K. R. (2000). Youth with mental health disorders: Issues and emerging responses. *Juvenile Justice, 7*, 3–12.

Cosden, M., Ellens, J., Schnell, J., & Yamini-Diouf, Y. (2005). Efficacy of a mental health treatment court with assertive community treatment. *Behavioral Sciences & the Law, 23*, 199–214.

Ditton, P. M. (1999, July). *Mental health and treatment of inmates and probationers.* Washington, DC: U.S. Department of Justice, Office of Justice Programs, Bureau of Justice Statistics. Retrieved January 24, 2011, from http://www.ojp.usdoj.gov/bjs/pub/pdf/mhtip.pdf.

Finkle, M. J., Kurth, R., Cadle, C., & Mullan, J. (2009). Competency courts: A creative solution for restoring competency to the competency process. *Behavioral Sciences & the Law, 27*, 767–786.

Goldkamp, J. S., & Irons-Guynn, C. (2000). *Emerging judicial strategies for the mentally ill in the criminal caseload: Mental health courts in Fort Lauderdale, Seattle, San Bernadino, and Anchorage.* Washington, DC: U.S. Department of Justice, Office of Justice Programs, Bureau of Justice Assistance. Retrieved April 14, 2011, from http://www.ncjrs.org/html/bja/mentalhealth/contents.html.

Griffin, P. A., Steadman, H. J., & Petrila, J. (2002). The use of criminal charges and sanctions in mental health courts. *Psychiatric Services, 53*, 1285–1289.

Grisso, T. (2004). *Double jeopardy: Adolescent offenders with mental disorders.* Chicago, IL: The University of Chicago Press.

Grudzinskas, A. J., Jr., & Clayfield, J. C. (2004). Mental health courts and the lesson learned in juvenile court. *Journal of the American Academy of Psychiatry & the Law, 32*, 223–227.

Hora, P. F. (2002). A dozen years of drug treatment courts: Uncovering our theoretical foundation and the construction of a mainstream paradigm. *Substance Use and Misuse, 37*, 1469–1488.

Lurigio, A. J. (2008). The first 20 years of drug treatment courts: A brief description of their history and impact. *Federal Probation, 72,* 13–17.

Miller, S. L., & Perelman, A. M. (2009). Mental health courts: An overview and redefinition of tasks and goals. *Psychology Review, 33,* 113–123.

Mirchandani, R. (2005). What's so special about specialized courts? The state and social change in Salt Lake City's domestic violence court. *Sociology Review, 39,* 379–417.

Mirchandani, R. (2006). "Hitting is not manly": Domestic violence courts and the re-imagination of the patriarchal state. *Gender & Sociology, 20,* 781–804.

Monahan, J., Bonnie, R. J., Applebaum, P. S., Hyde, P. S., Steadman, H. J., & Swartz, M. S. (2001). Mandated community treatment: Beyond outpatient commitment. *Psychiatric Services, 52,* 1198–1205.

New York State Office of Mental Health. (2002). Brooklyn Mental Health Court opens. *Office of Mental Health Quarterly, 8,* 11, 15.

Petrila, J., & Monahan, J. (2003). Introduction to this issue: Mandated community treatment. *Behavioral Sciences & the Law, 21,* 411–414.

Powell, J. (2003). Letter to the editor. *Issues in Mental Health Nursing, 24,* 463.

Poythress, N. G., Petrila, J., McGaha, A., & Boothroyd, R. (2002). Perceived coercion and procedural justice in the Broward mental health court. *International Journal of Law & Psychiatry, 25,* 517–533.

Rabasca, L. (2000). American Psychological Association, Public Information and Media Relations, Public Communications. (2000). A court that sentences psychological care rather than jail time. *Monitor on Psychology, 31,* 58–60.

Raines, J., & Laws, G. (2008). Mental health court survey (2008). Retrieved from http://ssrn.com/abstract=1121050.

Schaefer, M. N., & Bloom, J. D. (2005). The use of the insanity defense as a jail diversion mechanism for mentally ill persons charged with misdemeanors. *Journal of the American Academy of Psychology & the Law, 33,* 79–84.

Sipes, G. P., Schmetzer, A. D., Stewart, M., & Bojrab, S. (1986). A hospital-based mental health court. *Community Mental Health Journal, 22,* 229–237.

Skowyra, K., & Powell, S. D. (2006). Juvenile diversion: Programs for justice-involved youth with mental health disorders. Retrieved January 24, 2011, from http://www.ncmhjj.com/pdfs/publications/DiversionRPB.pdf.

Smith, D. (2002). $4 million for mental health courts: A new federal program will provide much needed services for nonviolent offenders with mental health problems. *Monitor on Psychology, 33,* 20.

Steadman, H. J., Davidson, S., & Brown, C. (2001). Mental health courts: Their promise and unanswered questions. *Psychiatric Services, 52,* 457–458.

Thompson, M., Osher, F., & Tomasini-Joshi, D. (2008). Improving responses to people with mental illness: The essential elements of a mental health court. Retrieved January 24, 2011, from http://www.ojp.usdoj.gov/BJA/pdf/MHC_Essential_Elements.pdf.

Trupin, E., & Richards, H. (2003). Seattle's mental health courts: Early indicators of effectiveness. *International Journal of Law & Psychiatry, 26,* 33–53.

Tyuse, S. W., & Linhorst, D. M. (2005). Drug courts and mental health courts: Implications for social workers. *Health & Social Work, 30,* 233–240.

U.S. Department of Justice, Bureau of Justice Statistics (2006). Mental health problems of prison and jail inmates, 2006. Retrieved January 24, 2011, from http://bjs.ojp.usdoj.gov/index.cfm?ty=pbdetail&iid=447.

Watson, A., Hanrahan, P., Luchins, D., & Lurigio, A. (2001). Mental health courts and complex issue of mentally ill offenders. *Psychiatric Services, 52,* 477–481.

Winick, B. J. (1997). The jurisprudence of therapeutic jurisprudence. *Psychology, Public Policy and Law, 3,* 184–206.

Winick, B. J. (1999). Redefining the role of the criminal defense lawyer at plea bargaining and sentencing: A therapeutic jurisprudence/preventive law model. *Psychology, Public Policy, and Law, 5*, 1034–1083.

Winick, B. J. (2003). Outpatient commitment: A therapeutic jurisprudence analysis. *Psychology, Public Policy, and Law, 9*, 107–144.

Winick, B. J. (2008). A therapeutic jurisprudence approach to dealing with coercion in the mental health system. *Psychiatry, Psychology, & Law, 15*, 25–39.

Wolff, N. (2002). Courts as therapeutic agents: Thinking past the novelty of mental health courts. *Journal of the American Academy of Psychiatry & the Law, 30*, 431–437.

CHAPTER 20

Violent Offenders: A Perspective on Dynamic Federal Supervision Practices

Alan J. Drury
Iowa State University and United States Probation Office

Some of the highest-risk offenders whom U.S. probation officers supervise are those who have long criminal histories with varied offending patterns that are predominately violent in nature. These offenders are categorized as high risk because they present significant risk factors to the safety of the community and the well-being of others. Moffitt (1993) has identified these types of offenders as "life-course persistent offenders." Frequently, their offending patterns do not stop while they are incarcerated. When these offenders are released from the Bureau of Prisons (BOP) facilities, many have institutional misconduct records that include disciplinary actions taken for various infractions, including continued patterns of victimization. When these high-risk offenders are released from their term of imprisonment, U.S. probation officers are responsible for their supervision.

In the federal probation system, U.S. probation officers are tasked with supervising offenders who have received any of the following sentences: conditional release, juvenile supervision, parole, probation, or supervised release. For purposes of this chapter, only sentences of adults under supervised release (18 U.S.C. § 3583) and

Disclaimer: The content in this chapter is not formal representation of the opinion or policies of the U.S. courts or the U.S. probation system.

probation (18 U.S.C. § 3561) are discussed. For more information on conditional release, see (18 U.S.C. §§ 4243 and 4246); for juvenile supervision, see (18 U.S.C. § 5035); and, for parole and mandatory release, see (18 U.S.C. § 4205 [U.S.C., 2010]). While many different forms of supervision are possible in the federal system, the two most common are supervised release and probation. Recently, the U.S. Sentencing Commission (2010) explained the difference between supervised release and probation sentences, describing probation sentences as a form of punishment which is applicable to the nature and severity of a crime that has been committed, and supervised release sentences as focusing on the reentry of offenders back into the community (Sessions et al., 2010). Each type of sentence is formally defined below:

Probation sentences are applicable to offenders who committed their offenses on or after November 1, 1987, based on the Sentencing Reform Act of 1984. Probation is a sentence in its own right, which is meant to reflect the seriousness of the crime committed, to promote respect for the law, and to provide just punishment for the offense committed. In the federal system, an offender may be sentenced to probation in lieu of imprisonment when this punishment is thought to provide adequate deterrence from criminal conduct and to protect the public from further crimes of the defendant. A sentence of probation also seeks to provide the defendant with needed education or vocational training, medical care, or other correctional treatment in the most effective manner (18 U.S.C. § 3561).

Supervised release is a sentence to a term of community supervision to follow a period of imprisonment. It is available only for offenders who committed their crimes on or after November 1, 1987. Supervised release is not an early release from prison or an additional punishment; instead, it is a separate sentence imposed in addition to the sentence of imprisonment, which assists offenders in re-entering the community after a term of imprisonment. Supervised release seeks to ensure there is adequate deterrence to criminal conduct, the public is protected from further crimes of the defendant, and the defendant is provided with needed education or vocational training, medical care, or other correctional treatment in the most effective manner (18 U.S.C. § 3583).

Since the late 1980s, in the United States, approximately one million federal offenders in the United States have been sentenced to a term of supervised release following a term of imprisonment (Sessions et al., 2010). While violent offenders represent a small percentage of the total number of offenders sentenced to a term of supervised release, some offenders have the potential to present an elevated risk to the community (Collins, 2010; Elliott, 1994). As depicted in **Table 20.1**, from 2005 to 2009, there were

TABLE 20.1 Offense Type for Offenders Sentenced to Supervised Release, Fiscal Years 2005–2009

		Prison Term Imposed	Supervised Release Imposed	Average Prison Term	Average Supervised Release Term
Offense Type	**Total Cases**	**Number**	**Number**	**Months**	**Months**
Murder	386	385	365	250	53
Manslaughter	280	271	269	53	35
Kidnapping	230	230	225	204	54
Sexual Abuse	1,845	1,782	1,771	97	71
Assault	2,981	2,510	2,336	43	33
Robbery	5,365	5,241	5,220	88	40
Arson	309	298	296	78	38
Burglary	199	175	173	22	35
Extortion	3,139	2,870	2,840	96	38
Total (Violent)	**14,734**	**13,762**	**13,495**	**103**	**44**
Total (All)	**356,339**	**313,811**	**297,955**	**60**	**41**

Note: The "Total (Violent)" line represents the total number of violent cases and the "Total (All)" line represents the total number of cases in the federal system for fiscal years 2005–2009.

Source: Sessions et al., 2010; U.S. Sentencing Commission, 2005–2009 data files, FY2005–FY2009.

approximately 13,495 (out of a total of 13,762) federal offenders who were sentenced to a term of supervised release following a term of imprisonment for a violent offense. Such violent offenses include: murder, manslaughter, kidnapping, sexual abuse, assault, robbery, arson, burglary of a dwelling, and extortion. During that same period, 267 offenders did not receive a sentence of supervised release (Sessions et al., 2010).

Table 20.1 shows the average number of months to which offenders were sentenced, separated categorically by violent offense type. Supervised release terms for violent offenses range from an average of 33 months to 71 months. On the low end, the 33-month supervised release term reflects the average length of a supervised release term imposed for offenders convicted of assault. The 71-month supervised release term reflects the average length of the term imposed for offenders convicted of sexual abuse. While these supervised release sentences are averages, some offenders are sentenced to a term of lifetime supervised release, whereas others do not have any term of supervised release imposed.

When offenders who have a history of violent behaviors are released from the BOP facilities to begin their terms of supervised release or for those violent offenders who receive a sentence of probation, there are several procedures in place to ensure they are properly supervised. This chapter examines the assessment of risk, needs, and responsivity factors; classification leading to levels of supervision; supervision and caseload management strategies; and supervision styles employed by federal probation officers.

ASSESSMENT AND CLASSIFICATION

A significant component of community corrections is the need to protect the public from future harm, a consideration that is especially important in the supervision of offenders who have been convicted of a violent crime or those who have a history of violence. In addition to ensuring offenders abide by the conditions of release set forth by the U.S. district courts, U.S. probation officers are responsible for managing the risk of their clients to ensure the safety of the community. Managing risk requires probation officers to intervene with strategies that provide offenders with the tools and resources needed to access social services that target specific criminogenic needs (*Supervision of Federal Offenders*, 2010).

Criminogenic needs are dynamic risk factors that, when addressed, change the likelihood that an offender will engage in criminal behavior (Latessa & Lowenkamp, 2006). Criminogenic needs include: antisocial attitudes, antisocial associates, antisocial behaviors, antisocial personality, dysfunctional family, unemployment or lack of education, lack of productive leisure and recreation time, and substance abuse (**Table 20.2**). Targeting criminogenic needs and influencing an offender's conduct by improving the circumstances that are linked to his or her criminal behavior is an effective way to reduce an offender's risk level and will impact the likelihood that an offender will engage in violent behavior (Andrews & Bonta, 2003, Bucklen & Zajac, 2009; Gendreau, French, & Taylor: 2002; *Supervision of Federal Offenders*, 2010).

Conducting a risk assessment assists officers in determining which offenders pose an elevated risk to the community. A risk assessment measures the risk, needs, and responsivity factors associated with each client. The goal of the risk assessment, when working with violent offenders, is to accurately identify those offenders who represent the greatest threat to commit another violent crime and, in turn, to employ supervision strategies that reduce the risk that these offenders will engage in future violent behaviors. Accordingly, risk assessments are individualized and address the respective risk factors associated with each offender.

Risk assessments are designed to measure risk. Risk is defined as the probability of re-arrest within a given time period. For example, offenders who are high risk are those

TABLE 20.2 Criminogenic Needs: The "Big Eight"

Antisocial attitudes	Offenders do not abide by the values and norms of the community in which they reside. They do not engage in prosocial interactions with others. They maintain that negative experiences such as being arrested are the result of external circumstances and not their own behaviors.
Antisocial associates	Associating with individuals who engage in criminal behaviors increases the likelihood of an offender committing new crimes.
History of antisocial behaviors	Such a history includes retreating from prosocial interactions to be alone with their thoughts and ideas; the thought that no one can be trusted and everyone is out to get them; and having a history of criminal behaviors.
Antisocial personality pattern	Offenders with an antisocial personality do not care how their actions affect others. As a result, they are not able to express remorse or empathy for others for what they have done. They act impulsively to obtain immediate gratification that is of the greatest benefit to themselves with disregard for others.
Family/marital: problematic circumstances at home	Offenders who come from a dysfunctional family are more likely to be in a setting where they have learned or engaged in criminal behaviors.
Education/employment: problematic circumstances at school or work	Offenders who are unemployed or underemployed and those who have low levels of education are more likely to engage in criminal behaviors.
Leisure and recreation: problematic circumstances with leisure and recreation activities	Offenders who do not have prosocial interests or those who do not use their time productively to develop prosocial networks are more likely to engage in criminal behaviors.
Substance abuse	Substance use impairs judgment and lowers inhibitions, which could lead to a propensity to engage in criminal behavior. Controlled substance use without a doctor's prescription is an illegal act in itself.

Source: Andrews and Bonta, 2003.

offenders who present a higher probability of being re-arrested for a new criminal offense within a specific time period (Latessa & Lowenkamp, 2006). In this sense, risk is conceptualized as an offender's likelihood to negatively impact community safety through his or her own intentions and actions.

Risk assessments utilize three concepts when assessing this probability. First, the *risk principle* asserts that the highest-risk offenders need to be targeted with the most intensive services. High-risk offenders present the greatest likelihood of engaging in violent criminal behavior and should be targeted with multiple interventions in an effort to reduce this risk. This approach promotes the most efficient use of resources by specifically targeting offenders who are in the most need of resources in an effort to change behaviors and protect the community.

Second, the *need principle* asserts that specific areas or identified criminogenic behaviors need to be targeted with specific intervention strategies to effectively change behavior and control risk (Latessa & Lowenkamp, 2006). Cognitive behavioral treatment programs, for example, have been shown to be effective at targeting antisocial attitudes and correcting antisocial behaviors in some offenders (Andrews & Bonta, 2003; Landenberger & Lipsey, 2005; Latessa & Lowenkamp, 2002; Lipsey, 2009; Petersilia, 1999; Taxman, 1999; Wilson, Bouffard, & McKenzie, 2005). Cognitive behavioral treatment programs focus on changing thinking processes, which in turn, changes behavioral outcomes. These treatment programs target many criminogenic needs, including antisocial attitudes, antisocial personality, and criminal thinking processes.

Third, *responsivity factors* are barriers that minimize the effectiveness of treatment or some other intervention. These factors include, but are not limited to: mental health problems, lack of transportation, lack of motivation, inability to read or write, language barriers, hygiene issues, and homelessness. Responsivity factors represent basic barriers that must first be addressed for an offender to obtain maximum benefit from the treatment or other interventions aimed at reducing risk and targeting criminogenic needs. Addressing responsivity factors with high-risk offenders provides probation officers with an optimal opportunity to build rapport and establish a positive relationship. Often, overcoming small barriers and working with offenders demonstrates to them that probation officers seek to establish a firm but fair relationship with the goal of helping offenders with successful reentry into society. Overcoming small barriers demonstrates to the offender that the probation officer is committed to help the offender be successful. This is important because most high-risk offenders view the criminal justice system as a negative institution that is out to get them.

Risk, needs, and responsivity factors interact dynamically. While there are a variety of risk assessments that capture these concepts, U.S. probation officers use the Post Conviction Risk Assessment (PCRA) to assess the risk, needs, and responsivity factors of each offender. The PCRA, which was developed by the Office of Probation and Pretrial Services (2010), is a risk assessment instrument that seeks to measure seven domains: criminal history, education/employment, substance abuse, social networks, cognitions (thinking processes of offenders), other (housing, finances, recreation), and responsivity factors. The 55 items on the PCRA are scored, rated, and classified into seven domains. The PCRA provides probation officers with an objective, quantifiable instrument that uses a valid method of predicting the risk of re-arrest and reconviction for any case plan period (Latessa & Lowenkamp, 2006; Office of Probation and Pretrial Services, 2010). The accurate assessment and classification of offenders based on risk, needs, and responsivity factors is essential to developing a case plan and supervision strategy that is effective in controlling risk and influencing behaviors (Vose, Cullen, & Smith, 2008).

While the PCRA is not designed to predict which offenders will engage in specific subsequent violent offenses or other types of specific offending patterns, this instrument does identify which persons are at a greater risk to engage in subsequent criminal behavior. If an officer identifies an offender who has patterns of violent behaviors and has reason to believe the offender presents a greater risk than the risk category produced from using the assessment, the officer may use a professional override to address the situation.

The professional override provides officers with the ability to gauge the accuracy of the assessment information in comparison to the resources used in computing the assessment. When using the professional override, it is emphasized that if officers notice patterns of violent behaviors that were not captured in the overall classification score, they override the assessment. When overriding the assessment, officers are required to pay special attention to areas of the assessment in which they believe their classification scores are either too high or too low and to provide explanations and documentation as to why. Using the professional override changes the results of the assessment, thereby affecting the validity and the reliability of the assessment, and in turn the risk management techniques that the officer uses during the course of supervision.

As a result, the risk assessment tool aids in identifying which offenders fall into which risk categories, while leaving a "safety valve" to exert for professional discretion. The resulting risk categories used in the assessment include: low, low/moderate, moderate, and high. While there are quantitative differences between each risk category, typically, probation officers are able to see qualitative differences as well. For instance, the following descriptions are examples of the qualitative differences that may be observed by a probation officer between risk categories when assessing offenders.

High Risk

Offenders who fall into this category can be characterized as callous, impulsive, criminally versatile, having poor social skills, restless or aggressive, having poor problem-solving skills, self-centered, and violent. They exhibit antisocial attitudes, criminal thinking patterns, and often have antisocial personalities. Approximately 5% to 10% of offenders fall into this category.

High-risk offenders present multiple risk factors, such as: lengthy criminal histories, including both violent and non-violent crimes; possible diagnoses of antisocial personality disorder; and narcissistic and predatory thought processes. These offenders may have extensive substance abuse problems, suffer from chronic unemployment, have disadvantaged families with long criminal histories, suffer from significant marital issues, do not engage in pro-social recreational activities, and have peer networks that are predominantly criminal in nature.

Moderate Risk

Moderate-risk offenders can be characterized as displaying antisocial attitudes and criminal thinking patterns that may be narcissistic in nature. They are restless and at times aggressive, have poor problem-solving skills, and have problems effectively communicating with others. These individuals will often blame others for their circumstances and fail to accept responsibility for their behaviors. They may or may not have lengthy criminal histories, but usually do have varied offending patterns consisting of drug, property, and violent crimes. These offenders usually have substance abuse problems, have long periods of unemployment or sporadic employment histories, they may be in an unhealthy marriage or have other ongoing family disputes, usually they do not engage in pro-social recreation but may from time to time, and have peer networks that include both criminal and non-criminal associates.

Low/Moderate Risk

Offenders in the low/moderate risk category internalize their behaviors and cognitively process that their circumstances are a result of their own actions. They display emotional traits such as empathy and remorse for their actions. They usually take responsibility for their behaviors. Although they may have criminal histories that consist of violent, drug, and property crimes, usually their criminal histories are not lengthy. These offenders have support networks that are largely pro-social in nature. They may have had prior substance abuse problems that are being effectively managed through treatment programs, voluntary Alcoholics Anonymous (AA) programs, or community support. They are usually employed or consistently seeking employment, are motivated to live a conventional lifestyle, and use community resources effectively. They may or may not be in a healthy marriage or have healthy family circumstances. They engage in activities that are community or family oriented and pro-social in nature. Although they have peer networks that consist primarily of non-criminal associates, they may have some criminal associates or criminal family members.

Low Risk

Low-risk offenders have extremely limited criminal histories, usually consisting of only one or two prior convictions, if any. These offenders have minimal criminal thinking tendencies and are pro-social in nature. They effectively internalize thought processes aligned with thought processes in conventional society. They have strong family support or have reestablished an identified support network. They have been employed consistently for long periods of time. They may not have ever had a substance abuse problem or the problem has been in remission for several years, and they have developed an effective relapse prevention plan for themselves that they can articulate. They have peer networks that do not engage in criminal behaviors, and they proactively engage in community-based activities on a regular basis that are positive in nature.

While these categories are presented here as static descriptions, it is important to note that an offender's risk, needs, and responsivity factors are dynamic and change over time. Therefore, offenders fluidly move from one risk category to the next as a result of life circumstances. For instance, obtaining employment, reestablishing family ties, or completing a drug treatment program, are all accomplishments that are likely to affect an offender's risk level.

When conducting a risk assessment, it is important to understand that certain qualitative differences, when observed, may require a reassessment that will measure the quantitative differences, changing the offender's risk category and the level of intensity of supervision. This consideration is especially important with violent offenders who experience changes in their risk factors that may lead to re-offending or subsistence from offending.

SUPERVISION AND CASELOAD STRATEGIES

Certain statutory duties are required of U.S. probation officers in the supervision process of offenders. During the supervision of all offenders, especially violent offenders, probation officers are expected to employ the principles of good supervision. These principles include utilizing an individualized supervision approach tailored to the specific risks, needs, strengths, and responsivity factors presented by the offender.

In addition, the supervision must be proportional and purposeful. "Proportional and purposeful" means that the correctional tools used in the supervision process must be the least intrusive means necessary to achieve specific supervision goals—e.g., targeting domestic violence patterns through anger management or cognitive behavioral treatment and using specified dosages of various types of treatment that are consistent with the research to maximize the likelihood of bringing about positive changes.

Case plans are an effective way of ensuring probation officers use principles of good supervision. A case plan is an individualized strategic supervision plan that identifies the specific supervision approach that will be employed with the offender. It reflects identified risks, needs, strengths, and responsivity factors associated with the offender. Case plans are an important planning tool used with all offenders. However, when dealing with violent offenders the case plans become even more important. Case plans allow officers to identify both the correctional and controlling strategies that will be used to protect the community and assist the offender in changing behaviors. The case plan allows probation officers the ability to methodically go through and identify any mandatory conditions imposed by the court on an offender, such as association restrictions, domestic violence treatment, anger management treatment, sex offender treatment, location monitoring, and any identified third-party risk issues that may need to be resolved.

Third-party risk is defined as the "duty to warn specific third parties of a particular prospect of harm (either financial or physical), which the officer reasonably foresees the probationer may pose." (*Supervision of Federal Offenders*, 2010). Probation officers pay special attention to patterns of violence when assessing third-party risk. Officers assess: the offender's current employment status, substance abuse, and current mental state; the instant offense; and the offender's prior criminal history and conduct. If a third-party risk is identified, the supervision case plan must include some controlling method to reduce or eliminate the third-party risk. Such methods may include increasing the intensity of supervision, increasing the number and frequency of home contacts and office appointments, warning third parties of the risk, or obtaining court approval to notify employers, even if it might result in loss of the offender's employment. If a third-party risk is identified with an offender who has demonstrated patterns of violent behavior, probation officers must take action to reduce the risk to the third party and to protect the community.

When supervising violent offenders, a two-pronged approach is used to effectively manage risk (Taxman, 2002). This dual approach consists of correctional and controlling supervision strategies.

Correctional Strategies

Correctional strategies include strategies that seek to provide the offender with skills, resources, and treatment to affect positive behavioral change both during the term of supervision and after the individual is released from supervision. Some of these skills, resources, and treatments include various types of programs such as employment programs, workforce/skill development, reentry programs, cognitive behavioral treatment, domestic violence treatment, and anger management programs.

Some offenders may be resistant to participating in such correctional strategies. In these cases, the probation officer may petition the court to modify the offender's conditions of release, at which time a hearing will be held to examine if the offender should be required to participate in the prescribed treatment regimen. When used in this sense, the correctional strategies become risk-controlling strategies aimed at correcting cognitive processes that are criminogenic in nature and are potentially associated with violent behaviors.

Correctional strategies used in supervising high-risk violent offenders may also include referrals for mental health and substance abuse evaluations. In some cases, an evaluation for a co-occurring disorder may be necessary; in co-occurring disorders, both a mental disorder and a substance abuse problem are present. Given the degree of risk and the needs associated with offenders who suffer from co-occurring disorders, these individuals are likely to require significant amounts of treatment. Different types of treatment may include individual co-occurring counseling, group counseling for co-occurring disorders, treatment readiness groups, family counseling, cognitive behavioral groups, and educational groups that assist offenders in obtaining their General Educational Development (GED) degree or basic computer skills.

When placing high-risk offenders in treatment programs, probation officers must be aware of the characteristics of the treatment group where the offender will be placed. Latessa and Lowenkamp conducted a halfway house study in Ohio that included approximately 15,000 offenders who were receiving intensive treatment services. They found that including high-risk offenders in treatment programs with low-risk offenders increased criminal conduct among the low-risk offenders. Latessa and Lowenkamp suggest this outcome was the result of low-risk offenders establishing negative peer networks and developing antisocial behaviors similar to those demonstrated by the high-risk offenders, resulting from their association in the treatment group. While high-risk offenders generally benefit from the treatment groups, and their recidivism rates are decreased through such interventions, mixing high-risk offenders in with a general treatment group has been shown to have negative effects on treatment rehabilitation of other currently low-risk offenders (Latessa & Lowenkamp, 2002;

Lowenkamp & Latessa, 2004, 2005; Lowenkamp, Latessa, & Holsinger, 2006). As a result, probation officers must target high-risk offenders with either individual treatment or establish treatment groups for offenders with specific risk levels in an effort to minimize the negative consequences associated with mixing high-risk and low-risk offenders.

Controlling Strategies

Controlling strategies focus on controlling an offender's behavior so as to protect the community from further harm. Controlling strategies enable probation officers to maintain awareness of an offender's activities while ensuring compliance with the conditions of release. Strategies that control behaviors include: location monitoring technologies such as global positioning systems, electronic monitoring, and electronic alcohol monitors; drug testing; in-person and written reports to the probation officer; home, community, collateral, and employment contacts; conducting criminal record checks; residential reentry center placements; and, in some cases, conducting a search of the offender's residence. Controlling strategies are employed to control the behaviors of offenders who have a history of non-compliance, those who exhibit violent behaviors, and those who violate or have violated the conditions of release.

Controlling strategies aimed to control risk among high-risk violent offenders require probation officers to increase the intensity of their supervision. Increasing the intensity of supervision can consist of a variety of methods. For example, probation officers may have frequent contact with other law enforcement agencies that have information about the activities of certain offenders. These agencies are not limited to, but may include, the original arresting agency, federal task forces, local intelligence agencies, Postal Service personnel, and the various agencies whose representatives attend community policing meetings. These agencies are all good sources of information that may provide probation officers with information regarding patterns of criminal activity. Other agencies may also provide probation officers with information on specific offenders and their current associations and peer groups, as well as general patterns of behavior of residents located in neighborhoods throughout the city. This information provides valuable assistance in monitoring an offender's activities while he or she remains under supervision.

Using information provided by other law enforcement agencies assists probation officers in managing high-risk violent offenders. This information becomes very important when the activities of the probation officer require community observation. Community observation consists of conducting fieldwork in a manner that does not involve any direct contact with the offender. This method seeks to unobtrusively

FIGURE 20.1 The correctional management of risk.
Source: Monograph 109. (2010). Supervision of Federal Offenders. Washington, D.C.

monitor compliance with specific conditions of supervision, such as association restrictions or location monitoring restrictions. It is an effective method to confirm the truthfulness of an offender. For example, if an offender claims that he is attending a domestic violence program offered through his church, the probation officer may use this method to drive by the church to verify that the offender's car is parked in the parking lot. Such methods are important to confirm the truthfulness of an offender, should the probation officer have suspicions to the contrary.

The goal of using the two-pronged approach to manage risk is to protect the community from further harm and to facilitate the successful reentry of offenders back into the community (**Figure 20.1**). Throughout the supervision process, it is expected that the offenders will refrain from committing any new crime, be held accountable for their victim(s), family(s), community(s), and other court-imposed responsibilities, and will prepare for continued success through improvements in their personal conduct and conditions beyond the term of supervision imposed by the court (*Supervision of Federal Offenders*, 2010).

RISK LEVELS AND IMPLEMENTATION OF CASE STRATEGIES

Implementation of case management strategies based on risk is not a "cookie cutter" operation approach. Case management strategies will inevitably vary based on the type of risks each case presents, the needs to be addressed, and the responsivity factors that need to be dealt with early on in the term of supervision. Case management strategies are individualized and explicated in the case planning process for each offender. This section summarizes the general risk categories and "typical" qualitative differences that may be seen in the different levels of supervision intensities. The categories identified here are just one example of the varying degrees of intensity found in a supervision process. Depending on the risk level of the case, the supervision strategies are likely to use a blended approach of both correctional and controlling strategies (Taxman, 2002).

High Risk

Conduct a home contact twice per month, and make one contact during non-traditional working hours. Conduct office contacts twice per month. Follow up with employers to verify employment. Contact any collateral contacts as needed for purpose-driven issues. Require the offender to participate in self-reflective journaling exercises with the probation officer, targeting specific risk areas based on the results of the risk assessment.

Make treatment referrals consistent with any special conditions imposed by the court or consistent with any output produced by the risk assessment. Direct the offender to attend a specified course of treatment for a specified time and duration, based on the recommendation from the initial treatment assessment. Follow up with the treatment provider on a regular basis to verify the offender's status and discuss treatment progress reports.

Review and compare written monthly reports. Review and compare documents accompanying written reports to verify residence, employment, and payment of outstanding financial penalties. Obtain criminal record checks at 6 months and annually thereafter, as well as 30 days prior to the expiration of the term of supervision. Perform credit checks, and require and review updated net worth and cash flow forms every 6 months. Follow up on any changes in circumstances, including third-party risk assessments upon a change in circumstances.

Moderate Risk

Conduct a home contact once per month during non-traditional working hours. Conduct office contacts once per month.

Make treatment referrals consistent with any special conditions imposed by the court or consistent with any output produced by the risk assessment. Based on the evaluation recommendation from the treatment referral, have the offender attend a specified course of treatment for a specified time and duration. Follow up with the treatment provider on a regular basis to verify the offender's status and discuss treatment progress reports.

Review and compare written monthly reports. Review and compare documents accompanying written reports to verify residence, employment, and payment of outstanding financial penalties. Obtain criminal record checks at 6 months and annually thereafter, as well as 30 days prior to the expiration of the term of supervision. Perform credit checks, and require and review updated net worth and cash flow forms every 6 months. Follow up on any changes in circumstances, including third-party risk assessments upon a change in circumstances.

Low Moderate Risk

Conduct a home contact once every 2 to 3 months. Conduct office contacts once every 2 to 3 months. Review and compare written monthly reports. Make treatment referrals consistent with the special conditions imposed by the court, and follow up with treatment providers. Review and compare written monthly reports. Review and compare documents accompanying written reports to verify residence, employment, and payment of outstanding financial penalties. Obtain criminal record checks at 6 months and annually thereafter, as well as and 30 days prior to the expiration of the term of supervision. Perform credit checks, and require and review updated net worth and cash flow forms every 6 months. Follow up on any changes in circumstances, including third-party risk assessments upon change of circumstances.

Low Risk

Conduct a home contact once every 3 to 6 months. Conduct office contacts as needed. Review and compare written monthly reports. Review and compare documents accompanying written reports to verify residence, employment, and payment of outstanding financial penalties. Obtain criminal record checks at 6 months and annually thereafter, as well as 30 days prior to the expiration of the term of supervision. Perform credit checks, and require and review updated net worth and cash flow forms every 6 months. Follow up on any changes in circumstances, including third-party risk assessment upon change of address or employment.

The supervision strategies outlined here are templates that should be modified on a case-by-case basis to ensure that each offender is receiving the proper level of supervision. Special conditions or modifications of supervision (e.g., use of location monitoring device), may alter the level of intensity of the supervision practice. During the course of supervision, the risk of each case will necessarily change as a result of changes in life circumstances, including both accomplishments and failures. When non-compliant behavior occurs, it is important for probation officers to consistently intervene early and effectively. Non-compliant behavior could be an indicator of a change in risk category and must be responded to with an intervention consisting of both a correcting and a controlling strategy and reported to the court. Non-compliant behavior, depending on its type and severity, may also require the probation officer to conduct a re-assessment to ensure an effective supervision strategy is being used and to determine the degree and nature of any controlling and/or correctional interventions that are to be imposed (Taxman, 2002; Vose et al., 2008).

Controlling interventions hold offenders accountable for their non-compliant behaviors. Such measures include reprimands, warnings, or the imposition of more intrusive or restrictive requirements to serve as a consequence for the non-compliant behavior. Correctional interventions, by comparison, promote corrective behaviors by providing offenders with resources, information, education, and treatment to change the circumstances that led to the non-compliance. Probation officers must consistently and swiftly respond to non-compliant behaviors to effectively protect the community and to deter future non-compliant behavior. Use of multidimensional approaches consisting of correctional and controlling strategies has been shown to be more effective than using either strategy (Petersilia, 1998; Taxman, 2002) over the other.

Appropriate strategies adopted in response to non-compliant behaviors usually consist of community-based interventions. Community-based interventions can be used in response to technical violations, for example, failure to appear for a urine test, failure to appear as directed, or not being employed. In contrast, revocation responses are encouraged for high-risk violent cases when the offender poses a significant risk to the community, for example, an arrest for domestic violence or a new felonious criminal behavior, a behavior that results from chronic or serious non-compliance, or failure of other interventions, such as not abiding by a location monitoring curfew, walking away from a halfway house, obstructing a urine screen, continued substance use, or other non-compliant behavior that requires revocation by statute.

CONCLUSION

Supervising high-risk violent offenders who have long criminal histories and varied offending patterns is a dynamic process. As has been discussed in this chapter, supervising these offenders with the appropriate level of supervision as dictated by the offender's risk category is paramount to efficiently using resources and targeting those high-risk individuals who need and will benefit from them the most. A blended approach of incorporating both controlling and correctional supervision strategies allows officers access to various resources to assist offenders in their reentry. Using a combination of controlling and correcting supervision approaches provides probation officers with effective methods to manage risk and presents a good opportunity to correct behaviors.

Employing strategies such as targeting antisocial attitudes and antisocial personalities with cognitive behavioral treatments forces offenders to reflect on and analyze their thinking processes and means to manage their behaviors. These programs provide offenders with skills that can help them to slow down and effectively manage their impulsive tendencies. When high-risk violent offenders are exposed to pro-social peers—probation officers, treatment counselors, mentors, or others in the community—and have the opportunity to practice pro-social responses to criminogenic stimuli, they can effectively learn thought processes and responses to manage their behaviors. Enhancing the supervision strategies dictated by offenders' risk categories provides structure and expectations for offenders through the supervision process. This structure attempts to minimize the likelihood that an offender will reconnect with people, places, and things that encourage criminal behavior. If such contact does occur, it tries to ensure that offenders will likely have the tools and resources necessary to manage their thoughts and behaviors to avoid negative antisocial outcomes.

Case management of high-risk violent offenders is a dynamic process. Probation officers must in turn consistently use many different approaches and resources to effectively supervise these offenders and to manage risk. Probation officers are responsible for identifying offenders who have a history of violence or those offenders who are likely to engage in violent behavior by using their professional judgment in conjunction with risk assessment. The PCRA assesses the general risk, need, and responsivity factors related to reoffending. While violent offenders have other factors that may need to be assessed when predicting violence, the Administrative Office for the U.S. Courts is in the process of developing additional assessments that would follow up on the PCRA results to assist the PCRA in targeting specific areas including violence, domestic violence, sex offenses, and white-collar crime. Once these assessments are developed, they will better assist the PCRA in assessing specific types of offending behaviors more accurately.

REFERENCES

Andrews, D. A., & Bonta, J. (2003). *The psychology of criminal conduct.* (3rd ed.). Cincinnati, OH: Anderson.

Bucklen, K. B., & Zajac, G. (2009). But some of them don't come back (to prison!): Resource deprivation and thinking errors as determinants of parole success and failure. *The Prison Journal, 89,* 239–264.

Collins, R. (2010). The effect of gender on violent and non-violent recidivism: A meta-analysis. *Journal of Criminal Justice, 38*, 675–684.

Elliott, D. S. (1994). Serious violent offenders: Onset, developmental course, and termination: The American Society of Criminology 1993 presidential address. *Criminology, 32*, 1–21.

Gendreau, P., French, S. A., & Taylor, A. (2002). *What works: (What doesn't work).* Invited submission to the International Community Corrections Association monograph series project.

Landenberger, N. A., & Lipsey, M. (2005). The positive effects of cognitive-behavioral programs for offenders: A meta-analysis of factors associated with effective treatment. *Journal of Experimental Criminology, 1*, 451–476.

Latessa, E. J., & Lowenkamp, C. T. (2002). *Evaluation of Ohio's community-based correctional facilities and halfway house programs.* Cincinnati, OH: University of Cincinnati, Division of Criminal Justice Research.

Latessa, E. J., & Lowenkamp, C. T. (2006). What works in reducing recidivism? *University of St. Thomas Law Journal, 3*, 521–535.

Lipsey, M. W. (2009). The primary factors that characterize effective interventions with juvenile offenders: A meta-analytic overview. *Victims & Offenders, 4*, 124–147.

Lowenkamp, C. T., & Latessa, E. J. (2004). Understanding the risk principle: How and why correctional interventions can harm low-risk offenders. *Topics in Community Corrections*, 3–6.

Lowenkamp, C. T., Latessa, E. J., & Holsinger, A. M. (2006). The risk principle in action: What we have learned from 13,676 offenders and 97 correctional programs? *Crime and Delinquency, 52*, 77–89.

Lowenkamp, C. T., & Latessa, E. J. (2005). Increasing the effectiveness of correctional programming through the risk principle: Identifying offenders for residential placement. *Criminology and Public Policy, 4*, 263–277.

Moffitt, T. E. (1993). Adolescence-limited and life-course-persistent antisocial behavior: A developmental taxonomy. *Psychology Review, 100*, 674–701.

Office of Probation and Pretrial Services (OPPS). (2010). *Federal Post Conviction Risk Assessment.* Washington, DC: Author.

Petersilia, J. (1998). Intermediate sanctions: What have we learned? In *Perspectives on Crime and Justice: 1997–1998 Lecture Series.* National Institute of Justice.

Petersilia, J. (1999). A decade with experimenting with intermediate sanctions: What have we learned? *Perspectives, 23*, 39–44.

Sessions, W. K. III., Castillo R., Carr W. B. Jr., Jackson, K. B., Hinojosa R. H., Howell, B. A., . . . Wroblewski, J. J. (2010). United States Sentencing Commission. *Federal offenders sentenced to supervised release.* Washington, D.C: United States Sentencing Commission.

Monograph 109. (2010). *Supervision of federal offenders.* Monograph 109. (2010). Washington, DC.

Taxman, F. S. (1999). Unraveling "what works" for offenders in substance abuse treatment services. *National Drug Court Institute Review, 2*, 93–134.

Taxman, F. S. (2002). Supervision: Exploring the dimensions of effectiveness. *Federal Probation, 66*, 14–27.

U.S.C. (2010). *Federal criminal code and rules.* Washington, D.C.

Vose, B., Cullen, F. T., & Smith, P. (2008). The empirical status of the level of service inventory. *Federal Probation*, 72, 22–29.

Wilson, D. B., Bouffard, L. A., & McKenzie, D. L. (2005). Quantitative review of structured group-oriented cognitive-behavioral programs for offenders. *Criminal Justice and Behavior, 32*, 172–204.

CHAPTER 21

Criminal Predatory Behavior in the Federal Bureau of Prisons

Glenn D. Walters
Federal Correctional Institution, Schuylkill, Pennsylvania

We have all heard the stories about inmates being raped or assaulted in prison; some of us may even have been deterred from committing a crime after hearing one of these stories. Rape and assault undoubtedly occur in U.S. prisons and correctional facilities, but the vast majority of inmates are neither raped nor beaten senseless. Most, however, have been victims or perpetrators of predation in prison. The seeming paradox of high rates of predation coupled with low rates of sexual and nonsexual assault can be explained by the fact that predation comes in many forms, ranging from the overt and obvious to the covert and discreet. Predation assumes planning, yet the level of planning varies from inmate to inmate and from one predatory act to the next. Given the heterogeneous nature of the federal prison population, variety is the rule when it comes to predation. This chapter offers a model of inmate predation in the federal Bureau of Prisons (BOP) that calls attention to two related but distinct strains of criminal predatory behavior: proactive and reactive.

THE SCOPE OF THE PROBLEM

Back in 1984 when I began working in the BOP, there were 32,000 inmates, 12,000 staff, and 44 different institutions in the federal prison system. Ronald Reagan was running for re-election, Los Angeles was hosting the summer Olympics, and TV shows like *Dynasty* and *Dallas* were at the top of the Nielsen ratings. By 2011, the BOP has grown to 210,093 inmates, 37,544 staff, and 113 institutions. Hence, between 1984 and 2011 there was a 557% increase in the inmate population of the BOP, but only a 213% increase in staffing and a 157% increase in institutions. The density of institutions and the inmate-to-staff ratio have more than doubled in the past 23 years. To put it succinctly, BOP staff is increasingly being asked to do more with less.

What might account for the dramatic increase in the BOP inmate population since 1984? The Sentencing Reform Act of 1984, which established determinate sentencing, abolished parole, and reduced the amount of good time inmates could earn, and in combination with mandatory sentencing charters enacted in 1986, 1988, and 1990, both increased the number of inmates entering prison and decreased the number of inmates leaving prison. However, these were not the only factors responsible for the rapid growth of the federal prison population since 1984. Although the United States has been engaged in a "war on drugs" since the early 1900s, this effort gained new momentum in the late 1980s and early 1990s with the advent of several presidential initiatives. The focus on prosecution of drug-related crimes not only increased the BOP inmate population but also changed the population's composition. In 1984, 29% of the BOP inmate population was serving time for drugs; by 2011, that the percentage had risen to 55%.

Federal statutes on gun control, immigration, child pornography, and carjacking, as well as the Revitalization Act of 1997 in which the BOP took custody of all District of Columbia (DC) code felony offenders, not only increased the overall prison population but also introduced a new class of violent offender to the BOP. In 1984, the only violent crime that the United States Attorney's Office was prosecuting to any significant extent was bank robbery; today, murderers, rapists, assaulters, burglars, and child molesters are entering the prison system in record numbers.

Changes in both the size and the composition of the federal prison population may have contributed to a rise in violence within the BOP, as exemplified by a 155% increase in the rate of inmate assaults on staff and a 206% increase in the rate of inmate assaults on other inmates since 1984. In trying to make sense of these changes, I came up with a two-dimensional theory of criminal predatory behavior as a means to assess the character of today's prisoners.

A TWO-DIMENSIONAL MODEL OF CRIMINAL PREDATORY BEHAVIOR

In an attempt to understand criminal predatory behavior in the BOP, I developed several working theoretical models of criminality. I published my initial findings in a book on the criminal lifestyle and created an 80-item self-report measure, the Psychological Inventory of Criminal Thinking Styles (PICTS), to assess the eight thinking styles that I believed supported this lifestyle (Walters, 1990). A continuing program of research and clinical observation has shown that these eight thinking styles can be grouped into two primary factors: proactive criminal thinking and reactive criminal thinking. In turn, I constructed PICTS composite scores to measure each factor. I also began to notice similarities between the eight thinking styles measured by the PICTS and such crime-related psychological constructs as psychopathy and antisocial personality disorder. These observations eventually led to the formation of a general theory of criminal predatory behavior comprised of two interrelated factors: proactive criminality and reactive criminality.

The terms *proactive* and *reactive* were popularized by Kenneth Dodge in his work on aggression in children. Dodge and his colleagues observed that childhood aggression assumes one of two forms: instrumental (proactive) aggression, and hostile (reactive) aggression (Dodge, 1991; Dodge & Coie, 1987). Proactive aggression is a goal-oriented behavior designed to achieve a tangible reward or objective, such as intimidating another child into giving up a prized possession. Reactive aggression, by contrast, is committed in retaliation for a perceived injustice or provocation, such as pushing a child who has bumped into you under the assumption that the bump was intentional. In comparing these two forms of aggression, Dodge held that while proactive aggression is cold-blooded, goal-directed, and calculated, reactive aggression is hot-blooded, affective, and unplanned. Although only a handful of studies have addressed the issue of proactive and reactive aggression in adults, results from this line of research studies suggest that Dodge's proactive–reactive model may apply to adults as well as to children (Cornell et al., 1996; Kockler et al., 2006).

Two studies conducted on PICTS measures of proactive and reactive criminal thinking provide preliminary support for the construct validity of the proactive–reactive model in adults. Research on children, for instance, demonstrates that proactive aggression is associated with positive outcome expectancies for aggression (Crick & Dodge, 1996), whereas reactive aggression is associated with hostile attribution biases (Dodge & Newman, 1981). In a study of incarcerated male federal prisoners, Walters (2007) determined that the PICTS Entitlement scale, a strong correlate of proactive criminal thinking, predicted positive outcome expectancies for crime but not hostile attribution biases, while the PICTS Cutoff scale, a strong correlate of reactive criminal thinking, predicted hostile attribution biases but not positive outcome expectancies for crime. In a second study, Walters, Frederick, and Schlauch (2007) discovered that the PICTS proactive composite scale postdicted prior arrests for instrumental crimes such as robbery and burglary, but not reactive crimes such as assault and domestic violence;

conversely, the PICTS reactive composite scale postdicted prior arrests for reactive crimes but not prior arrests for proactive crimes.

Several important questions relating to proactive and reactive aggression in children appear to have been answered in recent years. First, there is growing consensus that proactive and reactive aggression are dimensions rather than categories (Dodge et al., 1987). Second, rather than falling at opposite ends of the same dimension, proactive and reactive aggression occupy different dimensions (Poulin & Boivin, 2000). Third, although the proactive and reactive dimensions give rise to different sets of correlates, such as the outcome expectancy–proactive and hostile attribution bias–reactive relationships described earlier, these two dimensions correlate .41 to .90 with one another (Price & Dodge, 1989), thus giving rise to a correlated two-dimensional model (**Figure 21.1**). A two-dimensional model of criminal predatory behavior was consequently introduced using Dodge's work with childhood aggression as a theoretical guide. This developmentally informed theory linked proactive and reactive aggression in children to proactive and reactive aggression and criminal thinking in adults (Walters, 2005).

Criminal predation is a broad concept that encompasses proactive–reactive differences in behavior, affect, and cognition (**Table 21.1**). This chapter discusses whether adult proactive–reactive criminal predatory behavior parallels childhood proactive–reactive aggression, a purpose that will be addressed in three questions:

1. Does criminal predation, like childhood proactive–reactive aggression, form dimensions (quantitative distinctions) rather than categories (qualitative distinctions)?
2. If adult criminal predatory behavior, like childhood aggression, is dimensional in nature, then does it also partition into the proactive and reactive forms that have been observed in childhood aggression?

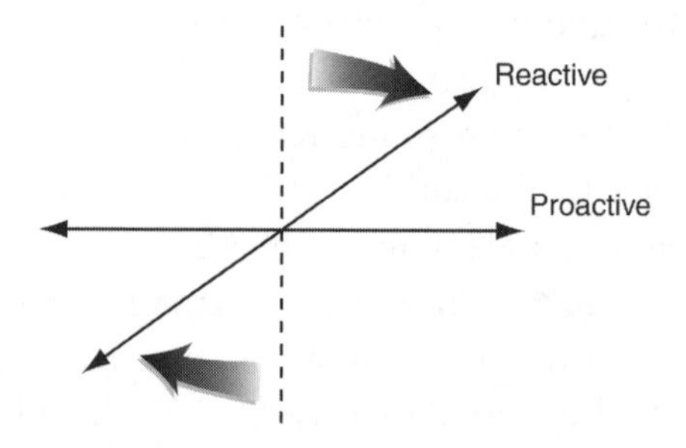

FIGURE 21.1 Relationship between reactive aggression and proactive aggression.

TABLE 21.1 Behavioral, Affective, and Cognitive Features of Proactive and Reactive Criminal Predation

Feature	Proactive	Reactive
Behavioral	Planned/premeditated	Spontaneous/impulsive
	Scheming/cold-blooded	Emotional/hot-headed
	Calculated/manipulative	Changeable/capricious
Affective	Anticipation (of positive outcomes)	Anger (at perceived injustices)
Cognitive	Attitude of privilege	Loss of focus
	Rationalization	Weak personal control

3. In the event proactive and reactive criminal predation are identified, is there evidence that these two strains of criminal behavior and thinking correlate with one another, as do proactive and reactive childhood aggression?

DIMENSIONS OR CATEGORIES?

An ongoing debate continues to rage over whether crime-related concepts such as psychopathy, antisocial personality, and criminal lifestyle are best construed as dimensions or as categories. The dimensional view of criminal predatory behavior holds that offenders fall along one or more dimensions of predatory behavior and that individual differences between offenders are quantitative (differences in degree) rather than qualitative (differences in kind). The taxonomic view of predatory behavior maintains that offenders fall into discrete categories of individuals who do and do not engage in predatory behavior.

Paul Meehl and his colleagues performed a valuable service to researchers interested in distinguishing between dimensional and taxonomic structure by creating the taxometric method. Taxometric procedures, such as mean above minus below a cut (MAMBAC), maximum covariance (MAXCOV), and maximum eigenvalue (MAXEIG), were constructed from statistical algorithms to assess the latent structure of psychopathological constructs such as schizophrenia, depression, and anxiety, and have been applied to crime-related constructs such as psychopathy, antisocial personality, and criminal lifestyle (Meehl, 2004; Meehl & Golden, 1982; Meehl & Yonce, 1994, 1996; Waller & Meehl, 1998; Ruscio, Haslam, & Ruscio, 2006).

In the first study to investigate the latent structure of psychopathy with the taxometric method, Harris, Rice, and Quinsey (1994) collected Psychopathy Checklist–Revised (PCL-R) data on 653 mentally disordered offenders and observed taxonomic latent structure in items from Factor 2 of the PCL-R, and in the eight individual items that correlated had the highest correlations with the total PCL-R score. Edens, Marcus, Lilienfeld, and Poythress (2006) have criticized Harris et al.'s choice of indicators (dichotomized), participants (mentally disordered forensic patients), and procedures (distributional analysis). Conducting their own study on a group of 876 prison inmates and non-incarcerated patients in court-ordered substance abuse programs, Edens and his colleagues found evidence of dimensional structure on the PCL-R, a finding that was replicated in several subsequent investigations (Edens et al., 2006; Guay, Ruscio, Knight, & Hare, 2007; Walters, Duncan, & Mitchell-Perez, 2007; Walters et al., 2006). Two self-report measures designed to assess psychopathy, the Psychopathic Personality Inventory (PPI) and the Levenson Self-Report Psychopathy scale (LSRP), also display dimensional structure when subjected to taxometric analysis (Marcus, John, & Edens, 2004; Marcus, Lilienfeld, Edens, & Poythress, 2006; Walters, Brinkley, Magaletta, & Diamond, 2008).

Ratings obtained from the *Diagnostic and Statistical Manual of Mental Disorders*, fourth edition (*DSM-IV*), diagnostic criteria for antisocial personality disorder showed signs of taxonicity when Skilling, Harris, Rice, and Quinsey (2002) studied a sample composed largely of participants from the earlier Harris et al. investigation. Several more recently conducted studies, however, have produced results more consistent with a dimensional interpretation of the latent structure of antisocial personality. First, Marcus, Lilienfeld, Edens, and Poythress (2006) administered the Structured Interview for *DSM-IV* Axis II Personality Disorders (SCID-II) and the Personality Diagnostic Questionnaire-4 (PDQ-4) to 1146 state inmates and non-incarcerated patients in court-ordered substance abuse programs and found evidence of dimensional structure on both measures. Second, Walters, Diamond, Magaletta, Geyer, and Duncan (2007) observed dimensional results when the three subscales of the Antisocial Features (ANT) scale of the Personality Assessment Inventory (PAI) were subjected to taxometric analysis, regardless of participant gender (male, female), race (white, black, Hispanic), or security level (low, medium, high).

Taxometric research on two measures designed to assess features of a criminal lifestyle, a chart audit procedure known as the Lifestyle Criminality Screening Form (LCSF) and a self-report measure known as the Psychological Inventory of Criminal Thinking Styles (PICTS), showed consistent support for dimensional latent structure in a group of 771 male federal inmates (Walters, 2006). A follow-up study determined that even when extreme scoring groups were employed (a situation conducive to the formation of pseudotaxa), the PICTS continued to exhibit dimensional structure (Beauchaine & Waters, 2003; Walters & McCoy, 2007). Research conducted over the past several years on multiple measures of three crime-related constructs—psychopathy, antisocial personality, and criminal lifestyle—indicate that these constructs exist along one or more continua. The question that needs to be addressed next is whether these constructs form a single dimension or several different dimensions.

HOW MANY DIMENSIONS?

Now that it has been established with a reasonable degree of certainty that the latent structure of criminal predatory constructs such as psychopathy, antisocial personality, and criminal lifestyle is dimensional, the next order of business is to calculate the number of individual dimensions that support criminal predation. The technique that researchers most often use to determine whether a set of variables falls along a single dimension or partitions into several different dimensions is factor analysis. In situations where a theory is available to guide our research efforts, confirmatory factor analysis is the recommended approach. The theory that will guide a structural equation modeling (SEM) confirmatory factor analysis of psychopathy, antisocial personality, and criminal lifestyle is that advanced by Walters, in which criminality is divided into two dimensions, proactive and reactive. It is accordingly reasoned that individual measures of psychopathy, antisocial personality, and criminal lifestyle should fit a two-dimensional model (proactive and reactive criminality) significantly better than the alternative one-dimensional and three-dimensional models.

The one-dimensional model of criminal predation holds that psychopathy, antisocial personality, and criminal lifestyle converge on a single common trait or dimension. In contrast, the two-dimensional model of criminal predation maintains that psychopathy, antisocial personality, and criminal lifestyle form two overlapping dimensions, which in the current research context, are referred to as proactive and reactive criminality. The three-dimensional model considered in the present set of analyses holds that psychopathy, antisocial personality, and criminal lifestyle form three separate dimensions, one for each construct or measure. The three-dimensional model can be viewed as a control model in the sense that it proposes greater commonality between different constructs from the same measure than between related constructs from different measures. This model may also be referred to as the measures model because it predicts that each measure constitutes a separate dimension.

In selecting measures for a confirmatory factor analysis of psychopathy, antisocial personality, and criminal lifestyle, it is imperative that we avoid comparing measures derived from different methods (self-report versus rating scale) because how a construct is measured (method) can have a profound effect on its pattern of correlation with other measures (Podsakoff, MacKenzie, Lee, & Podsakoff, 2003). Thus, if we were to combine a behavioral measure of psychopathy (e.g., passive avoidance errors), with an interview-based measure of antisocial personality (e.g., SCID-II), and a self-report measure of criminal lifestyle (e.g., PICTS), this multiplicity of methods would skew the data in favor of the three-dimensional or measures model because of common method variance. In other words, measures of different constructs utilizing the same method are more likely to correlate than measures of the same construct utilizing different methods. Consequently, self-report measures of psychopathy (LSRP), antisocial personality (ANT), and criminal lifestyle (PICTS) that had previously yielded dimensional results were used to test the one-, two-, and three-dimensional models in the analyses described next.

In investigating the number of dimensions that support criminal predatory behavior, the one-dimensional model regressed the 13 LSRP, ANT, and PICTS scales onto a single latent factor (**Table 21.2**). The two-dimensional model, by comparison, regressed the LSRP Primary Psychopathy scale, the ANT Egocentricity subscale, and the five PICTS scales that, according to the results of a study by Egan, McMurran, Richardson, and Blair (2000), constitute a "willful criminality" factor (Mollification, Entitlement, Power Orientation, Sentimentality, and Superoptimism) onto a proactive latent factor and the LSRP Secondary Psychopathy scale, the ANT Antisocial Behaviors and Stimulus Seeking subscales, and the three PICTS scales constitute a "lack of thoughtfulness" factor (Cutoff, Cognitive Indolence, Discontinuity) onto a reactive latent factor. The three-dimensional or control model regressed each of the individual scales onto their respective measures: the LSRP Primary and Secondary Psychopathy scales onto a LSRP latent factor, the three ANT subscales onto an ANT latent factor, and the eight PICTS thinking style scales onto a PICTS latent factor.

SEM analysis of the three models in a preliminary sample of 139 male, medium-security federal inmates generated support for the absolute fit of the two-dimensional model as measured by the comparative fit index (CFI) and root mean square error of approximation (RMSEA) criterion (Arbuckle, 1999). In addition, relative fit analyses revealed that the two-dimensional model furnished a significantly better fit for the data than either the one-dimensional or three-dimensional models. The results of these analyses are suggestive—but far from conclusive—given the fact that reliable results for an SEM analysis require a sample size twice that of the present investigation to allow for at least 20 participants per measured variable ($20 \times 13 = 260$). Assuming that these initial findings hold up in a larger sample, they offer preliminary support for the argument that—like childhood aggression—adult criminal predatory behavior partitions into two dimensions, proactive and reactive.

TABLE 21.2 Descriptions and Factor Assignments for the Thirteen Indicators

Measure	Indicator	Description	1D	2D	3D
Levenson Self-Report Psychopathy scale (LSRP)	Primary psychopathy	Selfish, callous, manipulative, impulsive	1	1	1
	Secondary psychopathy	Irresponsible, poor self-control	1	2	1
Personality Assessment Inventory Antisocial Features scale (ANT)	Antisocial behaviors	History of behavior and conduct problems	1	2	2
Psychological Inventory of Criminal Thinking Styles (PICTS)	Egocentricity	Self-centered, callous, remorseless	1	1	2
	Stimulus seeking	Low tolerance of boredom	1	2	2
	Mollification	Blaming, justifying, rationalizing	1	1	3
	Cutoff	Rapid elimination of deterrents to crime	1	2	3
	Entitlement	Ownership, privileged status	1	1	3
	Power orientation	Exerting power and control over others	1	1	3
	Sentimentality	Performing good deeds to excuse crime	1	1	3
	Superoptimism	Unrealistic appraisal of criminal success	1	1	3
	Cognitive indolence	Short-cut thinking, lack of critical reasoning	1	2	3
	Discontinuity	Poor follow-through, lack of consistency	1	2	3

Abbreviations: 1D, one-dimensional model; 2D, two-dimensional model; 3D, three-dimensional model.
Numbers under each dimensional model represent the dimension to which indicator was assigned.

The next step is to determine whether these two dimensions are correlated, as with proactive and reactive childhood aggression, or whether they are independent of one another.

ARE THE DIMENSIONS CORRELATED OR INDEPENDENT?

The third question posed in this chapter is whether the putative proactive and reactive dimensions of criminal predatory behavior are correlated or independent.

The proactive and reactive dimensions of childhood aggression have been found to correlate .41 to .90 with one another. Likewise, research studies conducted on the PCL-R Factor 1 and Factor 2 scores, the LSRP Primary and Secondary Psychopathy scales, and the PICTS proactive and reactive criminal thinking composite scales denote that the two dimensions correlate .40 to .72 with one another (Hare, 2003; Levenson, Kiehl, & Fitzpatrick, 1995; Walters & Mandell, 2007). Therefore, adult predatory behavior parallels childhood aggressive behavior in the sense that both appear to be subsumed by two correlated dimensions referred to in this chapter as proactive and reactive criminality/aggression. One might therefore ask, what are the implications of this two-dimensional model of adult criminal predatory behavior?

IMPLICATIONS OF THE TWO-DIMENSIONAL MODEL

One implication of the dimensional nature of the two-dimensional model is that criminal predatory behavior is not a categorically distinct form of behavior, but is rather an extension of normal aggressive behavior in both adults and children. Consequently, labeling someone as a proactive or reactive predatory criminal may be inappropriate, in that people differ in degree rather than in kind on the proactive and reactive dimensions of predatory criminal behavior. It would be more appropriate to identify a person's current level of proactive and reactive predatory criminal behavior and the situational factors that moderate his or her expression of one or both of these forms of criminal predation.

The dimensional structure of criminal predatory behavior has policy implications as well. Popular measures of criminality such as the PCL-R and *DSM-IV* criteria for antisocial personality disorder are sometimes included in death penalty and parole evaluations. In light of the dimensional latent structure of criminal predatory behavior, it would seem advisable that instead of relying on a single cutoff score, the entire range of scores or perhaps several cutoff scores on measures such as the PCL-R should be considered in making death penalty and parole determinations.

Preliminary analysis denotes that criminal predatory behavior is supported by two different dimensions, proactive and reactive. One implication of this research finding is that different intervention strategies may be required to address these two forms of criminal predatory behavior. Although reactive criminality is best managed with skill development techniques such as anger and stress management, proactive criminality requires that outcome expectancies for crime and criminal goals be explored and re-evaluated. Many correctional programs do a good job of addressing the skill deficits and overriding issues that support reactive criminality, but few programs are equipped to handle the manipulative features of proactive criminality. Furthermore, parole/release decision-making generally takes into account the more observable aspects of reactive criminality but often overlooks the more subtle aspects of proactive criminality. Hence, effort needs to be directed at designing correctional programs that address both reactive and proactive criminality as well as creating guidelines for parole/release decision makers that consider both reactive and proactive issues.

The fact that the proactive and reactive dimensions of criminal predatory behavior are correlated also has important theoretical, practical, and policy implications. Given the moderately high correlation that exists between proactive and reactive criminal predatory behavior, it will be relatively rare to find someone who rates extremely high on one dimension and extremely low on the other dimension (i.e., a "pure" proactive or reactive criminal). Instead, we are more apt to find someone who rates high on both dimensions or low on both dimensions. Shifting from the individual criminal offender to the individual criminal act, the evidence suggests that most criminal acts involve both proactive and reactive aggression/criminality, although the degree of each tends to vary from one criminal event to the next. Therefore, rather than be concerned with whether a particular offender is a proactive or reactive criminal, or whether a specific criminal act is a proactive or reactive criminal event, a more productive approach would be to consider the degree of proactive and reactive criminality currently being displayed by an individual offender or contained in an ongoing criminal event.

CONCLUSION

This chapter describes a two-dimensional model of criminal predatory behavior that links adult criminal behavior to childhood aggressive behavior. As mentioned at the beginning of this chapter, this model summarizes how I view criminal predatory behavior in the BOP as well as in general. Analysis of data with the taxometric method reveals that adult criminal predatory behavior is dimensional in nature; using factor analysis leads to the discovery that adult criminal predatory behavior partitions into two dimensions, proactive and reactive; and using correlational analysis indicates that these two dimensions are correlated rather than independent. To the extent that the proactive and reactive dimensions apply to both criminal and non-criminal human behavior, and given the dimensional nature of these constructs, additional research is required to understand the transition from proactive and reactive non-criminal behavior to proactive and reactive criminal behavior.

REFERENCES

Arbuckle, J. L. (1999). *AMOS user's guide* (Version 4.0). Chicago, IL: SmallWaters.

Beauchaine, T. P., & Waters, E. (2003). Pseudotaxonicity in MAMBAC and MAXCOV analyses of rating scale data: Turning continua into classes by manipulating observers' expectations. *Psychological Methods, 8*, 3–15.

Cornell, D. G., Warren, J., & Hawk, G. (1996). Psychopathy in instrumental and reactive violent offenders. *Journal of Consulting and Clinical Psychology, 64*, 783–790.

Crick, N. R., & Dodge, K. A. (1996). Social information-processing mechanisms in reactive and proactive aggression. *Child Development, 67*, 993–1002.

Dodge, K. A. (1991). The structure and function of reactive and proactive aggression. In D. Pepler, & K. Rubin (Eds.), *The development and treatment of childhood aggression* (pp. 201–218). Hillsdale, NJ: Erlbaum.

Dodge, K. A., & Coie, J. D. (1987). Social-information processing factors in reactive and proactive aggression in children's peer groups. *Journal of Personality and Social Psychology, 53*, 1146–1158.

Dodge, K. A., Lochman, J. E., & Harnish, J. D. (1997). Reactive and proactive aggression in school children and psychiatrically impaired chronically assaultive youth. *Journal of Abnormal Psychology, 106*, 37–51.

Dodge, K. A., & Newman, J. P. (1981). Biased decision-making processes in aggressive boys. *Journal of Abnormal Psychology, 90*, 375–379.

Edens, J. F., Marcus, D. K., Lilienfeld, S. O., & Poythress, N. G. (2006). Psychopathic, not psychopath: Taxometric evidence for the dimensional structure of psychopathy. *Journal of Abnormal Psychology, 115*, 131–144.

Egan, V., McMurran, M., Richardson, C., & Blair, M. (2000). Criminal cognitions and personality: What does the PICTS really measure? *Criminal Behavior and Mental Health, 10*, 170–184.

Guay, J. P., Ruscio, J., Knight, R. A., & Hare, R. D. (2007). A taxometric analysis of the latent structure of psychopathy: Evidence for dimensionality. *Journal of Abnormal Psychology, 116*, 701–716.

Hare, R. D. (2003). *The Hare Psychopathy Checklist—Revised Manual* (2nd ed.). Toronto, Ontario, Canada: Multi-Health Systems.

Harris, G. T., Rice, M. E., & Quinsey, V. L. (1994). Psychopathy as a taxon: Evidence that psychopaths are a discrete class. *Journal of Consulting and Clinical Psychology, 62*, 387–397.

Kockler, T. R., Stanford, M. S., & Nelson, C. S. (2006). Characterizing aggressive behavior in a forensic population. *American Journal of Orthopsychiatry, 76*, 80–85.

Levenson, M. R., Kiehl, K. A., & Fitzpatrick, C. M. (1995). Assessing psychopathic attributes in a noninstitutionalized population. *Journal of Personality and Social Psychology, 68*, 151–158.

Marcus, D. K., John, S. L., & Edens, J. F. (2004). A taxometric analysis of psychopathic personality. *Journal of Abnormal Psychology, 113*, 626–635.

Marcus, D. K., Lilienfeld, S. O., Edens, J. F., & Poythress, N. G. (2006). Is antisocial personality disorder continuous or categorical? A taxometric analysis. *Psychological Medicine, 36*, 1571–1582.

Meehl, P. E. (2004). What's in a taxon? *Journal of Abnormal Psychology, 113*, 39–43.

Meehl, P. E., & Golden, R. (1982). Taxometric methods. In P. Kendall &, J. Butcher (Eds.), *Handbook of research methods in clinical psychology* (pp. 127–181). New York, NY: Wiley.

Meehl, P. E., & Yonce, L. J. (1994). Taxometric analysis: I. Detecting taxonicity with two quantitative indicators using means above and below a sliding cut (MAMBAC procedure). *Psychological Reports, 74*, 1059–1274.

Meehl, P. E., & Yonce, L. J. (1996). Taxometric analysis: II. Detecting taxonicity using covariance of two quantitative indicators in successive intervals of a third indicator (MAXCOV procedure). *Psychological Reports, 78*, 1091–1227.

Podsakoff, P. M., MacKenzie, S. B., Lee, J.-Y., & Podsakoff, N. P. (2003). Common method biases in behavioral research: A critical review of the literature and recommended remedies. *Journal of Applied Psychology, 88*, 879–903.

Poulin, F., & Boivin, M. (2000). Reactive and proactive aggression: Evidence of a two-factor model. *Psychological Assessment, 12*, 115–122.

Price, J. M., & Dodge, K. A. (1989). Reactive and proactive aggression in childhood: Relations to peer status and social context dimensions. *Journal of Abnormal Child Psychology, 17*, 455–471.

Ruscio, J., Haslam, N., & Ruscio, A. M. (2006). *Introduction to the taxometric method: A practical guide.* Mahwah, NJ: Lawrence Erlbaum.

Skilling, T. A., Harris, G. T., Rice, M. E., & Quinsey, V. L. (2002). Identifying persistently antisocial offenders using the Hare Psychopathy Checklist and *DSM* antisocial personality disorder criteria. *Psychological Assessment, 14*, 27–38.

Waller, N. G., & Meehl, P. E. (1998). *Multivariate taxometric procedures: Distinguishing types from continua.* Thousand Oaks, CA: Sage.

Walters, G. D. (1990). *The criminal lifestyle: Patterns of serious criminal conduct.* Thousand Oaks, CA: Sage.

Walters, G. D. (2005). Proactive and reactive aggression: A lifestyle view. In J. P. Morgan (Ed.), *Psychology of Aggression* (pp. 29–43). Hauppauge, NY: Nova Science Publishers.

Walters, G. D. (2006). The latent structure of the criminal lifestyle: A taxometric analysis of the Lifestyle Criminality Screening Form and Psychological Inventory of Criminal Thinking Styles. *Criminal Justice and Behavior, 34*, 1623–1637.

Walters, G. D. (2007). Measuring proactive and reactive criminal thinking with the PICTS: Correlations with outcome expectancies and hostile attribution biases. *Journal of Interpersonal Violence, 22*, 1–15.

Walters, G. D., Brinkley, C. A., Magaletta, P. R., & Diamond, P. M. (2008). Taxometric analysis of the Levenson Self-Report Psychopathy scale. *Journal of Personality Assessment, 90*, 491–498.

Walters, G. D., Diamond, P. M., & Magaletta, P. R. (2007). Taxometric analysis of the antisocial features scale of the Personality Assessment Inventory in Federal prison inmates. *Assessment, 14*, 351–360.

Walters, G. D., Duncan, S. A., & Mitchell-Perez, K. (2007). The latent structure of psychopathy: A taxometric investigation of the Psychopathy Checklist—Revised in a heterogeneous sample of male prison inmates. *Assessment, 14*, 270–278.

Walters, G. D., Frederick, A. A., & Schlauch, C. (2007). Postdicting arrests for proactive and reactive aggression with the PICTS Proactive and Reactive composite scales. *Journal of Interpersonal Violence, 22*, 1415–1430.

Walters, G. D., Gray, N.S., & Jackson, R. L. (2006). A taxometric analysis of the Latent Structure of the Psychopathy Checklist: Screening version. *Psychological Assessment, 19*, 330–339.

Walters, G. D., & Mandell, W. (2007). Incremental validity of the Psychological Inventory of Criminal Thinking Styles and Psychopathy Checklist: Screening version in predicting disciplinary outcome. *Law and Human Behavior, 31*, 141–157.

Walters, G. D., & McCoy, K. (2007). Taxometric analysis of the Psychological Inventory of Criminal Thinking Styles in male and female incarcerated offenders and college students. *Criminal Justice and Behavior, 34*, 781–793.

CHAPTER 22

Institutional Misconduct Among Capital Murderers

Mark D. Cunningham
Private Practice Forensic Psychologist

The understanding of the prison behavior of capital offenders has expanded rapidly in recent years as a result of large-scale digitized correctional databases becoming available for study. This trend has yielded an increasingly detailed and reliable illumination of the comparative rates and correlates of prison misconduct among prison inmates in general, and more specifically among convicted murderers and capital offenders. Even so, the body of studies examining the prison misconduct of murderers and capital murderers remains limited. This paucity of data is surprising, given the national and international interest in U.S. capital jurisprudence and the implications that the prison misconduct of capital murderers have for attitudes regarding the death penalty, for juror determinations in capital sentencing cases, for life-without-parole sentencing, and for the confinement of death-sentenced inmates.

Studies providing inferential or direct data regarding the institutional misconduct of capital murderers have utilized six broad types of samples (**Table 22.1**):

1. Convicted murderers in the general prison population (Sorensen & Cunningham, 2007a, 2007b; Sorensen & Pilgrim, 2000).

2. Inmates sentenced to life-without-parole (Cunningham & Sorensen, 2006b; Cunningham, Sorensen, & Reidy, 2005; Sorensen & Wrinkle, 1996).
3. Capital offenders sentenced to life terms (i.e., never sentenced to death) (Cunningham & Sorensen, 2007; Marquart, Ekland-Olson, & Sorensen, 1989).
4. Capital offenders initially sentenced to death at trial, but who subsequently gained relief by commutation, retrial, or other remedy (Akman, 1966; Bedau, 1964; Edens et al., 2005; Marquart & Sorensen, 1988; Reidy, Cunningham, & Sorensen, 2001; Wagner, 1988).
5. Death-sentenced inmates on death row (Cunningham, Sorensen, & Reidy, 2004; Marquart, Ekland-Olson, & Sorensen, 1994; Reidy et al., 2001).
6. Death-sentenced offenders who are mainstreamed in the general prison population rather than being maintained on a segregated death row (Cunningham et al., 2005).

The findings of correctional research regarding these categories of murderers and capital murderers are illustrated in the sections that follow (Cunningham & Reidy, 1998, 2002). The implications of these findings for questions confronting correctional procedures or public policy are highlighted to illustrate the practical applications of this research. Finally, emerging research on the correlates of prison violence among murderers and capital murderers is considered.

ARE MURDERERS MORE LIKELY THAN OTHER OFFENDERS TO ENGAGE IN VIOLENCE IN PRISON?

Three large-scale studies have examined misconduct among convicted murderers in the general prison population, with only one of these providing comparative data with non-murderers. Sorensen and Pilgrim (2000) retrospectively reviewed the disciplinary records of 6390 murderers in Texas prisons who had been convicted between 1990 and 1998. During prison tenures averaging four-and-a-half years, 8.4% of these offenders were disciplined for violent acts. The prevalence rates of specific misconduct varied by the severity of the assault: 0%, homicide of staff; 0.1%, homicide of inmate; 0.5%, aggravated assault on staff; 4.4%, assault on inmate with a weapon; 4.2%, fight with a weapon; and 0.2%, other violence. From these prevalence rates, Sorensen and Pilgrim (2000) extrapolated that a convicted murderer is projected to commit serious violence during a 40-year term in prison at probabilities of: 16.4% likelihood of serious assault, 1% likelihood of aggravated assault on staff, and 0.2% likelihood of a homicide of an inmate.

In a subsequent study, Sorensen and Cunningham (2007b) examined the prison behavior of 1656 convicted murderers who had been admitted to the Texas prison

TABLE 22.1 Assaultive Rule Violations of Murderers, Capital Murderers, and Comparison Inmates

Rate of Assaults				
Study	**Sample**	**Follow-up Interval**	**Capital**	**Comparison**
Sorensen & Cunningham (2007a)	51,527 system-wide, Florida	2003 (12 months)		0.034 annual
	5010 first-degree murderers			0.032 annual
	3256 second-degree murderers			0.038 annual
	1320 lesser homicide			0.021 annual
	837 any homicide	(2002 admission)		0.042 annual
	13,251 no homicide	(2002 admission)		0.037 annual
	450 any homicide	(2002 admission close custody)		0.044 annual
	3663 no homicide	(2002 admission close custody)		0.082 annual
Sorensen & Cunningham (2007b)	1659 murderers, Texas	2001–2003 (initial) $M = 20$ months		
	223 lesser homicide			0.045 cum. preval.
	1108 murder			0.070 cum. preval.
	328 capital murder (life)		0.162 cum. preval.	
Sorensen & Pilgrim (2000)	6390 murderers, Texas	1990–1999 (initial) $M = 4.5$ years		.024 annual .084 cum. preval. (serious assault)
Life-Without-Parole Inmates				
Cunningham et al. (2005)	149 MS-DS, Missouri	1991–2002 ($M = 6.7$ years)	0.076 annual	
	1054 LWOP	($M = 4.3$ years)		0.096 annual
	2199 parole eligible	($M = 1.5$ years)	0.425 annual	

(Continued)

TABLE 22.1 (Continued)

Study	Sample	Follow-up Interval	Capital	Comparison
Rate of Assaults				
Cunningham & Sorensen (2006b)	9044 long-term inmates in close custody, Florida	1998–2003 (initial)		
	1897 LWOP	$M = 3.4$ years		0.074 cum. preval.
	1985 30+ year sentence	$M = 3.4$ years		0.061 cum. preval.
	1726 20–29 year sentence	$M = 3.2$ years		0.072 cum. preval.
	1469 15–19 year sentence	$M = 3.3$ years		0.097 cum. preval.
	1967 10–14 year sentence	$M = 3.2$ years		0.117 cum. preval.
Sorensen & Wrinkle (1996)	648 murderers, Missouri	1977–1992		0.218 cum. preval.
	93 death row	$M = 6.62$ years	0.237 cum. preval.	
	323 LWOP	$M = 6.66$ years	0.176 cum. preval.	
	232 LWP (second degree)	$M = 7.13$ years		0.224 cum. preval.
Capital Murderers Sentenced to Life Terms				
Cunningham & Sorensen (2007b)	136 capital murderers, Texas	2001–2004 (initial)		0.094 annual
		$M = 2.37$ years	0.14 cum. preval.	
Marquart et al. (1989)	107 CLS murderers, Texas	1974–1988 ($M = 7.2$ years)		0.026 annual
	38,246 system-wide, Texas	1986	0.12 annual	
	1712 high security, Texas	1986	0.20 annual	
Sorensen & Wrinkle (1996)	See above			
Former Death-Sentenced Murderers				
Akman (1966)	69 FDR, Canada	1964–1965 (2 years)	0 cum. preval.	
	7447 system-wide, Canada	1964–1965 (2 years)		0.007 annual

Study	Sample	Period	Rate	
Bedau (1964)	55 FDR, New Jersey	1907–1960 (53 years)	0 cum. preval. (serious assault)	
Edens et al. (2005)	See below			
			(serious assault)	
Marquart et al. (1989)				
Marquart et al. (1994)			(serious assault)	
	156 LS, Texas (128 murderers/28 rapists)	1973–1988 ($M = 11$ years)		0.10 cum. preval. (serious assault)
Marquart & Sorensen (1988)	533 FDR, nationwide (453 murderers, 80 rapists)	1973–1988	0.031 cum. preval.	
Reidy et al. (2001)	On DR	$M = 6.7$ years	0.054 annual	
	Post-DR	$M = 9.3$ years (serious assaults)	0.028 annual	
Wagner (1988)	100 FDR, Texas	1924–1972 ($M = 12$ years)	0.20 cum. preval.	
Death Row Inmates				
Edens et al. (2005)	155 DR, expert predicted, Texas			
	65 DR executed	$M = 12$ years	0.046 cum. preval.	
	42 DR	$M = 8$ years	0.071 cum. preval.	
	48 DR/FDR	$M = 22$ years (serious assaults)	0.042 cum. preval.	
Marquart et al. (1996)	421 DR, Texas	1974–1988	0.107 cum. preval.	
Reidy et al. (2001)	See above			
Mainstreamed Death-Sentenced Inmates				
Cunningham et al. (2005)	See above			

Abbreviations: DR, death row; FDR, former death row; DR/FDR, tenure on death row and post death row in general prison population; MS-DS, mainstreamed death sentenced; LS, life sentence; CLS, capital life sentence; LWOP, life without parole; LWP, life with parole; cum. preval., cumulative prevalence rate; annual, annual frequency rate.

system in 2001–2003. Averaging 22 months in prison at the time of this study, 8.5% of the convicted murderers had engaged in an assault and 2.17% in an assault with serious injury, with annual frequency rates of 71.7 and 12.5 per 1000 inmates, respectively. Again, comparisons were not made with the misconduct rates of inmates who had been convicted of other offenses.

These two studies have impressive sample sizes and provide important base rate data illuminating the frequency and prevalence rates of prison violence during an initial phase of incarceration among convicted murderers. Neither, however, answers the question of whether convicted murderers are more likely to engage in violence in prison.

A large-scale study in the Florida Department of Corrections did address this issue, in addition to illustrating how sample selection in a study can be varied to control for inherent flaws. Sorensen and Cunningham (2007a) compared the prison misconduct of various overlapping cohorts of convicted murderers and other offenders in Florida prisons. Most broadly, the 2003 disciplinary records of all inmates serving the entire 2003 calendar year (*N* = 51,527) were compared in terms of the type of offense that had resulted in their conviction. This sample included 9586 inmates who had been convicted of some form of homicide, of whom 5010 had been convicted of first-degree murder. The first-degree murderers had better disciplinary records, and equivalent annual prevalence rates of assault, as compared to other offenders: assault, 2.6%; assault with injury, 0.6%; and assault with serious injury, 0.2%. This analysis, while informing how conviction offense was related to prison violence in any given year, did not control for the murderers being deeper in their sentences and, therefore, older. Both of these factors have been associated with lower rates of institutional violence and thus could potentially confound comparisons of murderers with other offenders.

Another analysis examined the 2003 disciplinary records of a cohort of inmates who entered prison in 2002 (*N* = 14,088). Even this analysis, however, did not control for whether the respective inmates were held at the same level of custody. Accordingly, a third sample consisted of inmates who entered prison in 2002 and were assigned to close custody (*N* = 4113). An analysis was also conducted to determine whether murderers were disproportionately represented in various forms of assault. Regardless of the sample, analysis, or the severity of violence specified, convicted murderers were not found more likely to be involved in institutional violence.

The emerging conclusions of these studies indicate that only a minority of convicted murderers are cited for violence in prison, with progressively lower prevalence rates for more serious prison violence. Thus it appears that convicted murderers are no more likely to be involved in assaultive misconduct in prison than offenders convicted of other crimes.

ARE OFFENDERS CONVICTED OF MORE SEVERE FORMS OF HOMICIDE MORE LIKELY TO BE VIOLENT IN PRISON THAN OFFENDERS CONVICTED OF LESS SEVERE FORMS OF HOMICIDE?

Whether offenders who have been convicted of more severe forms of homicide are more likely to be involved in prison violence than those convicted of less serious forms is unclear. Two studies referenced previously examined the misconduct of homicide offenders by the severity of the homicide (e.g., capital, first degree, second degree, manslaughter) with contradictory results. A severity-related effect was strongly observed in research on Texas homicide offenders, with capital murderers exhibiting higher rates of assault and assault with serious injuries than offenders who had been convicted of lesser forms of homicide. Among convicted murderers in Florida prisons, however, first-degree murderers and second-degree murderers had equivalent rates of assaults, regardless of the severity of their crime. In a third study in the Missouri Department of Corrections, detailed in a subsequent section, Sorensen and Wrinkle (1996) reported that life-sentenced capital offenders and second-degree murderers had equivalent rates of prison assaults.

Severity of homicide, therefore, has an inconsistent relationship with assaultive misconduct in prison, apparently varying by correctional department. This inconsistency potentially illustrates the role of "institutional" variables, as opposed to "personal" variables in the occurrence of prison violence (Gendreau, Goggin, & Law, 1997).

DO FEMALE HOMICIDE OFFENDERS ENGAGE IN PRISON VIOLENCE AT DIFFERENT RATES THAN MALE MURDERERS?

Data related to rates of misconduct among women were drawn from the general population of inmates and convicted homicide offenders in the general prison population. Whether female inmates, or more specifically female homicide offenders, exhibit different base rates of assaultive misconduct in prison is controversial. In a large-scale study not restricted to homicide offenders, Harer and Langan (2001) reported that serious assaults by female inmates in the federal Bureau of Prisons occurred at one-twelfth the rate of such assaults among males. Similarly, among convicted homicide offenders in the Texas Department of Criminal Justice (TDCJ) during an initial confinement period of 6 to 30 months, females exhibited equivalent rates of assault, but had no incidents of serious assaults. Somewhat inconsistent with both of these studies, however, females in the Florida Department of Corrections demonstrated similar prevalence rates of assault at all levels of severity during their first year in confinement as compared to male inmates (Cunningham & Sorensen, 2006a). Institutional homicides by female inmates are apparently extraordinarily rare, with none occurring in the history of the Federal Bureau of Prisons (Harer & Langan, 2001).

We conclude, therefore, that the data are mixed regarding whether female inmates or female homicide offenders are less frequently involved in prison violence. Some trends point to females being generally less likely to engage in serious assaults, particularly prison homicides.

ARE LIFE-WITHOUT-PAROLE INMATES UNMANAGEABLE AND PREDATORY BECAUSE THEY HAVE NOTHING TO LOSE?

> *"I think that life sentences without parole do create a segment of the prison population who have no hope. They know that they are going to be there for life, and they have nothing to lose. And I think it does create a terrific security problem for prison officials and for the staffs that work in prisons."*
>
> —Harris County Houston Assistant District Attorney Roe Wilson, in testimony before the Judicial Committee of the Texas Senate in opposition to Senate Bill 348 (2003 legislative session) (McInnis, 2003).

Comparisons of life-without-parole (LWOP) and parole-eligible inmates have been undertaken in three studies, each of which examined rates of prison misconduct among life-without-parole (LWOP) and parole-eligible inmates, to illuminate an assertion that LWOP inmates represent a particular prison security and management problem. In two of these studies, LWOP inmates had rates of prison misconduct that were equivalent to those of parole-eligible inmates. The largest of these investigations involved a substantial sample of long-term inmates admitted to the Florida Department of Corrections in 1998–2002, including 1897 inmates sentenced to LWOP terms and 7147 other close custody inmates serving sentences of 10 to 30-plus years (Cunningham & Sorensen, 1996). Retrospective review of the 1998–2003 disciplinary records of these inmates revealed that the likelihood and pattern of disciplinary infractions and potentially violent rule infractions among LWOP inmates were broadly similar to those of other long-term inmates. Also, during their initial years in their LWOP sentences (M = 3.3 years) when they would be considered to be most at risk of violence, only 0.6% of the LWOP inmates were cited for an assault with serious injury.

Consistent with these findings, Sorensen and Wrinkle (1996) found that convicted murderers in the Missouri Department of Corrections exhibited similar rates of assaultive misconduct, regardless of whether they had been sentenced to LWOP or parole-eligible life sentences. More specifically, this study compared the disciplinary records (1977–1992) of 323 LWOP inmates who had been convicted of capital murder and 232 inmates sentenced to life with parole for second-degree murder. Approximately 20% of these inmates were sanctioned for an assault in prison during this period of time. Two-thirds of these assaults were minor and one-third was serious. During the 15-year study period, approximately 1.2% of the LWOP inmates killed another inmate.

Interestingly, in another study LWOP inmates were less frequently involved in assaultive misconduct in prison. Cunningham, Sorensen, and Reidy (2005) compared 960 LWOP inmates to 1503 parole-eligible inmates in a high-security prison in the Missouri Department of Corrections (1991–2002). The LWOP inmates were half as likely to have been cited for violent misconduct as the parole-eligible inmates with whom they were serving time in the same correctional facility. As often occurs in research design, the methodology used in this study is double-edged: It provided for side-by-side comparisons within the same prison, but this restriction to a single facility limits generalization of the findings.

Thus, LWOP inmates are not a disproportionate source of violence in prison. This conclusion rests on data from an aggregate of more than 12,000 inmates, drawn from two correctional jurisdictions, and encompassing more than two decades of retrospective records review. Although a rationale of "nothing to lose" has intuitive appeal, it is not borne out when the comparative prison behavior of LWOP inmates is examined. The findings of these studies were presented to the Texas Legislature in 2005. Texas subsequently became the 38th state to provide a LWOP sentencing option at capital sentencing. Eleven states without the death penalty also have LWOP sentencing.

ARE CAPITAL OFFENDERS LIKELY TO PERPETRATE SERIOUS VIOLENCE IN PRISON?

> *"He absolutely will, regardless of whether he's inside an institution-type setting or whether he's outside. No matter where he is, he will kill again. . . . He would be a danger in any type setting, and especially to guards or other inmates. No matter where he might be, he is a danger."*
>
> —Sentencing phase testimony of James Grigson, M.D., in *State of Texas v. Rodriquez*, a death-penalty case (1980).

Only three studies have reported on the prison behavior of capital offenders who received life sentences rather than the death penalty at their capital trials, none of which were available during the era of Dr. Grigson's recurrent testimony in Texas capital cases (Dr. Grigson gave testimony in more than 100 death penalty cases in Texas). In 1989, Marquart, Ekland-Olson, and Sorensen reported on 107 Texas capital defendants who had been convicted in 1974–1988 and sentenced to capital life terms after their juries had rejected the Texas capital sentencing "special issue," i.e., "whether there is a probability that the defendant would commit criminal acts of violence that would constitute a continuing threat to society." (Marquart, Ekland-Olson, & Sorensen, 1989, 1994; *State of Texas v. Rodriquez*, 1980). Averaging slightly more than seven years in prison at the time of retrospective file review, 12% of these inmates had been sanctioned for violent misconduct. Interestingly, the annual frequency of violent misconduct among these 107 Texas capital life offenders (0.026) was substantially lower than that exhibited by inmates system-wide (0.117) or by non-capital inmates at a similar high security level (0.195).

Sorensen and Wrinkle (1996) examined assaultive misconduct rates among 323 LWOP-sentenced capital murderers (1977–1992), finding a prevalence rate of assaults of approximately 20%. This rate was similar to that exhibited by second-degree murderers. A broader comparative analysis of assaults among all inmates in the Missouri Department of Corrections was not reported.

A third study by Cunningham and Sorensen (2007) described rates of assaultive prison misconduct among 136 Texas capital offenders during the initial phase (M = 2.37 years) of their life sentences. As anticipated, prevalence rates decreased as the severity of the misconduct increased (i.e., any potentially violent act = 0.368; assaultive violations = 0.14; assaults with serious injury = 0.051; homicide = 0). Comparison data with an admission cohort of non-capital offender inmates was not available. The capital offenders did, however, exhibit higher rates of assaultive prison misconduct than an admission cohort of inmates who had been convicted of non-capital forms of homicide.

A handful of other studies have reported on the institutional conduct of inmates who had initially been sentenced to death, but whose sentences were subsequently revised to life terms by commutation, retrial and sentence to life, or capital case dismissal. These reports inform considerations of whether these offenders constitute a long-term threat to institutional safety, but only following the initial years of incarceration when violence is most likely to occur.

In 1972, the death penalty, as it was then being practiced, was declared unconstitutional by the U.S. Supreme Court in *Furman v. Georgia* (1972). Two of these studies examined capital offenders from the pre-*Furman* era. Though not providing specific misconduct rates, Bedau (1964) reported that among 55 New Jersey capital offenders released from death row to the general prison population in 1907–1960, none had institutional histories that adversely affected their parole determinations. Wagner (1988) reported that among 100 Texas offenders who obtained relief from their death sentences in 1924–1971, during general prison population tenures averaging 12 years, 80% were not sanctioned for serious institutional violence (i.e., murder, aggravated assault, sex by force, striking a guard, or escape) and none assaulted a correctional officer.

Four other studies have illuminated the post-relief institutional conduct of death-sentenced offenders who were removed from death row under *Furman* or during the post-*Furman* era. Among 533 capital offenders whose death sentences were commuted nationwide under *Furman*, 31.5% were subsequently sanctioned for violent misconduct in the general prison population. Marquart, Ekland-Olson, and Sorensen reported on 92 Texas offenders who had been sentenced to death under the "special issue," i.e., "whether there is a probability that the defendant would commit criminal acts of violence that would constitute a continuing threat to society," but who subsequently obtained relief from their capital sentences (Marquart & Sorensen, 1989). During ensuing tenures averaging 6 years in the general prison population, their annual rate of serious violent rule infractions among these former condemned prisoners was 0.0161, well below that exhibited by other general-population inmates. A smaller study of 39 former death-sentenced inmates in Indiana by Reidy, Cunningham, and Sorensen (2001) reported that 20.5% were involved in violent acts in the general prison population following their removal from death row, and only one-third of these acts resulted in serious injury. Edens and his colleagues (2005) reported on the violent prison misconduct of 48 former death-sentenced inmates where a mental health expert had testified at their death penalty trials in Texas predicting that they would be a "future danger." While on death row and following relief from their death sentences, 4% of these inmates were sanctioned for serious assaults.

Only a minority of capital offenders are disciplined for serious prison violence. Further, the trend is toward these offenders not representing a disproportionate risk of prison violence as compared to other inmates. This finding appears consistent whether these convicted capital murderers are on death row pending a death sentence, or in the general prison population as a result of a life sentence at trial or relief from their death sentences. Even the capital murderers for whom there was an expectation by mental health experts or jurors of future violence were unlikely to be involved in prison violence.

DO DEATH-SENTENCED INMATES REQUIRE SEGREGATED AND SUPER-MAXIMUM CONFINEMENT?

Studies of Inmates on Death Row

Four studies have reported on rates of institutional violence among death-row inmates (Marquart & Sorensen, 1989). In the largest of these investigations, Marquart, Ekland-Olson, and Sorensen (1989) retrospectively reviewed the disciplinary records of offenders who had passed through the Texas death row in 1974–1988 (*N* = 421). These researchers reported that 45 inmates (10.7% percent) assaulted correctional staff or other inmates during the 15 years encompassed by the retrospective review, equivalent to the prevalence rate of aggravated/weapons assaults demonstrated by convicted murderers and rapists in Texas prisons who were serving life sentences. Two death-row inmates (0.47% percent) killed another inmate. Given the much higher rate of inmate homicide in the general prison population during this era, it is unclear whether the prevalence rate of prison homicide exhibited by the death-row inmates was disproportionate. Obviously, the death-row inmates were under higher security and did not have the same opportunities for violence as inmates in the general population. However, during this era, many Texas death-sentenced inmates worked in a death-row garment factory with objects that could serve as weapons, and had routine meal and recreation contact with staff and other inmates.

Sorensen and Wrinkle (1996) reported on 93 inmates who had passed through death row in Missouri in 1977–1992. Twenty-four percent were cited for assaultive misconduct during this tenure. In another retrospective review, Reidy, Cunningham, and Sorensen (2001) examined the disciplinary records of 39 Indiana capital offenders who had gained relief from their death sentences between 1972 and 1999. During their tenures on death row, which averaged 6.7 years, 25.6% were involved in violent misconduct.

In a fourth study, Cunningham, Sorensen, and Reidy (2004) compared the rates of inmate and staff assaults among Arizona death-row inmates (*N* = 127) to those exhibited by the general prison population of the Arizona Department of Corrections in fiscal year 2003 (*N* = 30,000). The death-row inmates had a one-year rate of assault of 0.78 per 100 inmates, compared to a rate of 3.26 per 100 inmates among all inmates in that prison system. Because it was possible that the death-sentenced inmates had "aged out" of serious prison misconduct by virtue of their averaging a longer time in prison on their current conviction than other inmates in Arizona Department of Corrections facilities DOC (119 months versus 36.9 months, respectively), their rate of assault since admission to death row was analyzed. Neither their average annual rate of 3.96 assaults per 100 inmates, nor the average annual rate of 3.42 assaults per

100 inmates if a psychotic outlier inmate was excluded, were significantly different than that exhibited by non-death-sentenced inmates. The prevalence rate of serious violent misconduct among Arizona condemned inmates was 17.3% during their entire tenure at risk on death row. Also informative, 52% of the death row inmates had three or fewer disciplinary infractions of any sort since admission to death row, and 16.5% had never had a disciplinary write up. Cunningham et al. (2005) characterized this finding as "a rather remarkable adaptation by a significant proportion of these inmates—particularly given their time at risk."

Studies of Mainstreamed Death-Sentenced Inmates

Rather than maintaining death-sentenced inmates on a segregated death row, since 1991 Missouri has maintained an innovative policy of making death-sentenced inmates eligible for all housing and programming assignments in a high-security prison. In other words, these individuals are intermingled in their cell and unit assignments, work roles, programming, recreation, and visitation with non-death-sentenced inmates. As with any other inmate in this high-security prison, assignments and activities are determined by inmate conduct and not sentence.

Cunningham, Reidy, and Sorensen (2005) reported that the 149 death-sentenced inmates who had been mainstreamed in the Missouri prison in the risk period of this study (1991–2002) exhibited an annual rate of violent misconduct of 0.076. This rate was equivalent to LWOP inmates serving life-without-parole (0.096) and substantially lower than that observed among parole-eligible inmates (0.425) within the same correctional facility. In fact, holding other factors constant, the death-sentenced and life-without-parole inmates were about half as likely to be sanctioned for violent misconduct. These reports by Cunningham and colleagues represented the first "apples to apples" comparison of the prison conduct of death-sentenced and non-death-sentenced inmates under the same conditions of confinement. This research also provided a quantitative validation of Missouri's ground-breaking policy in mainstreaming rather than segregating death-sentenced inmates. Administrators in the Missouri Department of Corrections have attributed the success of this policy to staff initiative and a "web of incentives" developed to influence inmate behavior (Lombardi, Sluder, & Wallace, 1997).

The security-driven rationale that death-sentenced inmates require segregated, super-maximum conditions of confinement to deter assaults against inmates and staff is not supported by available research. If, as research appears to demonstrate, such conditions do not serve a legitimate penal interest, they are arguably in violation of the Eighth Amendment, which bars cruel and unusual punishment (Lyon & Cunningham, 2006; *Turner v. Safley*, 1987).

WHICH FACTORS CORRELATE WITH PRISON VIOLENCE AMONG CAPITAL OFFENDERS?

Validation studies of risk assessment instruments and actuarial models comprise a rapidly expanding body of literature that is examining factors correlated with violence in prison and among capital offenders. This complex arena of inquiry could easily fill a chapter in its own right. Much of this research has focused on broad samples of prison inmates, with inferential application to capital offenders. My colleagues and I have examined correlates of prison misconduct among admission cohorts of inmates, prisoners in higher security classifications, and homicide offenders (Cunningham, 2006).

As the "counterintuitive" findings reviewed in this chapter suggest, murderers and capital offenders appear not to be distinctive in their prison violence proclivity; and thus findings generated from broader inmate samples are likely to generalize to them as well. Among these predictive factors from broader inmate samples, increasing age is associated with decreased rates of inmate misconduct of all severities (DeLisi, Berg, & Hochstetler, 2004; Harer & Steffensmeier, 1996; Lemieux, Dyeson, & Castiglione, 2002). Inmates who have earned a high school diploma or General Educational Development (GED) certificate have lower rates of disciplinary violations in general as well as assaultive misconduct in prison (Cooper & Werner, 1990). A prior prison term has been associated with an increased likelihood of assaultive institutional misconduct in some studies but not others (Cao, Zhou, & Van Dine, 1997; Cunningham, Sorensen, & Reidy, 2005). Membership in a prison gang is a significant risk factor for prison violence (Gaes, Wallace, Gilman, Klein-Saffran, & Suppa, 2002). Convicted murderers with a contemporaneous robbery or burglary demonstrate an increased prevalence rate of prison assault.

Two studies have examined the correlates of prison violence among samples that include capital offenders (Cunningham, Sorensen, & Reidy, 2005; Sorensen & Wrinkle, 1996). These first investigations on the correlates of prison violence among capital offenders in the general prison populations have produced findings consistent with the general trends emerging from research on broader inmate samples. Mainstreamed death-sentenced inmates were included, constituting a small minority of the Missouri high-security inmates (132 of 2595 prisoners) studied, in developing an actuarial instrument measuring risk of prison violence. Predictive factors for violent misconduct included age, type and length of sentence, education, prior prison terms, prior probated sentences, and years served.

In the only predictive study restricted to capital offenders, Cunningham and Sorensen (2006a) examined correlates of assaultive prison misconduct during the initial phase of incarceration ($M = 2.37$ years) among 136 Texas capital murderers sentenced to life terms. A simplified scale was developed (i.e., RASP-Cap), utilizing weightings for age, contemporaneous robbery or burglary, and prior prison term to identify three levels of risk. No inmate scoring at the lowest point totals (i.e., level of risk) had engaged in an assault, as compared to inmates at the highest score level, where 25% had been disciplined for an assault and 11.5% had committed an assault with serious injury. Though promising at varying severities of inmate violence (AUC = .715 to .766), this instrument remains experimental. Further, although higher scores reflected a comparatively higher risk, even at the highest risk classification there remained an overwhelming improbability of violent misconduct.

Correlates or predictive factors for prison violence among capital offenders appear to be consistent with those identified with other inmate groups. These factors are useful for classification, security, programming, and resource allocation. None, however, identify a "more likely than not" probability of violent misconduct. Though knowledge of the rates and correlates of prison conduct among capital offenders and other convicted murderers is rapidly expanding, these emerging trends require confirmation and elaboration with samples from diverse jurisdictions. The role of arrest history as a predictive factor for institutional misconduct among these offenders remains largely unexplored. There is also a need to go beyond "importation" factors in explaining the occurrence of prison violence among convicted murderers (Jiang & Fisher-Giorlando, 2002; McCorkle, Miethe, & Drass, 1995; Patrick, 1998). This research should include studies examining the contribution of situational and deprivation factors as well.

CONCLUSION

Research findings regarding the comparative prison conduct of murderers and capital murderers, even when they are sentenced to death or life-without-parole, demonstrate the importance of obtaining and relying on data rather than intuitive expectations. Despite the severity of their offenses and the bleakness of their institutional futures, the majority of these offenders do not continue on a trajectory of serious violence following their admission to prison. These findings suggest a reexamination of longstanding public policies and correctional procedures directed toward these offenders. For example, data regarding rates and correlates of prison violence among capital offenders raises grave concerns with whether "future dangerousness" can be reliably applied at capital sentencing to determine who lives and who dies (American Psychological Association, 2005). Equally notable, over more than a decade of data from the Missouri Department of Corrections, as well as from other studies of death-row and former-death-row inmates, challenges assumptions that death-sentenced inmates require housing in the segregated, super-maximum units that typify the confinement of these offenders in U.S. prisons.

Much of the public policy and correctional mores regarding capital murderers was conceived in an era that did not have the benefit of studies examining the prison behavior of these offenders. Our better understanding of this issue provides an unparalleled opportunity for criminal justice research and the associated illumination of science to prompt more enlightened perspectives.

REFERENCES

Akman, D. D. (1966). Homicides and assaults in Canadian penitentiaries. *Canadian Journal of Corrections, 8*, 284–299.

American Psychological Association (2005). Brief of amicus curie in support of defendant-appellant, *U.S. v. Sherman Lamont Fields*, in the United States Court of Appeals for the Fifth Circuit.

Bedau, H. A. (1964). Death sentences in New Jersey, 1907–1960. *Rutgers Law Review, 19*, 1–64.

Cao, L., Zhou, J., & Van Dine, S. (1997). Prison disciplinary tickets: A test of the deprivation and importation models. *Journal of Criminal Justice, 25*, 103–113.

Cooper, R., & Werner, P. (1990). Predicting violence in newly admitted inmates. *Criminal Justice and Behavior, 17*, 431–477.

Cunningham, M. D. (2006). Dangerousness and death: A nexus in search of science and reason. *American Psychology, 61*, 828–839.

Cunningham, M. D., & Reidy, T. J. (1998). Integrating base rate data in violence risk assessments at capital sentencing. *Behavioral Sciences & the Law, 16*, 71–95.

Cunningham, M. D., & Reidy, T. J. (2002). Violence risk assessment at federal capital sentencing: Individualization, generalization, relevance, and scientific standards. *Criminal Justice and Behavior, 29*, 512–537.

Cunningham, M. D., & Sorensen, J. R. (2006a). Actuarial models for assessment of prison violence risk: Revisions and extensions of the Risk Assessment Scale for Prison (RASP). *Assessment, 13*, 253–265.

Cunningham, M. D., & Sorensen, J. R. (2006b). Nothing to lose? A comparative examination of prison misconduct rates among life-without-parole and other long-term high security inmates. *Criminal Justice and Behavior, 33*, 683–705.

Cunningham, M. D., & Sorensen, J. R. (2007). Predictive factors for violent misconduct in close custody. *Prison Journal, 87*, 241–253.

Cunningham, M. D., Sorensen, J. R., & Reidy, T. J. (2004). Revisiting future dangerousness revisited: Response to DeLisi and Munoz. *Criminal Justice Policy Review, 15*, 365–376.

Cunningham, M. D., Sorensen, J. R., & Reidy, T. J. (2005). An actuarial model for assessment of prison violence risk among maximum security inmates. *Assessment, 12*, 40–49.

DeLisi, M., Berg, M. T., & Hochstetler, A. (2004). Gang members, career criminals and prison violence: Further specification of the importation model of inmate behavior. *Criminal Justice Studies, 17*, 369–383.

Edens, J. F., Buffington-Vollum, J. K., & Keilen, A. (2005). Predictions of future dangerousness in capital murder trials: Is it time to "disinvent the wheel"? *Law and Human Behavior, 29*, 55–86.

Gaes, G. G., Wallace, S., Gilman, E., Klein-Saffran, J., & Suppa, S. (2002). The influence of prison gang affiliation on violence and other prison misconduct. *Prison Journal, 82*, 359–385.

Gendreau, P., Goggin, C. E., & Law, M. A. (1997). Predicting prison misconducts. *Criminal Justice and Behavior, 24*, 414–431.

Harer, M. D., & Langan, N. P. (2001). Gender differences in predictors of prison violence: Assessing the predictive validity of a risk classification system. *Crime and Delinquency, 47*, 513–536.

Harer, M. D., & Steffensmeier, D. J. (1996). Race and prison violence. *Criminology, 34*, 323–350.

Jiang, S., & Fisher-Giorlando, M. (2002). Inmate misconduct: A test of the deprivation, importation, and situational models. *Prison Journal, 82*, 335–358.

Lemieux, C. M., Dyeson, T. B., & Castiglione, B. (2002). Revisiting the literature on prisoners who are older: Are we wiser? *Prison Journal, 82*, 440–458.

Lombardi, G., Sluder, R. D., & Wallace, D. (1997). Mainstreaming death-sentenced inmates: The Missouri experience and its legal significance. *Federal Probation, 61*, 3–11.

Lyon, A. D., & Cunningham, M. D. (2006). Reason not the need: Does the lack of compelling state interest in maintaining a separate death row make it unlawful? *American Journal of Criminal Law, 33*, 1–30.

Marquart, J. W., Ekland-Olson, S., & Sorensen, J. R. (1989). Gazing into the crystal ball: Can jurors accurately predict dangerousness in capital cases? *Law Society Review, 23*, 449–468.

Marquart, J. W., Ekland-Olson, S., U Sorensen, J. R. (1994). *The rope, the chair, the needle: Capital Punishment in Texas, 1923–1990.* Austin, TX: University of Texas Press.

Marquart, J. W., & Sorensen, J. R. (1988). Institutional and post release behavior of *Furman*-commuted inmates in Texas. *Criminology, 26*, 677–693.

Marquart, J. W., & Sorensen, J. R. (1989). A national study of the *Furman*-commuted inmates: Assessing the threat to society from capital offenders. *Loyola Los Angeles Law Review, 23*, 5–28.

McCorkle, R. C., Miethe, T. D., & Drass, K. A. (1995). The roots of prison violence: A test of the deprivation, management, and "not so total" institution models. *Crime and Delinquency, 41*, 317–331.

McInnis, J. (2003, April). Senate panel pushes no-parole sentencing option. *Houston Chronicle*, April 2, 2003, p. A-19.

Patrick, S. (1998). Differences in inmate—inmate and inmate—staff altercations: Examples from a medium security prison. *Social Science Journal, 35*, 253–263.

Reidy, T. J., Cunningham, M. D., & Sorensen, J. R. (2001). From death to life: Prison behavior of former death row inmates. *Criminal Justice and Behavior, 28*, 67–82.

Sorensen, J. R., & Cunningham, M. D. (2007a). Conviction offense and prison violence: A comparative study of murderers and other offenders. *Crime and Delinquency, 56*, 103–125.

Sorensen, J. R., & Cunningham, M. D. (2007b). Operationalizing risk: The influence of measurement choice on the prevalence and correlates of violence among incarcerated murderers. *Journal of Criminal Justice, 35*, 546–555.

Sorensen, J. R., & Pilgrim, R. L. (2000). An actuarial risk assessment of violence posed by capital murder defendants. *Journal of Criminal Law and Criminology, 90*, 1251–1270.

Sorensen, J. R., & Wrinkle, R. D. (1996). No hope for parole: Disciplinary infractions among death-sentenced and life-without-parole inmates. *Criminal Justice and Behavior, 23*, 542–552.

State of Texas v. Rodriquez, Texas, Tex. Crim. App., 597 S.W.2nd 917 (1980).

Texas Code of Criminal Procedure. Article 37.071 Procedure in capital case. This special issue was affirmed by the U.S. Supreme Court in *Jurek v. Texas* (1976).

Turner v. Safley, 482 U.S. 89 (1987).

Wagner, A. (1988). *A commutation study of ex-capital offenders in Texas, 1924–1971.* Unpublished dissertation, Sam Houston State University, Huntsville, TX.

CHAPTER 23

Civil Commitment Laws for Sexual Predators

Roxann M. Ryan
Iowa Department of Public Safety

The research on violence against women in the last two decades has led to new insights into the impact of sex offending on both victims and offenders. Despite extensive empirical research indicating that the vast majority of sexual assaults are committed by persons who either know the victim or are related to the victim, the stranger-in-the-bushes stereotype of a serial rapist remains strong. Perhaps it was that vivid stereotypical view that generated policy makers' interest in addressing the problem of sexually violent predators in the 1990s and the early 21st century. Alternatively, it may have been the new research on innovative treatment methodologies for sex offenders that prompted this consideration, or it may have been the research indicating that sex offenders are no different from other offenders, in that a small percentage of offenders are responsible for a disproportionate amount of the offending. For the politicians who enacted these laws, the motivation may well have been recognition that the strong, get-tough-on-crime campaign theme often wins over voters. In all likelihood, there were many motivations behind the grand experiment in enacting laws that permit the civil commitment of sexually violent predators. Make no mistake: It is an *experiment*, and an expensive one at that.

It is not the first time that such experiments have occurred. In the 1930s, many states enacted sexual psychopath laws to provide an alternative disposition for offenders who were thought to suffer from unique mental or emotional disorders that caused them to become sexual offenders (Becker & Murphy, 1998). Those laws fell out of favor when it appeared that the treatment methods were ineffectual in "curing" the mental illnesses. In the 1970s and 1980s, a great deal of research was devoted to sex offender causation and treatment. Many experts believed that this in-depth study of sex offender treatment could lead to a more refined and broad-based approach. In addition, the focus changed from elimination of recidivism to reduction of recidivism.

In the 1990s, the treatment response changed to focus on specialized treatment programs for violent sex offenders, designed to reduce the risk of reoffending using a variety of treatment modalities. Singling out sex offenders for special treatment was justified on several grounds. First, sex offenders are one of the few types of offenders who tend to specialize in their criminal activity. Although other offenders are as likely to commit a different type of offense as they are to commit the same offense a second time, sex offenders who commit another crime are more likely to commit another sex offense (Miethe & McCorkle, 1998). Second, sex offenders tend to escalate the seriousness of their offenses over the course of their criminal lifetimes. Many sex offenders commit increasingly violent sex offenses as they continue their criminal careers. The effects of sex offenses on victims are particularly profound and often directed toward children (Koss & Harvey, 1991). Third, a variety of treatment programs have been developed to address behavioral patterns, cognitive thinking skills, biological basis, and social skills behavior in sex offenders. Finally, social scientists have refined their ability to make risk predictions with respect to sex offenders. Advances in research design and statistical capacity have improved, so that the actuarial prediction of reoffense risk outperforms clinical predictions (Becker & Murphy, 1998).

The result of these research and policy considerations was a movement in the 1990s to adopt a specialized civil commitment program designed specifically for sexually violent predators—that is, those sex offenders who are most likely to commit further sex offenses against victims who are unrelated to them. Nineteen states had developed some form of sexually violent predator civil commitment laws by 2007. Civil commitment laws are quite different from criminal laws. Although both civil commitment and criminal conviction can affect fundamental liberty interests, the differences are stark. The goals are different, the processes are quite different, and the constitutional rights at stake are very different.

GOALS

The goals of criminal law are manifold: retribution/revenge, incapacitation, or rehabilitation. Public policy makers do not necessarily agree on which of the many goals is paramount, and a criminal law may serve more than one goal. Regardless of the goal, the result is that the criminal offender can be incarcerated upon conviction and incur the loss of liberty.

Although civil commitment also results in a loss of liberty, the goals of civil commitment are more circumscribed than those of criminal punishment (Appelbaum, 1992). The focus is on protecting both mentally ill persons from themselves and others who might be targeted by these individuals. The goal is treatment of the mental health condition that will provide protection of persons. If the only means to obtain that protection is civil commitment, then a loss of liberty for that individual is considered justified, at least until the risk of danger has been resolved. The assumption in a civil commitment action is that the involuntary commitment will last for a short time, often a matter of days or weeks.

Thus, both criminal conviction and civil commitment can result in the loss of liberty for an individual, but the reasons for the loss of liberty may be very different. The constraints on the institutions housing the individuals also are quite different. In a civil commitment action, there is a presumption that the person will be committed for purposes of treatment. Although the treatment is usually not characterized as "rehabilitation," as it is in the criminal context, it bears certain similarities to criminal rehabilitation programs. The assumption is that civil commitment is a temporary situation and that the committed person will be released as soon as he or she no longer poses a danger to self or others. There is no set time limit for commitment, and the constraints on the person are based entirely on the committed individual's risk to self or others.

In criminal incarceration, however, the requirements for treatment are quite different. Although corrections officials are constitutionally required to provide basic health care to inmates, the treatment requirements are far more limited. The standards for required care are set by the Eighth Amendment's "cruel and unusual punishment" standard. If failure to provide health care would be considered "cruel and unusual punishment," then correctional officials must provide that health care. This is a minimal standard in comparison to the standards used in civil commitment.

These differing goals translate into contrasting perspectives between civil commitment and criminal conviction. Correctional officials generally focus their attention on maintaining order in the institution. Individuals are viewed as part of the whole population, as people who need to be controlled, and their treatment programs are a secondary concern. In contrast, mental health care facility administrators who oversee persons who are civilly committed are focused on the individual treatment plans for each person committed. Their concerns for maintaining order are important, too, but their primary focus is on individual treatment plans. From the individual's perspective, the contrast between a prison environment and a civil commitment environment can be stark. Although the motivations of policy makers who enacted sexually violent predator laws certainly have been questioned, the use of civil commitment laws in addition to or in place of criminal laws necessarily implicates these contrasting goals.

PROCESSES

Legal processes are based largely on the underlying goal of the system, so it is not surprising that the process of criminal conviction is dramatically different from the process of civil commitment. The criminal process is more widely understood. First and foremost, it is an adversarial process in which the accused criminal defendant faces formal action by the government. The action is entitled, "State v. [defendant's name]" or "Commonwealth v. [defendant's name]." The government determines whether to charge a person with a crime based on the government's investigation of facts. A trial is held in which fact-finders determine beyond a reasonable doubt whether the person is guilty of the offense charged. If the person is found guilty, then a judge imposes the sentence prescribed by statute. Throughout the criminal process, the criminal defendant has the panoply of constitutional rights to protect against overreaching by the government.

A traditional civil commitment action generally is less adversarial. That is not to say that there is no animosity between parties involved in a civil commitment action. As a legal process, there are no "adversaries" in a civil commitment action. The action is entitled "In re Commitment of [person's name or initials]." The focus is on the individual facing civil commitment and whether that individual suffers from a mental health condition that poses a danger to self or others. Proof required for traditional civil commitment is a "clear and convincing evidence" standard, which is less demanding than "beyond a reasonable doubt," but more demanding than the civil standard of a preponderance of the evidence.

Civil commitment of sexually violent predators is somewhat different from traditional civil commitment, but it is more like civil commitment than criminal prosecution and conviction. States that have adopted sexually violent predator variations of civil commitment use standards that parallel traditional civil commitment. The primary difference is that sexually violent predator laws do not require proof of a type of mental health disorder that would support a traditional civil commitment action. For example, the Iowa legislature adopted a policy similar to that of many states:

> *The [Iowa] general assembly finds that a small but extremely dangerous group of sexually violent predators exists which is made up of persons who do not have a mental disease or defect that renders them appropriate for involuntary treatment pursuant to the treatment provisions for mentally ill persons under chapter 229, since that chapter is intended to provide short-term treatment to persons with serious mental disorders and then return them to the community. In contrast to persons appropriate for civil commitment under chapter 229, sexually violent predators generally have antisocial personality features that are unamenable to existing mental illness treatment modalities and that render them likely to engage in sexually violent behavior. The general assembly finds that sexually violent predators' likelihood of engaging in repeat acts of predatory sexual violence is high and that the existing involuntary commitment procedure under chapter 229 is inadequate to address the risk these sexually violent predators pose to society. The general assembly further finds that the prognosis for rehabilitating sexually violent predators in a prison setting is poor, because the treatment needs of this population are very long-term, and the treatment modalities for this population are very different from the traditional treatment modalities available in a prison setting or for persons appropriate for commitment under chapter 229. Therefore, the general assembly finds that a civil commitment procedure for the long-term care and treatment of the sexually violent predator is necessary (Iowa Code § 229A.1, 2011).*

The goal of sexually violent predator civil commitment is similar to the goal of traditional civil commitment: treatment of a mentally disordered person who poses a risk to public safety. The process for achieving the goal is quite different, however.

Who Qualifies?

When lawmakers set the standards for civil commitment of sexually violent predators, they usually begin by defining their terms. The person who is the subject of the civil commitment has been characterized in several ways: as a sexually violent predator, as a sexually dangerous person, or as a sexually violent person. The term "sexually violent predator" generally refers to a person who has been convicted of or charged with a sexually violent offense and who suffers from a mental abnormality that makes the person likely to engage in predatory acts constituting sexually violent offenses if not confined in a secure facility. Generally, states have established two broad categories of sex offenders who may be civilly committed as sexually violent predators. The first group is *persons who are guilty of sex offenses*. This may include persons who were found guilty by a judge or jury, pleaded guilty, found incompetent to stand trial, or were acquitted by reason of insanity after being charged with designated sex offenses. This provision usually applies to offenders who are currently incarcerated at the time that the sexually violent predator civil commitment action is initiated. The second group is *persons who commit a "recent overt act" warranting commitment*. Some sexually violent predator statutes may include offenders who are not currently incarcerated, but who commit a "recent overt act," that is, an act that has either caused harm of a sexually violent nature or creates a reasonable apprehension of such harm.

Most sexually violent predator statutes also require proof of a history of "predatory" offenses. These definitions usually include more than serial rapists who stalk their victims. The term "predatory" refers to acts directed toward a person with whom a relationship has been established or promoted for the primary purpose of victimization. This definition generally excludes incest offenses, unless the offender has established a relationship with a child's caretaker for the primary purpose of sexually molesting the stepchild.

Standard of Proof

The standard of proof often is higher for a sexually violent predator civil commitment action than for a traditional civil commitment action. Many predator statutes require proof beyond a reasonable doubt rather than proof based on clear and convincing evidence.

Mental Abnormality

The predator statutes also require proof of a "mental abnormality," rather than a "mental disorder." Much like the legal term "insanity," which has no equivalent term in the mental health profession, a "mental abnormality" is a legal term rather than a mental health term. The term "mental disorder" is used by the American Psychiatric Association to describe conditions that are included in the *Diagnostic and Statistical Manual of Mental Disorders (DSM)*. A "mental abnormality," in contrast, is defined as an abnormality that either predisposes a person to commit sex offenses or makes it more likely than not that the person will commit a sexual offense. For example, the Iowa statute states that a "mental abnormality" is "a congenital or acquired condition affecting the emotional or volitional capacity of a person and predisposing that person to commit sexually violent offenses to a degree which would constitute a menace to the health and safety of others." (Iowa Code § 229A.2(3), 2011).

Who Decides?

States set up a variety of ways to prosecute the cases involving civil commitment. Some states give sole authority over the matter to the state attorney general. Other states invest sole authority in the local county prosecutor. Some states allow for dual jurisdiction. Generally, the prosecutor has very broad discretion in making decisions about civil commitment, just as the prosecutor has broad discretion in bringing criminal charges. Many states set up an elaborate process of screening to help prosecutors decide which civil commitment cases to bring. Because the sexually violent predator statutes are designed to identify only the most dangerous repeat sex offenders, a screening process helps to winnow the large pool of eligible sex offenders to a much smaller number who may qualify for the "sexually violent predator" designation.

Most statutes allow for two different types of screening processes. First, a prosecutor may initiate a civil commitment process if a previously convicted sex offender commits a recent overt act. The standards vary among jurisdictions, but generally provide a great deal of discretion to the prosecutor to decide which civil commitment actions to pursue. Second, an institutional review can result in the screening of potential commitment candidates. When an institutional review process is used, many states develop an initial screening process that applies specific baseline requirements for civil commitment, for example, a minimum number of predatory sex offenses and no treatment gain shown. This initial screening process may include more than one step. For instance, one screening may occur within a prison, and a second screening may be performed by a multidisciplinary group. In addition, preliminary screening may include some type of assessment by mental health professionals, who use actuarial risk assessment procedures to assess whether the person qualifies as being more likely than not to re-offend if not placed in a secure facility. Some states also include a prosecutor review committee that makes the final recommendation for filing a civil commitment action.

Several tools are available to assess the risk that an individual offender will re-offend. These tests are one means by which the data needed for decison-making can be generated.

- The Rapid Risk Assessment for Sexual Offense Recidivism (RRASOR) is a four-item test that is designed to measure a paraphilic component. It was validated by examining reoffense by 2900 sex offenders in the United States, Canada, and England. The highest possible score is 6. A high risk of reoffending is shown with a score of 5.

- The Minnesota Sex Offenders Screening Tool (MnSOST) and the Minnesota Sex Offender Screening Tool Revised (MnSOST-R) were developed by Doug Epperson in conjunction with the Minnesota Department of Corrections. These instruments are designed to measure sex offense recidivism. The MnSOST was first validated by examining reoffense by more than 250 sex offenders in Minnesota; these results have since been replicated in other studies. A high risk of reoffense is shown by a score of 47 or more on the MnSOST and by a score of 8 or more on the MnSOST-R.
- The Static Risk Assessment 1999 (Static 99 or SRA 99) is a combination of two other actuarial instruments that have been validated. The Static 99 measures both sex offense recidivism and violence recidivism. A high risk of reoffense is shown by a score of 6 or more.

These tests are one means by which the data needed for decision-making can be generated.

Preliminary Procedures

When the screening process is completed, the civil commitment action begins with the filing of a petition, which alleges that the person is a sexually violent predator and must state sufficient facts to support the allegation (Iowa Code § 229A.4(1), 2011). Every legal action must begin with some formal filing. Generally, the petition can be filed if the institutional review process results in a recommendation to file a civil commitment action, or if a prosecutor determines that the person has committed a "recent overt act." The person who is being civilly committed is called the "respondent." Because the liberty interest is at stake, a civil commitment respondent is entitled to have an attorney; if the respondent is indigent, an attorney can be appointed (Iowa Code § 229A.6(1), 2011).

Within three days after the petition is filed, the district court makes a preliminary determination about whether there is probable cause to believe that the person named in the petition is a sexually violent predator (Iowa Code § 229A.5, 2011). The rules of evidence do not apply, and the state may rely solely on the petition or add documentary evidence or live testimony to make its case. If probable cause is found, then the respondent remains incarcerated until the trial is over. Once probable cause is found, the respondent also is ordered to be evaluated by a qualified professional. In some states, the mental health expert appears as a witness for the court. In other states, the expert is a prosecution witness and the defense is permitted to hire a different expert witness to evaluate the respondent.

Evaluation of Respondent

In the full evaluation of the respondent, the court-appointed expert supplements a review of the records by conducting a personal interview with the respondent in the case. This personal interview is designed to assess the potential factors that would reduce the risk of reoffending. This includes the respondent's treatment gains, indicating that the respondent has gained insight into his sex offense pattern and recognizes potential risks for future offending; and the respondent's relapse prevention plans, including any official supervision or informal supervision or support systems available to the respondent. The respondent can retain experts or professionals to perform independent examinations, at state expense (Iowa Code § 229A.6(1), 2011).

The rules of governing civil procedures, including pretrial discovery rules, apply in sexually violent predator civil commitment actions. As a consequence, prosecutors are able to discover a great deal of information about respondents, while respondents are able to discover all of the prosecution's evidence before the trial as well. Application of civil rules of discovery also means that prosecutors can depose respondents in advance of trial—a procedure that is never available in a criminal case because it would violate the criminal defendant's right against self-incrimination. Civil commitment respondents do not lose their Fifth Amendment rights, but some courts have ruled that the respondent may be required to answer questions about previous crimes that either have been prosecuted or are no longer eligible for prosecution (either because the statute of limitations has run out or because the prosecutor has given the respondent immunity from criminal prosecution). Respondents can be required to testify about the cases in which they were convicted, because they no longer have a Fifth Amendment privilege against self-incrimination in those cases.

Treatment Before Trial

There is no constitutional or statutory right to specified pretrial treatment. In Iowa, the Iowa Supreme Court has suggested that it will not recognize a right to specified treatment (*In re C.S.*, 1994), and courts in other jurisdictions have rejected claims of a right to specified treatment (*James v. Wallace*, 1974; *Apodaca v. Ommen*, 1991). The U.S. Supreme Court has recognized only a substantive due process right to "minimally adequate treatment" (*Youngberg v. Romeo*, 1982) and has noted that even when adequate treatment may not be available, public safety may demand some action by the government (*Powell v. Texas*, 1968).

Trial

Many states establish a short time frame for trials to occur after the probable cause hearing, usually 90 to 120 days. Iowa law, for example, allows the respondent or the state to request a continuance based on good cause by the court in the due administration of justice. Because of lengthy delays in getting cases to trial, the statute now provides that "[i]n determining what constitutes good cause, the court shall consider the length of the pretrial detention of the respondent." (Iowa Code § 229A.6(2), 2007). Given the lengthy pretrial discovery permitted under the rules of civil procedure, it is not unusual to have the trial date moved far beyond the statutory time limit.

The fact-finder in the trial may be either a judge or a jury. Most states allow a jury to decide the question of civil commitment, although some states have left the decision solely to the judge. Some states allow the prosecution, the respondent, or the judge to ask for a jury trial.

At the trial on the merits of the case, the fact-finder must determine whether the respondent is a sexually violent predator—that is, whether the respondent is more likely than not to commit another sexually violent offense if not confined in a secure facility. Many states use a "beyond-a-reasonable-doubt" standard, although a few states apply a "clear and convincing evidence" standard. The U.S. Supreme Court ruled in the *In re Winship* (1970) case that a criminal conviction requires proof beyond a reasonable doubt, but a civil commitment action is not a criminal case, so a lower standard of proof is permissible.

The "reasonable doubt" definition varies among jurisdictions. For example, the Eighth Circuit Court of Appeals defines reasonable doubt as follows:

> *A reasonable doubt is a doubt based upon reason and common sense, and not the mere possibility of innocence. A reasonable doubt is the kind of doubt that would make a reasonable person hesitate to act. Proof beyond a reasonable doubt, therefore, must be proof of such a convincing character that a reasonable person would not hesitate to rely and act upon it. However, proof beyond a reasonable doubt does not mean proof beyond all possible doubt (Eighth Circuit Model Instruction No. 3.11, 1992).*

The Iowa definition is similar:

> *A reasonable doubt is one that fairly and naturally arises from the evidence or lack of evidence produced by the State. If, after a full and fair consideration of all the evidence, you are firmly convinced of the defendant's guilt, then you have no reasonable doubt and you should find the defendant guilty. But if, after a full and fair consideration of all the evidence or lack of evidence produced by the State, you*

are not firmly convinced of the defendant's guilt, then you have a reasonable doubt and you should find the defendant not guilty (Iowa Code § 229A.7(3), 2007).

The definition of "clear and convincing" evidence also varies among jurisdictions. In Iowa, the term "clear and convincing" is defined as "Evidence is clear, convincing and satisfactory if there is no serious or substantial uncertainty about the conclusion to be drawn from it."

In most states, when a jury decides whether to civilly commit the defendant, the decision must be by unanimous verdict of the jury (Iowa Code § 229A.7(3), 2007). If the fact-finder (whether it is the judge or the jury) is not satisfied that the respondent is a sexually violent predator, the court must order the respondent to be released from custody.

When a respondent is found to be a sexually violent predator, then the respondent is admitted to a treatment program. Those treatment programs generally are separate from other types of mental health treatment programs, because sexually violent predators are different from most civil commitment patients. The finding that the respondent is a sexually violent predator necessarily requires proof that the person is sexually dangerous, so it is inappropriate to house sexually violent predators with other civil commitment patients. In addition, the purpose of sexually violent predator treatment programs differs from that of traditional civil commitment programs. The sexually violent predator may suffer from a "mental abnormality" that is not treated in the same way as the mental disorders that warrant traditional civil commitment. The focus of the sexually violent predator treatment program is sex offending, which is not the primary focus of most civil commitment treatment programs.

WHAT HAPPENS IN COMMITMENT PROGRAMS?

Most sexually violent predator programs have dual purposes: of protection of society and treatment of the committed person. The U.S. Supreme Court has not specifically ruled that treatment programs are required for sexually violent predator programs. In *Kansas v. Hendricks*, (1996), the United States Supreme Court said, "We have never held that the Constitution prevents a State from civilly detaining those for whom no treatment is available, but who nevertheless pose a danger to others." The Court noted that persons with untreatable, highly contagious diseases might be involuntarily confined, or that confinement of dangerously insane persons who are untreatable also may be permitted.

The pledge to treat sexually violent predator patients varies among states, and there is no universally accepted treatment program for all sex offenders. Thus the content of treatment programming varies among jurisdictions. Some states have placed a stronger emphasis on the treatment programs than others, and some states have seen rapid growth in their sexually violent predator programs, which makes effective treatment programs far more difficult to implement. Treatment is far more expensive than incarceration without treatment.

A standard, comprehensive treatment program consists of a five-phase treatment program incorporating several treatment modalities that can be adapted to the individual needs of each patient. The standard treatment program can be completed in 3 to 5 years if the patient is cooperative and motivated to change. Individual treatment and group classes involving the first patient and the treatment staff begin immediately. Patients are instructed regarding the requirements for advancement through the program. During the first phase, patients complete classes on cognitive skills, victim empathy, relapse prevention, relationship skills, human sexuality, anger management, personal victimization, and other topics that are universally accepted as critical components of a comprehensive treatment program for sex offenders.

Most programs develop treatment goals for each patient based upon the many factors identified in the treatment literature that are believed to contribute to sexual offending. These goals may include the following criteria:

1. Thoroughly disclose sexual history
2. Gain insight into risk factors
3. Resolve victimization issues
4. Develop victim empathy
5. Develop solitary and interactive social skills
6. Develop strong cognitive coping skills
7. Modify deviant arousal
8. Complete the relapse prevention plan
9. Demonstrate relationship and intimacy skills
10. Modify negative self-concept
11. Develop problem-solving skills
12. Demonstrate motivation to change

In most programs, patients are evaluated periodically (often review occurs every 90 days). Three objective, physiological measures may be included in the assessment process to assess treatment progress, and these results are included in the patient's quarterly reports:

- Polygraph exams may assess patient honesty about the numbers and types of victims as well as types of sexual behaviors performed.

- Penile plethysmographic exams may be performed to determine each patient's sexual arousal patterns. This is important because research has demonstrated that sexual arousal to children is the number one predictor of recidivism by sexual offenders.
- An Abel assessment provides a physiological measure of the patients' sexual interests as measured by Visual Reaction Time technology.

For the treatment program staff, a therapeutic interaction model involving seven components to therapeutic interactions can help to set the stage for a therapeutic environment. The seven components of the treatment staff model, using the acronym MEDICAL, include:

- **M**odeling appropriate behavior
- **E**mpathizing with the patients
- **D**eescalating agitated patients
- **I**nstructing patients in alternative behaviors
- **C**onfronting patients with firmness and compassion
- **A**ccepting and Affirming patients as worthwhile individuals, and
- **L**istening actively to understand patients

The primary advantage of an indefinite civil commitment program for sex offenders is that the sex offenders must consistently demonstrate progress and insight into their own offending in order to be seriously considered for release. In a prison (or in a commitment of a defined time), the sex offenders can simply bide their time until the release date. In an indefinite commitment, the progress must be real and sustained for a lengthy time period before any form of release is possible. Treatment of psychopathic sex offenders is different, and some experts suggest that psychopathic offenders should be separated from non-psychopathic patients in treatment because the psychopathic offender may appear to be benefiting from the treatment, when in fact the individual is learning better techniques to avoid detection or responsibility.

As a matter of constitutional law, civilly committed patients must be evaluated periodically. In most states, the review of sexually violent predator patients occurs at least annually, with the results provided to the court for a determination as to whether the commitment should continue. In some states, the patient can petition for discharge at other times, but the courts have discretion to summarily deny a patient's request for discharge.

If the annual review hearing indicates that the patient's mental abnormality has changed so that it may be safe for the person to be at large, then a final hearing is scheduled to determine whether the patient should be released. Unless the prosecution demonstrates that the patient's mental abnormality or personality disorder remains such that the person is not safe to be at large, and if discharged is likely to engage in acts of sexual violence, then the patient may be released. Statutes vary regarding the conditions for release, but virtually all states make provision for a transitional release, and set out the consequences for violations of conditional release. These conditions usually include recommitment for serious violations.

CONSTITUTIONAL ISSUES

Critics of civil commitment for sexually violent predators question the constitutionality of the process. The U.S. Supreme Court and most state courts have upheld the basic elements of the programs, however. The Supreme Court has examined several due process issues, for example, and rejected the constitutional challenges. States have rejected those same due process challenges, as well as other variations of due process challenges, such as claims of vagueness (a complaint that the law does not adequately describe the prohibited behavior) and suggested the use of less restrictive alternatives to incarceration. State courts have also rejected constitutional challenges based on various criminal rights—ex-post facto (passing a criminal law that took effect after the person had already committed the offense), double jeopardy (repeated criminal punishment for the same offense), cruel and unusual punishment, and equal protection (different treatment for sex offenders)—and jury trial rights (unanimous verdict, number of jurors), speedy trial, competency to stand trial, and self-incrimination. Sexually violent predators have also raised a broad variety of evidentiary issues and procedural issues in the various state courts. Most state statutes have withstood the various challenges, albeit with some modifications in each state.

Aside from various legal challenges, the two primary complaints about sexually violent predator civil commitment laws are (1) a general sense of fairness and concern about the selection of persons committed; and (2) prohibitive costs.

Fairness

It is often argued that regardless of any legal determinations of fundamental fairness under the due process clause, it is unfair to simply lock up sex offenders because of society's moral panic about offenders.

Why are sex offenders selected for indefinite commitment when other offenders are not included? Recidivism rates for sex offenders do not appear to be very different from recidivism rates for other offenders. Supporters of civil commitment, however, argue that research on recidivism rates of sex offenders is questionable, because sex offenses are severely underreported. In addition, most civil commitment laws require several convictions for sex offenses—which demonstrates a pattern of recidivism that indicates continued dangerousness. These repeat offenders are among the small percentage of offenders who are the most dangerous. Supporters of civil commitment also argue that the impact of sex offending on the victims is greater than for other types of offenses, which justifies a different response to sex offenders than to other types of offenders.

Why are sex offenders confined indefinitely for "treatment" when no successful treatment program has yet been established? It seems unfair to confine a person as part of an experiment, when there is no evidence to indicate that the person can be successfully treated. Supporters of civil commitment respond that the small group of high-risk repeat sex offenders includes those persons who are most likely to have more victims. Society has a strong interest in avoiding this further victimization. In addition, given the nature of sex offending, indefinite and long-term commitment to a treatment program is most likely to show success. Sex offender treatment requires lengthy treatment, and sex offenders who have a known end date to their confinement can conform to requirements during the time of confinement, but then continue their offending when they are released. Intense, long-term treatment is most likely to change the offenders' view of themselves and the world, and thereby reduce the likelihood of reoffending. In addition, indefinite-length treatment programs can benefit the offenders in addressing the serious emotional and psychological problems that led them to sex offending.

Why are some sex offenders civilly committed when others are not? The selection process for choosing the "most dangerous" sex offenders is not based on clear, empirically supported criteria. As critics of these programs note, human behavior is difficult to predict, yet it is these flawed predictions that form the basis for long-term civil commitment. Supporters of civil commitment respond that although human behavior is difficult to predict, proof of a long-term pattern of offending, coupled with proof of a mental abnormality that predisposes the person to commit those types of offenses, is sufficient to warrant commitment for purposes of treatment.

Cost

The cost of sexually violent predator civil commitment programming is estimated to average four times more than the cost for incarceration in prison (Davey & Goodnough, 2007). Critics question whether the additional cost for civil commitment programming is justified, given the lack of proof that sex offender treatment works.

Some sex offenders choose not to participate in programming, and there is no way to force their compliance, yet the costs of security and programming do not decrease for such individuals. Given that so few patients have been released from the programs, the costs are likely to continue to increase. Because the program costs are so high, there is a possibility that the quality of the programs may decline as a result of cost-cutting measures.

Monitoring of programs is limited—most are self-monitored—and standards for programming have not yet been established. Although informal associations have been formed among the directors of the programs in the various states, these groups are not authorized to evaluate treatment or other programs at the facilities. Moreover, public support for official monitoring is not as strong as the public support for the idea of confining dangerous sex offenders, so it may be difficult to provide formal monitoring mechanisms.

CONCLUSION

Societal response to sex offending has ebbed and flowed over the course of many decades. The recent return to civil commitment of sex offenders harkens back to the efforts in the early 20th century. The new sex offender treatment programs have survived most of the legal challenges brought by the patients in the programs, but policy questions remain. Although the research literature on sex offending has grown, the existing treatment programs have not been shown to eliminate sex offending. This dearth of evidence supporting the programs' effectiveness raises significant policy questions about involuntary commitment to a program that may not work. Yet treatment is only one of the goals of an involuntary civil commitment program for sexually violent predators. Public safety is also a legitimate societal concern. Victims of sexual assault often suffer long-term consequences. If programs are developed to address the needs of repeat sex offenders who suffer from abnormalities that predispose them to commit more sex offenses, then the costs of victimization may be reduced.

From a policy perspective, it is difficult to turn back once the decision is made to adopt a civil commitment program for sexually violent predators. Even if the treatment programs fail to work and a decision is made to disband them, public safety and political concerns make it difficult to release sex offenders who have been found to be dangerous. The sexually violent predator civil commitment statues are, indeed, a grand experiment with very high stakes.

REFERENCES

Apodaca v. Ommen, 807 P.2d 939 (Colo. 1991).

Appelbaum, P. S. (1992). Civil commitment from a systems perspective. *Law and Human Behavior, 16*, 61–74.

Becker, J., & Murphy, W. (1998). What we know and don't know about assessing and treating sex offenders. *Psychology and Public Policy Law, 4*, 116–137.

Davey, M., & Goodnough, A. (2007, March 4–7). Doubts rise as states hold sex offenders after prison. *New York Times, March 4–7, 2007.* Retrieved April 14, 2011, from http://www.nytimes.com/2007/03/04/us/04civil.html?ex=1174276800&en=309030f49f197ff6&ei=5070

In re C.S., 516 N.W.2d 851 (Iowa 1994).

In re Winship, 397 U.S. 358 (1970).

James v. Wallace, 382 F. Supp. 1177 (N.D. Ala. 1974).

Kansas v. Hendricks, 521 U.S. 346 (1996).

Koss, M., & Harvey, M. (1991). *The rape victim: Clinical and community interventions*, (2nd ed.). Thousand Oaks, CA: Sage.

Miethe, T., & McCorkle, R. C. (1998). *Crime profiles: The anatomy of dangerous persons, places and situations.* Los Angeles, CA: Roxbury.

Powell v. Texas, 392 U.S. 514, 88 S. Ct. 2145, 20 L. Ed. 2d 1254 (1968).

Youngberg v. Romeo, 457 U.S. 307, 102 S. Ct. 2452, 73 L. Ed. 2d 28 (1982).

Eighth Circuit Model Instruction No. 3.11 (1992).

Iowa Code § 229A.1 (2011).

Iowa Code § 229A.2(3) (2011).

Iowa Code § 229A.4(1) (2011).

Iowa Code § 229A.6(1) (2011).

Iowa Code § 229A.5 (2011).

Iowa Code § 229A.6(2) (2007).

Iowa Code § 229A.7(3) (2007).

CHAPTER 24

Sex Offender Registries and Criminal Predators

Shelley L. Reese
Dallas Country Sheriff's Department

Sex offender registries were developed to aid in tracking offenders convicted of criminal offenses against a victim who is a minor or any sexually violent offense. In 1947, California became the first state to develop a sex offender registry (California Department of Justice, Office of the Attorney General, 2003). California ranks first in the United States with more than 100,000 sex offenders registered and 81,000 listed on its public website. Until the passage of the Jacob Wetterling Crimes Against Children and Sexually Violent Offenders Act in 1994, sex offender registries were not required in each state. In many states, once sex offenders were released from a correctional institution and were no longer on parole, probation, or supervision, no records were maintained as to their whereabouts. These offenders were free to move from community to community without the knowledge of law enforcement. Today, the number of registered sex offenders in the United States changes daily due to convictions, deaths, and individual state requirements on the length of registration.

SEX OFFENDER LEGISLATION

In October 1989, Jacob Wetterling, 11 years old, telephoned his father and mother at the dinner party they were attending in their hometown of St. Joseph, Minnesota. Jacob had been left at the family home to watch over his siblings: Trevor, 10, and Carmen, 8. Jacob's friend Aaron Larsen, 11, had also come to the house. Jacob telephoned Jerry Wetterling to ask permission for the three boys to ride their bikes to the local convenience store 10 minutes away to rent a video. Trevor had attempted to get permission just minutes before by telephoning their mother, Patty Wetterling. Patty, concerned about the boys riding their bicycles on the dark roadway, told them no. Jacob, coming up with a revised plan, called his father stating that Trevor would carry a flashlight and Aaron was wearing a white sweatshirt. Jacob would wear his father's reflective vest, thus providing plenty of visibility while riding their bikes to the video store. A neighbor would babysit for Carmen until they returned. Jerry, knowing that Jacob was disappointed due to skating poorly at hockey tryouts, agreed for the first time to allow the boys to ride their bicycles after dark.

Jacob and Aaron, on bicycles, and Trevor on his push scooter were returning from their ride to the convenience store after renting *"Naked Gun,"* when they approached a dark area of the roadway. The boys heard a male voice call out, ordering them to stop and turn off the flashlight. A stocking-masked man brandishing a gun ordered the children into the ditch. He then asked each of the boys their ages. As they replied, he told Trevor and Aaron to run away. The last thing the fleeing boys saw was the armed man grabbing Jacob by the sleeve. Jacob is still missing, and his abductor has never been identified. Unknown to local residents and law enforcement, area halfway houses were occupied by sex offenders released from prison (Irsay, 2002).

In 1994, as part of the Federal Violent Crime Control and Law Enforcement Act, the Jacob Wetterling Crimes Against Children and Sexually Violent Offender Registration Act was passed. Included in this law was the requirement of each state to develop and maintain a registry for sex offenders and offenders of crimes against children. The Jacob Wetterling Crimes Against Children and Sexually Violent Offenders Act defines the term "criminal offense against a victim who is a minor" as (1) kidnapping of a minor, except by a parent; (2) false imprisonment of a minor, except by a parent; (3) criminal sexual conduct toward a minor; (4) solicitation of a minor to engage in sexual conduct; (5) use of a minor in a sexual performance; (6) solicitation of a minor to practice prostitution; (7) any conduct that by its nature is a sexual offense against a minor; (8) production or distribution of child pornography; or (9) an attempt to commit any of the offenses listed above in (1) through (8), if the state makes such an attempt a criminal offense, and chooses to include such an offense in its listing of criminal offenses against a victim who is a minor for the purposes of this legislation. The Jacob Wetterling Act went on to define a sexually violent offense to be any criminal offense in a "State law which is comparable to or which exceeds the range of offenses encompassed by aggravated sexual abuse or sexual abuse . . . or an offense that has its elements engaging in physical contact with another person with intent to commit aggravated sexual abuse or sexual abuse."

Protection of children was the primary intent of this law. Of those inmates who are convicted of rape and sexual assault, two-thirds of their victims are younger than the age of 18, and 58% of those children are younger than the age of 12. In California, one in four sex offenders listed on the public website were in violation of the state's registration laws. Sex offender registries are an aid to law enforcement to determine the movements of convicted sex offenders and to protect our children (Bureau of Justice Statistics, 2000).

Megan's Law, named after victim Megan Kanka, was implemented in 1996. Amending the Jacob Wetterling Act, this law allows the state sex offender registries to provide information on specific offenders to the public. Community notification systems are developed within each state. Also during 1996, the Pam Lychner Sexual Offender Tracking and Identification Act passed. The Pam Lychner Act created the National Sex Offender Registry. This act was developed to organize individual state registries and track the offender if he or she moves throughout the country. Minimum standards were set for state registries, and the FBI was mandated to register and substantiate sex offenders in those states not meeting the minimum standards. In addition, the Pam Lychner Act also defines a "sexually violent predator" as "a person who has been convicted of a sexually violent offense and who suffers from a mental abnormality or personality disorder that makes the person likely to engage in a predatory sexually violent offense."

Pam Lychner was a Houston real estate agent waiting at a vacant house for a prospective buyer. Unknown to her, a twice-convicted felon was at the home, and he brutally assaulted her. During the attack, Lychner's husband arrived and saved her life. Within two years, Lychner's attacker had his first parole hearing. Angered by his quick parole, Pam Lychner formed the "Justice for All" victim rights advocacy group and lobbied for tougher sentencing for violent criminals. In 1996, Lychner and her two daughters were killed in the explosion of TWA Flight 800 off the coast of New York. Congress passed the Pam Lychner Sexual Offender Tracking and Identification Act in her memory.

The Pam Lychner Act sets forth fines and prison sentences for registered sex offenders who failed to comply with the law. Any registrant who moves and fails to notify authorities, if convicted of a misdemeanor for the first offense, can receive up to one year in prison and a fine of as much as $100,000. The second offense, a felony, is punishable for up to 10 years in prison and a fine not to exceed $100,000 (Federal Bureau of Investigation, n.d.).

The Jacob Wetterling Act was again amended in 1998 by the Commerce, Justice, and State, the Judiciary, and Related Agencies Appropriations Act (CJSA) to include registration of federal and military offenders and nonresident students and workers. In addition, the CJSA requires each state to participate in the National Sex Offender Registry (Bureau of Justice Assistance, n.d.).

The last significant change to the Jacob Wetterling Act came in 2000, with the passage of the Campus Sex Crimes Act. Under this act, registrants are required to report to area law enforcement if they are attending or working at an institution of higher learning. Sex offenders must provide this information to both the law enforcement agency where they live and also the local police where they are attending school. According to the National Sex Offender Registry, in 2005, only nine states had implemented this legislation: California, Florida, Illinois, Iowa, Kentucky, Michigan, South Carolina, Tennessee, and Utah. The intent of the law is to increase awareness on campuses and to deter those criminals from targeting employment at colleges and universities in order to have access to students.

In 2006, President George W. Bush signed the Adam Walsh Child Protection and Safety Act. To encourage state compliance with these policies, incentives were created. States that failed to abide by the Adam Walsh Act regarding sex offenders received reduced amounts of federal monies distributed to each state through the Omnibus Crime Control and Safe Streets Act, also known as the Edward Byrne grant funds. Those agencies implementing the policies prior to the three-year deadline will be eligible for additional monies under the Sex Offender Management Assistance Program.

An important change to registries made by this law is the requirement for sex offenders to be classified by tiers. The act classifies offenders into three tiers.

A "Tier I sex offender" is any sex offender other than a Tier II or Tier III sex offender. Tier I represents the least dangerous class of sex offenders.

A "Tier II sex offender" is any sex offender (other than a Tier III sex offender) whose offense is punishable by imprisonment for more than a year and is comparable to, or more severe than, the following offenses against a minor: sex trafficking; coercion and enticement; transportation with intent to engage in criminal sexual activity; abusive sexual contact; attempt or conspiracy to commit any of those offenses; involves use in a sexual performance; solicitation to practice prostitution; or production or distribution of child pornography. Tier II also includes persons whose offense occurs after the offender has become a Tier I sex offender (e.g., a Tier I offender who reoffends is reclassified as Tier II).

A "Tier III sex offender" is any sex offender whose offense is punishable by imprisonment for more than a year and is comparable to, or more severe than, aggravated sexual abuse, sexual abuse, abusive sexual contact against an individual younger than 13, or an attempt or conspiracy to commit any of those offenses; involves kidnapping of a minor, unless committed by a parent or guardian; or occurs after the offender has become a Tier II sex offender.

Tier I offenders are required to register for 15 years, Tier II for 25 years, and Tier III for life. A reduction in the length of required registration applies to those sex offenders in Tier I who maintain a clear record for 10 years and to Tier II sex offenders whose records remain clean for 15 years. In addition, adjudicated delinquent juveniles are eligible for a reduction in their registration period by keeping a clear record and successful completion of any supervised release, including probation and parole, and completing a sex offender treatment program certified by the Attorney General or by the state where they reside.

The Adam Walsh Act mandates that all sex offenders must appear within 3 days at the designated local law enforcement agency to update their registration when any of the following changes occur: name, residence, employment, or student status. In addition, each registrant must appear in person to this agency to verify information and have a new photograph taken. Tier I registrants must appear once a year, Tier II registrants every six months, and Tier III registrants every three months. A sex offender must register prior to being released from a correctional institution.

Each registrant must now provide not only his or her address, employment, student status, and offense and conviction dates, but also a set of fingerprints and palm prints, a DNA sample, and a photocopy of his or her driver's license or identification card. Prior to the passage of the Adam Walsh Act, most states did not collect palm prints or DNA samples.

State registries are required to provide public access to sex offender information through the Internet, and the public must be able to access this information by ZIP code or a geographic radius. The Attorney General will continue to maintain the National Sex Offender Registry, tracking offenders traveling from state to state or internationally. State public websites are forbidden from listing any victim information or the offender's Social Security number. These sites also may not list any information about any arrest that did not result in a conviction. With respect to Tier I offenders, each state has discretion in listing information to indicate that the victim was an adult and to name the employer of the offender and/or any educational institution where he or she is a student.

According to the provision of the Adam Walsh Act, each state's public website must carry a warning indicating that information found on the site should not be used to "unlawfully injure, harass, or commit a crime" against any of the registrants. This statement must also incorporate a warning that if such actions occur, criminal or civil penalties may apply. Public websites must also provide for any errors in information to be reported. E-mail links enable the public to contact state law enforcement to report errors, omissions, and noncompliant offenders. Increased penalties for sex offenses against children are also part of the Adam Walsh Act.

The Adam Walsh Child Protection and Safety Act of 2006 is dedicated to 17 victims. The Declaration of Purpose in this act pays tribute to those victims. To protect the public from sex offenders and offenders against children, and in response to the vicious attacks by violent predators against the following victims, Congress in this Act established a comprehensive national system for the registration of those offenders:

1. Jacob Wetterling, who was 11 years old when he was abducted in 1989 in Minnesota, and who remains missing;
2. Megan Nicole Kanka, who at 7 years old was abducted, sexually assaulted, and murdered in 1994, in New Jersey;
3. Pam Lychner, who at 31 years old was attacked by a career offender in Houston, Texas;
4. Jetseta Gage, who was 10 years old when she was kidnapped, sexually assaulted, and murdered in 2005, in Cedar Rapids, Iowa;
5. Dru Sjodin, who at 22 years old, was sexually assaulted and murdered in 2003, in North Dakota;
6. Jessica Lunsford, who was 9 years old when she was abducted, sexually assaulted, buried alive, and murdered in 2005, in Homosassa, Florida;
7. Sarah Lunde, who at 13 years old, was strangled in 2005 by a convicted sex offender in Ruskin, Florida;
8. Amie Zyla, who was sexually assaulted in 1996 at the age of 8 by a juvenile offender in Waukesha, Wisconsin, and who has become an advocate for child victims and protection of children from juvenile sex offenders;
9. Christy Ann Fornoff, who at age 13 was abducted, sexually assaulted, and murdered in 1984, in Tempe, Arizona;
10. Alexandra Nicole Zapp, 30 years old, who was brutally attacked and murdered in a public restroom by a repeat sex offender in 2002, in Bridgewater, Massachusetts;
11. Polly Klaas, who at 12 years old was abducted, sexually assaulted, and murdered in 1993 by a career offender in California;
12. Jimmy Ryce, age 9, who was kidnapped and murdered in Florida in 1995;
13. Carlie Brucia, 11 years old, who was abducted and murdered in Florida in February, 2004;

14. Amanda Brown, 7 years old, who was abducted and murdered in Florida in 1998;
15. Elizabeth Smart, who at 14 years old, was abducted and sexually assaulted in Salt Lake City, Utah, in June 2002;
16. Molly Bish, 16 years old, who was abducted in 2000 while working as a lifeguard in Warren, Massachusetts, where her remains were found 3 years later;
17. Samantha Runnion, 5 years old, who was abducted, sexually assaulted, and murdered in California in July, 2002.

NATIONAL SEX OFFENDER REGISTRY

The National Sex Offender Registry (NSOR) was developed in 1997 as a result of the Pam Lychner Sexual Offender Tracking and Identification Act, which was signed into effect in 1996. This law mandated the Attorney General to establish a national database with the Federal Bureau of Investigation with two goals:

- To develop and maintain a national database to track the location and movements of all persons convicted of a criminal offense against a minor, convicted of a sexually violent offense, or determined to be a sexually violent predator; and
- To maintain a registry and verify the addresses of these offenders in those states that do not have a "minimally sufficient" sex offender registry program.

With the exception of sexually violent predators, address verification occurs once every year. For sexually violent predators, address verification takes place once every 90 days. Crimes Against Children (CAC) is in the division of the Federal Bureau of Investigation that oversees the National Sex Offender Registry. Using the National Crime Information Center, the National Sex Offender Registry is able to maintain a database of registered sex offenders, their offenses, and their addresses. The Pam Lychner Act strictly prohibits the FBI from releasing the identity of victims of a registered sex offender.

STATE SEX OFFENDER REGISTRIES

Due to funding allocated by the Jacob Wetterling Act, all 50 states and the District of Columbia have established a sex offender registry. Initial registration is completed upon release from any correctional facility, placement on parole, probation, or supervision for offenses stipulated by law. Most states require that the offender report any address changes to local law enforcement within 1 to 10 days of moving. Federal guidelines also require offenders to register in any state they are a student or have employment, if different from their residence.

Currently, registration forms vary from state to state, but all include the offender's address, date of birth, Social Security number, place of employment, and conviction information. Most states also require parents' names and addresses, vehicle descriptions, and physical identifiers such as tattoos, scars, and marks. After completing the registration form, the offender must sign and date it, and acknowledge the requirements of registration for this state. At the time of registering, the offender must be fingerprinted and photographed.

Each state sex offender registry is required to maintain its own database and participate in the national database. State registries forward the information they collect to the National Sex Offender Registry. All offenders will have their addresses verified annually by their state of residence. Many states conduct this verification through the mail by sending a letter to the registrant that must be signed and returned. If no response is received, the state will follow up on the verification, either in person or by contacting the local police for verification.

The length of registration was initially set forth in the Jacob Wetterling Act, although states may invoke stricter registration requirements. Under the Wetterling Act, all offenders meeting the requirements of the law must register for a minimum of 10 years upon release from prison, probation, parole, or supervision, or register for life if the offender has more than one conviction, has been convicted of an aggravated offense, or has been determined to be a sexually violent predator. A sexually violent predator, as defined in the Jacob Wetterling Act, is a "person who has been convicted of a sexually violent offense and who suffers from a mental abnormality or personality disorder that make the person likely to engage in predatory sexually violent offenses."

Megan's Law allows each state sex offender registry program to determine what information will be given to the public. Some states do not release such information, including addresses or photographs. Other states have most or all registrants available for public view on their websites (National Sex Offender Registry, 2005). The Dru Sjodin National Sex Offender Public website is maintained by the U.S. Attorney General and provides access to information by ZIP code. Named after college coed Dru Sjodin, a kidnap and murder victim, this website is the first national online listing providing the public with information on sex offenders and a link to each state's website. According to supporters of Dru's Voice (2006), an independent website, the development of this database "will aid police officers in finding more than 100,000 unaccounted for sex offenders."

By accessing the National Sex Offender Registry website at (www.nsopr.gov), anyone can review and access individual state registries. Some states may provide limited information or provide information on offenders whom they have determined to represent a high risk. Each state has specific parameters in regard to the public notification program.

PROBLEMS WITH PUBLIC NOTIFICATION

Critics of public notification and Internet websites listing information on registered sex offenders cite many problems with this system. Not all states are consistent regarding what information is posted. Some websites for state registries post all of their registered sex offenders, while others list only those offenders who have been determined to be high risks. This inconsistency creates a situation in which, for some states, a sex offender registered in that state may not be identified as such to the public.

Another problem is that not all sex offenders have been caught and convicted. Many offenders remain in the community but have not been apprehended. Critics say that the state registries that do not list all offenders promote a false sense of security for the community. Individuals searching the Internet may not find the registered sex offender living next door because he does not meet the criteria for the state's' website. The vast majority of sexual offenders know their victims, and many times are a family member or a friend. Only 14% of offenders are unknown to their minor victims. The use and promotion of programs for locating sex offenders using the Internet or public notification furthers the misconception of what constitutes a safe environment.

A different problem is that websites listing sex offenders may not distinguish among sex offenders based on the nature of the offense. Registries may list both the serial rapist who murders his child victims—clearly a dangerous person in the community—alongside the 19-year-old who had sex with his willing, but underage, girlfriend two weeks before her 16th birthday and was convicted of statutory rape. Unless the type of offense is made clear, the 19-year-old "rapist" may face social ostracism as a sex offender, when in reality his wrongdoing consists primarily of bad judgment and he is unlikely to harm anyone else in the community.

Sex offender registration occasionally has serious unintended consequences. Incidents of violence have occurred when sex offenders' home addresses were listed on registries' websites. In Maine, a 20-year-old dishwasher located the addresses of two registered sex offenders using the state's website. He then drove to their homes, in two different cities, and shot and killed the two men before committing suicide himself. Many other cases of harassment and discrimination have been reported due to public websites listing employment addresses and photographs of registrants, despite the fact that such actions are clearly identified as illegal on the websites (Cavanaugh, 2006).

Finally, as published on most state websites, confirmation of a registered sex offender's identity can only be determined through fingerprint examination and not by a description or photo on a website. Inaccurate identification of a person as a sex offender is possible due to similar names or appearance.

RESIDENCY RESTRICTION LAW

Residency restriction laws stipulate the minimum distance a registered sex offender may live from facilities where children are cared for or educated. The restriction stipulates a radius of a specific number of feet from a facility, usually a school, library, or daycare center, within which a sex offender is prohibited from living. In Iowa, the law applies to those sex offenders who had victims who were younger than the age of 18 and limits the offender from living within 2000 feet of "property comprising a public or nonpublic elementary or secondary school or a child care facility." Aside from state laws, cities have passed additional local ordinances banning offenders from living close to public libraries, parks, and youth athletic fields (Iowa Code §692A.2A, 2011). For instance, Dyersville, Iowa has created a city ordinance that bans sex offenders from living anywhere within the city limits. The constitutionality of this ordinance has not been determined, as no one has yet challenged the law. Does this law really protect the citizens? Or do sex offenders just go underground, registering with an address at a rest area and living next door without law enforcement and community knowledge (Rood, 2007)?

These types of laws are difficult to enforce and may force sex offenders into noncompliance with the law. For the offenders, residency restrictions eliminate many towns and cities as options of where they can live due to the presence of schools and childcare facilities. Larger metropolitan areas, in particular, have few housing options due to the high frequency of childcare centers and schools. Even smaller towns that have several daycare programs and a school may become off-limits altogether to offenders. As a consequence, offenders who require public transit may not have access to a community with bus lines, subways, or taxi services. In an attempt to abide by the law, sex offenders may end up living miles away from choice employment opportunities, their families, and treatment opportunities. This constraint isolates the offender from a normal support structure. Feelings of isolation or estrangement from society, in turn, can serve as a precondition to reoffend.

ENFORCEMENT ISSUES

Although the residency restriction laws have good intentions, they are difficult to enforce. Most states define a person's residence as where a person sleeps. This means that a sex offender who perpetrated his crime on a minor can only be convicted of violating residency restriction laws if law enforcement can prove he is sleeping at a location where he is not registered. Because most residency laws do not stipulate when a person must sleep, a sex offender can spend unlimited time at an unregistered home so long as law enforcement cannot prove he was asleep.

In an effort to comply with the law, sex offenders are now registering at rest areas or commuter parking areas and are seeking out other offenders to rent a room or apartment together in a location acceptable under the law. Many times, the offender simply uses the address but does not actually live at this location. Prior to the residency restriction law, many police agencies conducted home checks on offenders and were able to determine with whom they were living and at which location they were spending the majority of their free time. Law enforcement would note if children were living in the home and investigate to determine if this arrangement was a violation of the offender's terms of probation or parole. Under the current law, investigations are limited to where the offender is sleeping. If the offender is registered at a rest area, parking lot, or other public place, law enforcement must monitor this location 24 hours per day for several days to determine if the offender is violating the law. This type of surveillance takes enormous amounts of human resources, long hours, and expensive equipment that local police may not have.

Another problem for law enforcement is the ability to find the offender on short notice. If the offender is registering at a public location, law enforcement cannot leave a message or note for him or her to contact the probation officer or police department.

Arguments against the current residency restrictions include the following points:

1. There is no research showing a correlation between residency restrictions and improving the safety of children;
2. Research does not support the belief that children are more likely to be victimized by strangers at the locations listed in the residency laws;
3. Residency restrictions are intended to reduce sex crimes committed against children at places such as schools and childcare centers, but statistics show that 80% to 90% of these crimes are committed by either a relative or someone known to the child;

4. Law enforcement is finding that the residency laws are causing offenders to be homeless, register incorrect addresses, or simply fail to register. In all of these situations, both law enforcement and the community do not know where this offender actually lives; and
5. Residency restrictions have failed to demonstrate any type of protection intended but have caused a huge drain on law enforcement resources as police attempt to enforce these laws.

Many offenders are back living with or are married to their victims, causing the residency laws to penalize the victims. Offenders who have families may be forced to move their families, uprooting children from school, friends, and any type of current community support system (Iowa County Attorneys Association, 2006).

Many offenders have physical or mental disabilities but cannot live with family who might provide them with assistance. Due to the residency restrictions, attempting to find affordable housing may be difficult. Prosecutors are seeing more cases going to trial because the offender is unwilling to plead to a charge that would require residency restrictions.

States without residency restrictions are seeing an increase of sex offenders coming to their states. Wyoming is considering adopting residency laws similar to other states. Currently, they have no laws restricting where a registered sex offender can live, but 56% of the state's 1200 registered sex offenders moved to Wyoming after they were convicted. Wyoming also publishes information only on high-risk offenders, another reason the state appeals to registered sex offenders (Miller, 2007).

PARAMOUR LAW AND SAFE ZONES

Paramour laws are common in many states. These laws are designed to protect children from being victims of registered sex offenders in their own homes. Paramour laws are usually part of the state code directed at child endangerment. Although each state may differ slightly in its approach, most paramour laws make it illegal for a person to knowingly allow a registered sex offender to live in his or her home where children also reside. This restriction does not apply if the sex offender is married to the parent of the child, or if the child is his or her own offspring. Under this law, it is not the sex offender who breaks the law: It is the parent who allows the sex offender to live at the residence who is guilty of the crime.

The intention of this law is to prohibit sex offenders from selecting a partner based on the availability of children in the home. Punishment is not sanctioned out against the sex offender, but rather against the partner who allows the offender to move into the home. Prosecution of this law depends on proving first that the sex offender resides in the home, and second that the partner knows he or she is a registered sex offender. Law enforcement may find these points difficult to prove, especially if the living arrangements are not constant, but the offender is staying overnight only occasionally.

Safe zones are another type of restriction placed on registered sex offenders. These safe zones are designated areas specified by law as to the footage surrounding a location frequented by children where a sex offender must have prior permission to enter. Among areas that can be designated as safety zones include schools, parks, and childcare facilities. The locations and restrictions are determined by the state law.

OTHER CRIMINAL JUSTICE RESPONSES

Ankle or wrist electronic monitoring has been used for years by correctional officers to monitor all types of offenders. These units coordinate with a unit inside the home and monitor when the offender is within a specified distance to the unit. In addition, this type of monitoring can be programmed to require the offender to maintain that distance during certain hours of the day. If the wearer violates the home curfew, the device will notify the proper correctional authorities through the phone system.

Some states require high-risk sex offenders to carry a GPS monitoring device that allows constant tracking of the offender. Although it is accurate, GPS tracking is expensive due to the equipment and labor costs. Use of this type of tracking system requires the registrant to wear an ankle band and also to carry the GPS device when farther than a predetermined distance from the receiver unit located in the offender's home. If the offender moves farther away from the receiver than the allocated permitted distance without carrying the GPS device, a notification is sent to the monitoring system. This type of tracking also allows law enforcement to check the movements of an offender in regards to a crime that occurred.

Voice verification is a tracking system that uses little equipment and requires the least amount of human resources. This type of monitoring is generally used with sex offenders who are restricted to being at their residences at particular times. An example would be a registered sex offender who is required to be at his home between 6:00 p.m. and 6:00 a.m. The voice verification system makes automated, random telephone calls during these hours to the residence asking questions that only the offender could answer. If the answers are incorrect or the voice template does not match, the system notifies the appropriate law enforcement agency.

New technology in tracking and identifying sex offenders has been introduced in North Carolina. The Sex Offender Registry and Identification System (SORIS) is a biometric database that will record and document images of the sex offender's irises. For instance, The Mecklenburg County Sheriff's Office is using this technology in the tracking of registered sex offenders. All sex offenders have their irises scanned; these scans are placed in a database, enabling law enforcement to identify them using a device the size of a smart phone. The technology allows officers to scan an iris and immediately know if the person is a registered sex offender. Law officers hope that the technology will spread across the United States, so that all sex offenders will eventually be entered into the iris database (Waddell & Campo-Flores, 2006).

REFERENCES

The Adam Walsh Child Protection and Safety Act of 2006, 42 U.S.C. §16901. Retrieved on January 13, 2011, from www.govtrack.us/data/us/bills.text/109/h/h4472.pdf

Bureau of Justice Assistance. (n.d.). Overview and history of the Jacob Wetterling Act. Retrieved January 13, 2011, from http://www.ojp.usdoj.gov/BJA/what/2a1jwacthistory.html

Bureau of Justice Statistics. (2000). *Sexual assault of young children as reported to law enforcement: Victim, incident, and offender characteristics.* Washington, DC: U.S. Department of Justice.

California Department of Justice, Office of the Attorney General. (2003). Public access to information about sex offenders expires at end of 2003. Retrieved January 13, 2011, from www.ag.ca.gov/newsalerts/release.php?id=1137&year=2003&month=9

The Campus Sex Crimes Prevention Act, 20 U.S.C. §1092 (2000).

Cavanaugh, T. (2006). Deadly sex offender registries. *Reason, 38*, 6–7.

Dru's Voice. (2006, August). Dru's Law. Retrieved January 13, 2011, from www.drusvoice.com/law

Federal Bureau of Investigation. (n.d.). Investigative programs: Crimes Against Children. Retrieved January 13, 2011, from http://www.fbi.gov/hq/cid/cac/registry.htm

Iowa Code §692A.2A (2011). Residency restrictions: Child care facilities and schools.

Iowa County Attorneys Association. (2006). *Statement on sex offender residency restrictions in Iowa.* Des Moines, IA: Iowa County Attorneys Association Author.

Irsay, S. (2002). The search for Jacob. Retrieved January 13, 2011, from http://archives.cnn.com/2002/LAW/11/19/ctv.wetterling/index.html

The Jacob Wetterling Crimes Against Children and Sexually Violent Offenders Act, 42 U.S.C. §14071 (1994).

Miller, K. (2007, February 21). Wyoming worries loose rules luring out-of-state sex offenders. *Des Moines Register*, February 21, 2007, p. 16a.

National Sex Offender Registry. (2005). Megan's Law. Retrieved on January 13, 2011, from http://www.registeredoffenderslist.org/megans-law.htm

The Pam Lychner Sexual Offender Tracking and Identification Act, 42 U.S.C. §14072 (1996).

Rood, L. (2007, February 12). Keep Iowa's sex offender laws strict, one city says. *Des Moines Register*, February, 13, 2007, pp. 1A, 12A.

Waddell, L., & Campo-Flores, A., (2006). Iris scans keeping an eye on sex offenders. *Newsweek, 148*, 8.

CHAPTER 25

Running to Stand Still? Reentry and Violent Offenders

Shelley Johnson Listwan
University of North Carolina–Charlotte

Inmates reentering society from prison have become a bit of a thorn in the side of correctional administrators. In most circumstances, the formerly incarcerated bring a host of problems back to the communities from which they came. The recent interest in the released and their problems may be new, however, the process of inmates reentering the community post incarceration is not. Parole supervision has been in existence since the early 1900s. In fact, by 1925, 46 states had some form of parole supervision in place (Friedman, 1993).

The increased interest in this reentry process, and ultimately in parole supervision, during the last decade stems from the reality that the vast majority of offenders, including violent offenders, will eventually be released. More importantly, recent studies have found that parolee outcomes are not as positive as we might hope. More than two-thirds of people being released are "failing" in some capacity. Underneath these rates of failure, however, is the reality of people involved, and the costs associated with their care are staggering.

Today, more than 1.6 million adults are serving time in the U.S. prison system (West, Sabol, & Greenman, 2010), with an additional 767,620 in local jails (Minton, 2010) and nearly 5 million on probation or parole (Glaze, Bonczar, & Zhang, 2010). In terms of the types of offenders in prison, approximately 50% of the prison population was sentenced for a violent offense, and approximately 20% for both property and drug offenses. This population rate translates into one in every 100 Americans, with African American and Hispanic populations being consistently overrepresented in the prison population (Pew Charitable Trust, 2008).

As mentioned, the vast majority of prisoners are eventually released. Generally speaking, this translates into approximately 650,000 inmates returning to the community each year (Travis, Solomon, & Wahl, 2001). The cost of serving these individuals in the correctional system is overwhelming many states' budgets. According to the Pew Charitable Trust (2008), it is estimated that the money spent on corrections by states climbed from an estimated $12 billion in 1987 to nearly $50 billion in 2007, and was expected to increase. Moreover, California continues to lead the way with expenditures on corrections topping $8.8 billion. In terms of funding from state budgets, corrections inevitably competes with the costs of education, health care, and basic infrastructure needs.

HISTORICAL CONTEXT

It is difficult to understand how the United States reached this "crisis" without understanding the historical context of our prison and parole systems. As a country, the United States incarcerates more people than any other place in the world. In the 1980s and 1990s, the "get tough" movement led to this expansive growth in the prison system and subsequently the number of people on parole. The "get tough" movement is often discussed in the context of the "war on drugs," and it is not difficult to see why. The increase in drug offenders incarcerated in prison was grossly disproportionate during this time (Blumstein & Beck, 1999; Goerdt & Martin, 1989). Policies implemented during this era, such as mandatory minimum sentences, mandatory arrest, and various drug sweeps or crackdowns throughout the country, led to this increase in the prison population.

Statistics indicate that violent offenders made significant contributions to the growth of the prison population. In fact, 50% of the growth in the prison population between 1990 and 1997 was attributed to violent offenders (Gilliard & Beck, 1998). Violent offenders have always been a concern for the criminal justice system, but the "get tough movement" included a number of policy developments specifically geared towards these offenders. One example is the "three strikes and you're out" legislation designed to target repeat violent offenders. This legislation was introduced after the highly publicized murder in California of Polly Klass, who died at the hands of a repeat violent offender (Zimring, Hawkins, & Kamin, 2001). Ultimately, the policy did not realize its intended purpose. Nevertheless, it became widely popular because it was seen as not only tough on crime, but also as a mechanism for selectively incapacitating the "worst of the worst"—namely, those repeat criminals who were disproportionately affecting the crime rate.

In addition to the "three strikes" legislation, other policies sought to increase the severity and certainty of penalties. Certain offenses, particularly drug offenses, were given mandatory minimum penalties. Moreover, to ensure that offenders stayed in prison for most of their sentence, states began to enact "the truth in sentencing laws." Through Truth in Sentencing policies, which mandated that offenders serve at least 85% of their sentences, they felt the maximum punishment possible. Finally, the passage of the Violent Crime Control and Law Enforcement Act of 1994 (HR 4092) set out tougher rules and penalties related to the use of assault weapons, made more crimes eligible for the death penalty, increased penalties for domestic violence, and provided resources for the creation of new prisons.

THE REENTRY CRISIS

Nearly two decades later, people began to question the results of these policies—that is, whether the "get tough" movement "worked" to keep the public safe from violent repeat offenders. On one hand it depends who is asked. Some observers would argue that prevention is difficult to measure and it is possible that crime rates in the United States might be worse off had the country not engaged in the mass incarceration during this era. Statistics, however, seems to paint a different picture.

As mentioned earlier, one study in particular conducted by the Bureau of Justice Statistics led people to take note of the ineffectiveness of the current process. The 15-state study found that two-thirds of parolees released in 1994 were arrested during a three-year follow-up period, with 25% returning to prison. Importantly, 30% failed in the first 6 months after their release. The same study further examined the recidivism rates for violent offenders, finding that nearly 62% of the individuals in prison for a violent offense recidivated during the study period, although not necessarily for a violent offense (Langan & Levin, 2002). Further compounding the issue, researchers reviewed the success rates of parolees in general and found that "over the past 20 years, as the number of people sent to prison on new convictions has increased threefold, the number sent to prison for parole violations increased sevenfold. We now send as many people back to prison for parole violations as the total number of prison admissions in 1980" (Travis & Lawrence, 2002).

So why are so many former inmates coming back? Given the harsh penalties and the conditions of deprivation they encountered in prison, why were these people not being sufficiently deterred from committing subsequent crimes? One possible reason is that increased surveillance and control does little to change the factors that brought these individuals to prison in the first place. As reported by many sources, inmates reentering the community are facing a great number of barriers (Petersilia, 2003; Travis, 2000). These barriers are often interrelated. For example, housing is difficult to secure, particularly for violent offenders whose violent past precludes their ability to live in certain locations or dwellings. These housing barriers can have a direct impact on an individual's ability to find employment (Metraux & Culhane, 2004; Roman & Travis, 2004). Further, a lack of marketable skills and the offender's violent criminal record typically inhibits the person's employability once released from the institution (Piehl, 2003; Solomon, Johnson, Travis, & McBride, 2004). Finally, the offender's family life rounds out the picture of this difficult community adjustment phase. The families who might be able to accept the reentering inmate back into their homes often face emotional barriers or issues of poverty and housing problems of their own (Henggeler & Borduin, 1990; Mumola, 1999; Naser & Visher, 2006; Travis, Chincotta, & Solomon, 2003).

The situation is further exacerbated by the treatment gap that exists within prison and the community. For example, nearly 75% of inmates have a history of substance abuse, more than half have not received their high school diploma, and more than half have children younger than the age of 18, which may lead to child support debt issues (Klein, Alexander, & Parsons, 1977). Only 13% of inmates receive substance abuse treatment in prison. Moreover, running adequate reunification programs in prison is often difficult, if nearly impossible, given the remoteness of many prison locations, which makes family visits difficult.

Prisoners also have medical problems and mental health issues, both created by the environment of prison and those imported with them upon their incarceration. For example, rates of HIV, hepatitis B and C, and tuberculosis are substantially higher among inmates than in the general population (Robertson, 2003). According to the Pew Foundation, California estimates that it costs roughly $30,000 to treat an inmate with tuberculosis; on a national scale, it is estimated that 20% to 45% of all inmates are impacted by this disease. Moreover, it is estimated that more than half of prisoners have mental health issues that require substantial intervention services, and nearly three-fourths of these individuals met the criteria for substance dependence (James & Glaze, 2006).

In response to this growing problem, the federal government engaged in a significant policy shift. Instead of continuing the "get tough" rhetoric of the 1980s and 1990s, the response in more recent years has been geared towards rehabilitation and service delivery. This shift has not only taken place within the context of reentry (Listwan, Jonson, Cullen, & Latessa, 2008); but changes in the policies surrounding violence and reentry are also noteworthy.

THE SERIOUS AND VIOLENT OFFENDER REENTRY INITIATIVE

In 2003, legislation establishing the Serious and Violent Offender Reentry Initiative (SVORI) was passed to better serve violent offenders reentering the community. The SVORI initiative is unique in that it directly targets serious and violent offenders returning to the community. Recognizing that serious and violent offenders were at high risk for recidivism once released, the initiative provided states with federal funding to develop or enhance existing parole services. Some states had few services for offenders coming back to the community, while other states utilized the money to enhance the services or the processes already taking place. Ultimately, 69 agencies nationwide received over $100 million in funds to develop reentry programs.

The federal initiative gave states a certain degree of latitude in selecting their target populations and developing their service delivery components. Each program was given guidelines it had to follow. For example, the target populations were supposed to be serious and violent; however, the SVORI did not mandate that certain offenders be accepted. For example, some jurisdictions chose to exclude sex offenders or mentally ill, whereas other accepted a more general population. The programs were able to choose the services they provided, however, the programs were encouraged to begin service while the inmate was still in prison (approximately 6 months prior to release) and continue this process once the individual was released into the community (upwards of 12 months post release) (Lattimore et al., 2004).

Over the last several years, there have been a number of process and outcome evaluations of these programs. The results of a national evaluation of SVORI-funded programs have recently been published. Some interesting findings include:

- SVORI participants were fairly high risk and high need. For example, the vast majority had a lengthy criminal history. Forty-two percent had never held a job for more than 1 year. Eighty-four percent had friends who had been convicted of a crime, and seventy-eight percent had family members who had been convicted of a crime. The vast majority indicated that they needed help in areas such as employment assistance, financial assistance, job training, education assistance, and obtaining a driver's license (Visher & Lattimore, 2009).

- In many jurisdictions, vast expansive service delivery systems were developed where none had existed before, thereby helping communities develop more effective systems of service delivery (Lattimore, 2008; Listwan, 2006; Winterfield, Lattimore, Steffey, Brumbaugh, & Lindquist, 2006).
- Although programs improved their systems of service delivery, program participation declined after the inmate was released. According to the evaluation, programs nationally had difficulty providing sufficient services to meet the needs of returning inmates (Visher & Lattimore, 2009).
- In terms of outcomes, SVORI participants were slightly more likely to show improvements in some key areas such as housing and employment. However, the arrest and reincarceration data indicate that the program participants had similar outcomes to those who did not participate in SVORI programs (Visher & Lattimore, 2009).

There has been much discussion about why, at least on an aggregate level, these programs failed to "work." Evaluators have cited implementation barriers, funding, staff turnover, and the fact that these programs were new and did not have enough time to develop (Visher & Lattimore, 2009).

An evaluation of one SVORI program conducted by the author of this chapter echoed some of these concerns. For example, an investigation of Nevada's Going Home Prepared (GHP) program (Listwan, 2000, 2006) examined the extent to which the program met its intended target population and the barriers experienced by the program. The program defined "serious and violent offenders" as offenders convicted of the state's most severe Category A and B felonies, and having a history of such offenses, weapons use, gang involvement, or violent prison disciplinary incidents. Some interesting findings emerged. In particular, the participants were rated fairly high in terms of need. For example, **Figure 25.1** illustrates some of the needs cited by participants before release.

With regard to transportation, many of the released prisoners were concerned about how they would meet the obligations for parole or treatment if they did not have access to transportation. Moreover, the participants were unsure of how they would manage to care for their dependents and simultaneously fulfill the requirements of treatment. Concerns about debt focused primarily on child support payments. In terms of medical needs, treatment services for narcolepsy, HIV/AIDS, epilepsy, and diabetes were most often cited.

The staff reported that they felt participants had certain needs that might impair the parolees' chance at successful parole. **Figure 25.2** illustrates a different set of needs. Employment and vocational training emerged as two important needs.

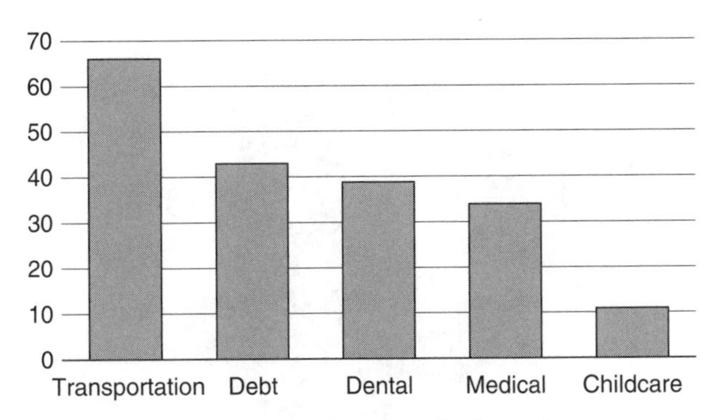

FIGURE 25.1 Percentage of participants by identified need.

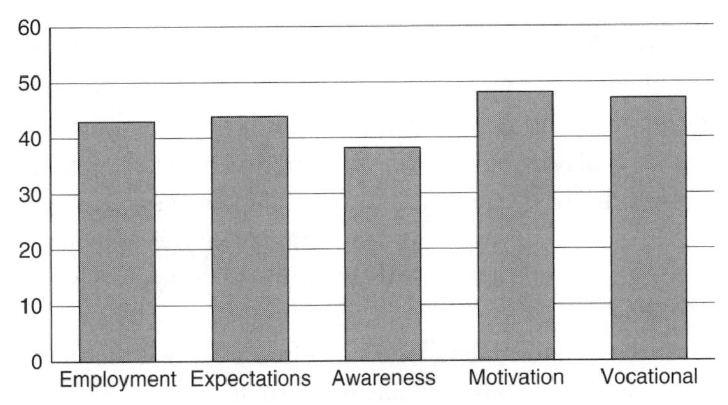

FIGURE 25.2 Staff identified needs among participants.

Caseworkers also identified problems in the areas of motivation, expectations for release, and awareness surrounding these offenders' situation and past. Specifically, these situations included expectations surrounding how to find a job and how much they needed to make to pay their bills and obligations. Staff also reported that the participants were unable to express a clear plan for how they would maintain a prosocial life in the community.

The evaluation examined the program completion rates among the participants. Of the 230 participants, the majority failed to successfully complete the program. Specifically, as shown in **Figure 25.3**, 69% of the program participants failed to complete the program during the follow-up period.

Barriers experienced by the participants were noted in the explanation for the low program graduation rates. For example, those who were unable to find a job, who did not live with family members, and who were higher risk were more likely to be unsuccessfully terminated. Most of the participants in Nevada's GHP program did not

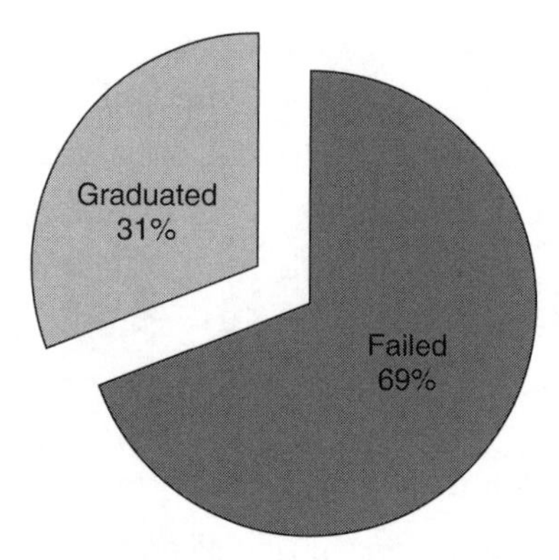

FIGURE 25.3 Status at termination.

work while in prison, thereby exacerbating their employment situation upon release. Forty percent of clients were unable to secure employment in the community upon release. In addition, housing represented a significant problem among participants. Given the restriction placed on felony offenders, many were unable to return to government-assisted homes, even when those homes were occupied by family members. Many clients were unable to raise the money for the down payment required for alternative housing, forcing the GHP program to provide the necessary funds. Further, many apartment complexes were unwilling to accommodate offenders who were identified as serious and violent. The staff also found that there were insufficient transitional or halfway house beds available for participants.

The authorities running the GHP program also found that other needs consumed considerable time and resources. For example, dental benefits, crisis intervention/mental health care, and transportation proved more costly than first imagined. The program developed a fairly extensive network of community agencies; however, resources to provide intensive services remained limited. Overall, the program staff felt they succeeded in developing an infrastructure needed to provide service delivery, but noted that the focus on external responsivity factors were often prioritized over criminogenic needs.

The concern for these SORVI-funded programs is not whether a treatment-focused approach can work, but rather whether such programs can be implemented as designed and with sufficient resources. While the money provided by the federal initiative was extensive, programs established in locations such as Nevada found themselves overwhelmed by the needs of the target population. It became clear fairly early in the process that a "one size fits all" approach to service delivery was insufficient for handling these offenders. By the end of the study, the stakeholders realized the need to further individualize treatment services. The federal initiative, however, was designed to provide states with temporary funding with which to create or expand these programs. After the initial funds were expended, states would be required to shoulder the entire financial burden to continue these programs. Ultimately, Nevada did not choose to continue the GHP program.

Another policy development worth noting is the Second Chance Act. This legislation, while not directly targeting serious and violent offenders, was signed into law in 2008 by President Bush. It provides federal grants to agencies to assist parolees with employment, housing, family services, and other needs in an effort to reduce recidivism. These federal initiatives are promising, however, as just discussed, the needs of parolees—particularly those labeled serious and violent—can quickly overwhelm a community's ability to handle them. While targeting housing and employment is necessary, the reentry problem requires a larger shift in our thinking about the structural barriers that many communities face. Research has found that high incarcerations rates contribute to the overall decline of urban areas and actually contribute to the crime rates in some areas both by destabilizing neighborhoods, breaking apart families, and increasing the overall risk of these criminals (Clear, 2007; Travis et al., 2003). A comprehensive approach to reentry should at least include a dialogue for the infrastructure of the inner cities across the United States.

While the issues and problems identified here often take center stage in the discussion surrounding reentry, one issue that receives less attention is the impact of the prison environment on those reentering the community. The previously mentioned data should at least prompt questions about whether prison makes people "better" once released. The real debate, however, is whether prison, at least those prisons that are coercive and depriving, might actually be making prisoners worse.

PRISON ENVIRONMENT AND REENTRY

The prison environment, and more specifically, prison violence, has been the subject of research and social media for many years. Prison violence takes many forms, as exemplified by the early research on deprivation by Hans Toch, to in-depth examinations of prison riots such as the New Mexico riot at Santa Fe (Colvin, 1992), to popular television shows such as *Oz*.

Unfortunately, it is difficult to know exactly how much violence actually occurs in prison. First, the rate depends on the type of violence measured. For example, some studies have found that between 20% and 35% of inmates have reported being seriously harmed while in prison (Blitz, Wolff, & Shi, 2008; Chen & Shapiro, 2007), whereas exposure to violence, such as witnessing a fight, might happen to nearly all prison inmates at some point in their incarceration experience. In terms of deaths in prison (McCorkle, 1993), reports indicate that in 2002 there were nearly 3000 deaths in U.S. prisons in 2002, although only 48 of them were classified as homicide. Finally, in an investigation of sexual victimization among prisoners, researchers found that 14% of inmates were the target of sexual aggression (Hensley, Tewksbury, & Castle, 2003). Perhaps more importantly, others suggest that many more inmates simply experience the emotional fear of living in a threatening, coercive prison environment (Toch, 1977).

Sexual violence in prison gained attention in 2003 with the passage of the Prison Rape Elimination Act (PREA). As part of this law, every state was required to adopt a set of standards and polices to reduce, prevent, and eliminate prison rape. The concern over the amount of rape occurring in prison (some argued more than 20% of the population was being raped) and the lack of national data and reporting led to the passage of this law. This law and the general interest in violence in prison, however, are not often discussed in the context of the reentry movement.

The impact of prison violence on inmate reentry highlights a logical relationship: What inmates experience in prison, regardless of whether they came into prison for nonviolent offenses or not, can potentially influence how they adjust to the community once released. While there is no universal prison environment experienced by each inmate, certain features of the prison are common. For example, the often discussed "'inmate code"' that emphasizes autonomy, hypermasculinity, and conflict provides the context for violence. The depriving nature of prison, in which inmates are given few opportunities for privileges or decision-making, often results in a loss of perceptua control among the inmates. This loss of control has been shown to lead to a host of emotional issues such as depression, anxiety, and anger (Ruback, Carr, & Hopper, 1986; Wright, 1991, 1993).

The degree to which prisons rely on coercive power matters as well. For example, studies have found that prison violence is more likely in prisons where staff rely on coercion for control (Colvin, 1992). When staff utilizes inconsistent discipline or rule enforcement, the inmates are more likely to feel that the correctional officers' means were unjust, which compromises their legitimacy. While the research in this area has produced mixed results, some studies have found that misconduct in prison is often greater among those who serve longer periods of time in prison and are subject to higher levels of "prisonization" (Kruttschnitt & Gartner, 2005).

In a study conducted by the author and colleagues (Listwan & Hanley, 2010), we examined whether prison victimization affected individuals as they came back into the community. We interviewed more than 1600 inmates who had recently been released from prison. While this study was not an examination of violent offenders specifically, nearly 50% of the sample had a prior record involving violence. We asked participants to indicate whether they had ever seen or been a direct victim of sexual assault, physical assault (fighting), verbal assault, or theft of property. The study was funded in the context of the interest in both PREA and the reentry movement. We were interested not only in sexual victimization, but also in violent victimization in general. **Figure 25.4** illustrates the types of victimization reported, both witnessed and direct.

Theft could include any item that belonged to the inmate, such as mail, clothes, food, and magazines, etc. Physical victimization was operationalized as fighting. Verbal victimization could include belittling, talking down to, or harassment. Finally, sexual victimization, which was rarely reported among inmates (1% reported direct victimization), but could include coercion or more traditional definitions of rape. The results regarding sexual violence are not completely out of the realm of what has subsequently been found by a national study of prison inmates. That study found that 2.1% of inmates reported sexual victimization by another inmate. We found that the physical victimizations could be particularly violent. Twenty percent of the direct victims felt their life was in danger, and the vast majority was sent to see the prison doctor or nurse as the result of their injuries. The type of harm ranged from stabbings, to broken bones or teeth, and to bleeding or bruising.

The prison code appears to be alive and well in prison. For example, 61% of those study participants who witnessed victimization and 82% of those who directly experienced it, did not report the incident to prison authorities. As seen in **Figure 25.5**, the most frequently cited reasons for not reporting included not wanting to be seen as a snitch and the desire to take care of the matter themselves. Finally, while the majority of former inmates indicated that they were told how to report rape, fewer indicated that they felt correctional officers would assist inmates who had been raped.

We examined whether experiencing victimization in prison (either witnessed or direct exposure) was related to psychological distress or poor community adjustment

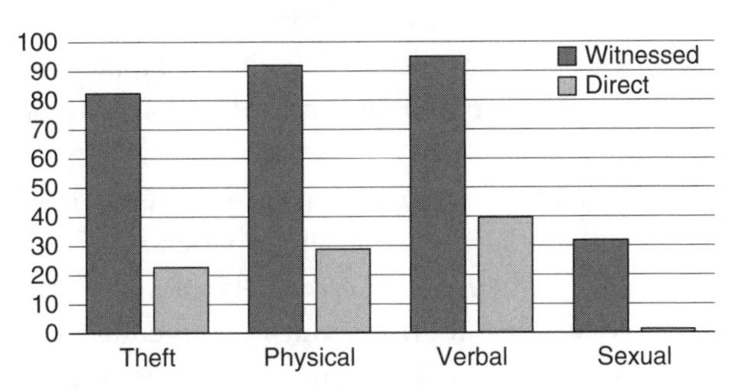

FIGURE 25.4 Victimization types reported.

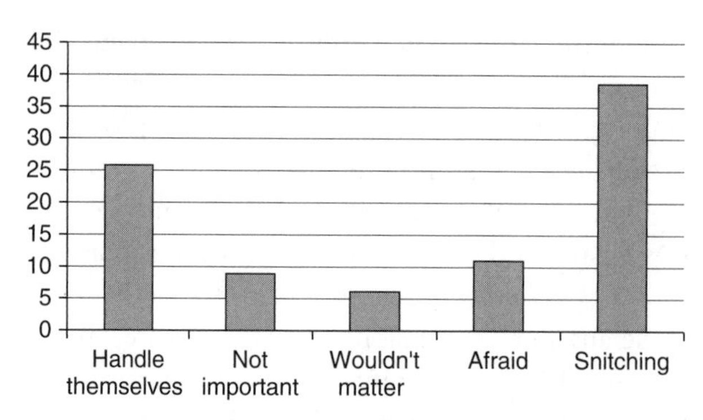

FIGURE 25.5 Responses to witnessed victimization.

such as arrest, arrest for violence, reincarceration, and parole termination. In our study, victimization in prison did matter when examining outcomes. First, victimized inmates were more likely to experience psychological distress, and with some variation were more likely to fail in the community. Specific victimization types did not predict violent arrests. However, when we combined victimization experiences with data describing how hostile the inmate felt the prison environment was, we found that the scale did predict violent arrests. In other words, people who reported experiencing victimization and felt that the prison environment was hostile and threatening were more likely to do worse on all outcomes that we measured.

CONCLUSION

Combining the results from these areas of inquiry reveals that many serious and violent inmates come back into the community with a great variety of needs. In many circumstances, communities, agencies, and families are ill equipped to handle these ex-inmates. Moreover, prisons that are hostile, violent, and coercive can contribute to the overall risk of criminal behavior in the community. While the policy developments in recent years are promising, they should supplement more concerted efforts to address the known structural and individual level factors associated with criminal behavior.

From a public safety standpoint, the concern is that violent offenders will remain unchanged in prison and potentially, after having experienced violence in prison, may be released to further continue their violent careers. Even with putting chronic violent career criminals aside, concerns still arise about how offenders are being let out of prison with few resources and skills, and how that situation is exacerbated by what they experience while they are inside prison. The deck is certainly stacked against many inmates as they reenter society. Not everyone released from prison attempts to change their previous criminal behavior. However, among those who do, it is apparent that many of them will hit the community with a number of problems.

Reentry programs should continue to take note of the research on effective programming. In fairness, the SVORI guidelines encouraged states to follow best practices when developing their programs (Gendreau, 1996). Those recommendations included assessing inmates prior to release, deliberately planning for their treatment service delivery in the community, and providing services for a sufficient length of time. The less than desirable results from the SVORI-funded programs can partly be blamed on the barriers experienced by both the programs and the participants. However, a program is only as good as its providers. Specifically, the implementation of effective service delivery requires that communities have treatment providers in the community who adhere to best practices. These best practices include, although not limited to, providing cognitive behavioral treatment at a duration and intensity commensurate with the individual's risks and needs, working with family members to support the inmate's transition, and reassessing clients during the process to determine whether those services have worked to reduce the risk and needs of the participants. The cognitive behavioral treatment must recognize that the attitudes encouraged in prison are exactly the type of attitudes that we hope inmates come back into the community without.

As programs and services mature, they must continue to expand their capacity to provide effective treatment services to those at highest risk for criminal behavior. This growth is clearly important for repeat violent offenders, given the known difficulty in breaking the cycle of violence. Nevertheless, those services cannot occur in a vacuum that ignores the experiences inmates have had in prison. As noted by Todd Clear (2008):

> *"In the United States, for example, there is no single study showing that people who leave prison are by and large (or even marginally) lucky to have had the experience. To the contrary, the effusive interest these days in the topic of reentry has as its foundational assumption that people who leave prison bring most of their problems back to the community, intact or amplified by what happened to them behind bars."*

The findings discussed in this chapter validate what others have suggested about prisons and the communities to which these inmates are returning: We need a policy shift at every level.

REFERENCES

Blitz, C. L., Wolff, N., & Shi, J. (2008). Physical victimization in prison: The role of mental illness. *International Journal of Law and Psychiatry, 31*, 385–393.

Blumstein, A., & Beck, A. J. (1999). Growth in the U.S. prison population: 1980–1996. *Crime and Justice, 26*, 17–61.

Chen, M. K., & Shapiro, J. M. (2007). Do harsher prison conditions reduce recidivism? A discontinuity-based approach. *American Law and Economic Review, 9*, 1–29.

Clear, T. (2007). *Imprisoning communities: How mass incarceration makes communities worse.* New York, NY: Oxford University Press.

Clear, T. R. (2008). Foreword. In J. M. Byrne, D. Hummer, & F. S. Taxman (Eds.), *The culture of prison violence* (p. viii). New York, NY: Pearson and Allyn Bacon.

Colvin, M. (1992). *Penitentiary in crisis: From accommodation to riot in New Mexico.* New York, NY: State University of New York.

Friedman, L. M. (1993). *Crime and punishment in American history.* New York, NY: Basic Books.

Gendreau, P. (1996). The principles of effective intervention with offenders. In A. Harland (Ed.), *Choosing correctional options that work* (pp. 117–130). Thousand Oaks, CA: Sage.

Gilliard, D. K., & Beck, A.J. (1998). *Prisoners in 1997.* Washington, D.C.: Bureau of Justice Statistics.

Glaze, L. E., Bonczar, T. P., & Zhang, F. (2010). *Probation and parole in the United States, 2009.* Washington, D.C.: Bureau of Justice Statistics.

Goerdt, J. A., & Martin, J. A. (1989). The impact of drug cases on case processing in urban trial courts. *State Court Journal, 13*, 4–12.

Henggeler, S. W., & Borduin, C. M. (1990). *Family therapy and beyond: A multi-systemic approach to treating the behavior problems of children and adolescents.* Pacific Grove, CA: Brooks/Cole.

Hensley, C., Tewksbury, R., & Castle, T. (2003). Characteristics of prison sexual assault targets in male Oklahoma correctional facilities. *Journal of Interpersonal Violence, 6*, 595–607.

HR 4092. Violent Crime Control and Law Enforcement Act of 1994. Retrieved from http://www.govtrack.us/congress/bill.xpd?bill=h103-4092

James, D. L., & Glaze, L. E. (2006). *Mental health problems of prison and jail inmates.* Washington, D.C.: Bureau of Justice Statistics.

Klein, N., Alexander, J., & Parsons, B. (1977). Impact of family systems intervention on recidivism and sibling delinquency: A model of primary prevention and program evaluation. *Journal of Consulting and Clinical Psychology, 45*, 469–474.

Kruttschnitt, C., & Gartner, R. (2005). *Marking time in the golden state: Women's imprisonment in California.* Cambridge: UK: Cambridge University Press.

Langan, P. A., & Levin, D. J. (2002). *Recidivism of prisoners released in 1994.* Washington, DC: Bureau of Justice Statistics.

Lattimore, P. K. (2008). SVORI programs: Positive impacts on housing, employment, and substance abuse. Retrieved on January 21, 2011, from http://www.svori.org/%5Cdocuments%5CPresentations%5C2008_06_WhiteHous e_FaithBased_Conference.pdf.

Lattimore, P. K., Brumbaugh, S., Visher, C., Lindquist, C. H., Winterfield, L., Salas, M., & Zweig, J. (2004). *National portrait of the Serious and Violent Offender Reentry Initiative.* Washington, DC: National Institute of Justice.

Listwan, S. J., (2000). Reentry for serious and violent offenders: An analysis of program attrition. *Criminal Justice Policy Review, 20*, 154–169.

Listwan, S. J. (2006). *Going Home Prepared in Nevada process evaluation findings.* Unpublished technical report. Kent, OH: Kent State University.

Listwan, S. J., & Hanley, D. (2010). *The prison experience and reentry: The impact of victimization on coming home.* Washington, D.C.: National Institute of Justice.

Listwan, S. J., Jonson, C. L., Cullen, F. T., & Latessa, E. J. (2008). Cracks in the penal harm movement: Evidence from the field. *Criminology & Public Policy*, 7, 423–465.

McCorkle, R. C. (1993). Fear of victimization and symptoms of psychopathology among prison inmates. *Journal of Offender Rehabilitation, 19*, 27–41.

Metraux S., & Culhane, D. P. (2004). Recent incarceration history among a sheltered homeless population. *Crime and Delinquency, 52*, 504–517.

Minton, T. D. (2010). *Jail inmates at mid-year 2009.* Washington, D.C.: Bureau of Justice Statistics.

Mumola, C. J. (1999). *Incarcerated parents and their children.* Washington, D.C.: Bureau of Justice Statistics.

Naser, R. L., & Visher, C. A. (2006). Family members' experiences with incarceration and reentry. *Western Criminology Review*, 7, 20–31.

Petersilia, J. (2003). *When prisoners come home: Parole and prisoner reentry.* New York, NY: Oxford University Press.

Pew Charitable Trust. (2008). *One in 100: Behind bars in America 2008.* Washington, D.C.: Author.

Piehl, A. (2003). *Employment dimensions of reentry: Understanding the nexus between prisoner reentry and work.* Washington, DC: Urban Institute Justice Policy Center.

Robertson, J. E. (2003). Rape among incarcerated men: Sex, coercion and STDs. *AIDS Patient Care and STDs, 8*, 423–431.

Roman, C. G., & Travis, J. (2004). *Taking stock: Housing, homelessness, and prisoner reentry.* Washington, DC: Urban Institute Justice Policy Center.

Ruback, R. B., Carr, T. S., & Hopper, C. (1986). Perceived control in prison: Its relation to reported crowding, stress, and symptoms. *Journal of Applied Social Psychology, 16*, 375–386.

Solomon, A. L., Johnson, K. D., Travis, J., & McBride, E. C. (2004). *From prison to work: The employment dimensions of prisoner reentry.* Washington, DC: Urban Institute Justice Policy Center.

Toch, H. (1977). *Living in prison: The ecology of survival.* New York, NY: The Free Press.

Travis, J. (2000). *But they all come back: Rethinking prisoners' reentry.* Washington, DC.: National Institute of Justice.

Travis, J., Chincotta, E. M., & Solomon, A. (2003). *Families left behind: The hidden costs of incarceration and reentry.* Washington, DC: The Urban Institute Justice Policy Center.

Travis, J., & Lawrence, S. (2002). *Beyond the prison gates: The state of parole in America.* Washington, DC: The Urban Institute.

Travis, J., Solomon, A., & Wahl, M. (2001). *From prison to home: The dimensions and consequences of prisoner reentry.*, Washington, DC: The Urban Institute.

Visher, C., & Lattimore, P. (2009). *Multisite evaluation of SVORI: Summary and synthesis.* Washington, DC: Urban Institute.

West, H. C., Sabol, W. J., & Greenman, S. J. (2010). *Prisoners in 2009.* Washington, DC: Bureau of Justice Statistics.

Winterfield, L., Lattimore, P. K., Steffey, D. M., Brumbaugh, S., & Lindquist, C. (2006). The serious and violent offender reentry initiative: Measuring the effects on service delivery. *Western Criminology Review*, 7, 3–19.

Wright, K. N. (1991). The violent and victimized in the male prison. *Journal of Offender Rehabilitation, 16*, 1–25.

Wright, K. N. (1993). Prison environment and behavioral outcomes. *Journal of Offender Rehabilitation, 20*, 93–113.

Zimring, F. E., Hawkins, G., & Kamin, S. (2001). *Punishment and democracy: Three strikes and you're out in California.* New York, NY: Oxford University Press

Name Index

NOTE: Page numbers followed by *t* refer to tables.

Subject Index

NOTE: Page numbers followed by *f* refer to figures; page numbers followed by *t* refer to tables.

D

M

N

O

T